Metaheuristi

J. Dréo   A. Pétrowski
P. Siarry   E. Taillard

Metaheuristics for Hard Optimization

Simulated Annealing, Tabu Search, Evolutionary
and Genetic Algorithms, Ant Colonies, ...

Methods and Case Studies

With 140 Figures

 Springer

Johann Dréo
Professor Patrick Siarry
Université Paris XII, Faculté des Sciences, LiSSi
61 avenue du Général de Gaulle, 94010 Créteil, France

Alain Pétrowski
Institut National des Télécommunications,
9 rue Charles Fourier, 91011 Evry, France

Professor Eric Taillard
EIVD, Ecole d'Ingénieurs du Canton de Vaud
route de Cheseaux 1, 1400 Yverdon-les-Bains, Switzerland

Translator: Amitava Chatterjee

Originally published in French by Eyrolles, Paris (2003) under the title:
"Métaheuristiques pour l'optimisation difficile"

Book coordinated by Patrick Siarry

ISBN-13 978-3-642-06194-3   e-ISBN-13 978-3-540-30966-6

Springer is a part of Springer Science+Business Media.

springeronline.com

© Springer-Verlag Berlin Heidelberg 2010
Printed in Germany

Cover design: de'blik, Berlin

Preface

Metaheuristics for Hard Optimization comprises of three parts.

The first part is devoted to the detailed presentation of the four most widely known metaheuristics:

- the simulated annealing method;
- the tabu search;
- the genetic and evolutionary algorithms;
- the ant colony algorithms.

Each one of these metaheuristics is actually a family of methods, of which we try to discuss the essential elements. Some common features clearly appear in most metaheuristics, such as the use of diversification, to force the exploration of regions of the search space, rarely visited until now, and the use of intensification, to go thoroughly into some promising regions. Another common feature is the use of memory to archive the best encountered solutions. One common drawback for most metaheuristics still is the delicate tuning of numerous parameters; theoretical results available by now are not sufficient to really help in practice the user facing a new hard optimization problem.

In the second part, we present some other metaheuristics, less widespread or emergent: some variants of simulated annealing; noising method; distributed search; Alienor method; particle swarm optimization; estimation of distribution methods; GRASP method; cross-entropy method; artificial immune systems; differential evolution.

Then we describe some extensions of metaheuristics for continuous optimization, multimodal optimization, multiobjective optimization and contrained evolutionary optimization. We present some of the existing techniques and some ways of research. The last chapter is devoted to the problem of the choice of a metaheuristic; we describe an unifying method called "Adaptive Memory Programming", which tends to attenuate the difficulty of this choice. The delicate subject of a rigorous statistical comparison between stochastic iterative methods is also discussed.

The last part of the book concentrates on three case studies:

- the optimization of the 3G mobile networks (UMTS) using the genetic algorithms. After a brief presentation of the operation of UMTS networks and of the quantities involved in the analysis of their performances, the chapter discusses the optimization problem for planning the UMTS network; an efficient method using a genetic algorithm is presented and illustrated through one example of a realistic network;
- the application of genetic algorithms to the problems of management of the air traffic. One details two problems of air traffic management for which a genetic algorithm based solution has been proposed: the first application deals with the en route conflict resolution problem; the second one discusses the traffic management in an airport platform;
- constrained programming and ant colony algorithms applied to vehicle routing problems. It is shown that constraint programming provides a modelling procedure, making it possible to represent the problems in an expressive and concise way; the use of ant colony algorithms allows to obtain heuristics which can be simultaneously robust and generic in nature.

One appendix of the book is devoted to the modeling of simulated annealing through the Markov chain formalism.

Another appendix gives a complete implementation in C++ language for robust tabu search method.

Créteil, Evry, Yerdon-les-Bains   *Johann Dréo*
September 2005   *Patrick Siarry*
                                                       Alain Pétrowski
                                                         Eric Taillard

Contents

Introduction 1

Part I Presentation of the Main Metaheuristics

1 **Simulated Annealing** 23
   1.1 Introduction 23
   1.2 Presentation of the method 24
        1.2.1 Analogy between an optimization problem and some
               physical phenomena 24
        1.2.2 Real annealing and simulated annealing 25
        1.2.3 Simulated annealing algorithm 25
   1.3 Theoretical approaches 27
        1.3.1 Theoretical convergence of simulated annealing 27
        1.3.2 Configuration space 28
        1.3.3 Rules of acceptance 30
        1.3.4 Annealing scheme 30
   1.4 Parallelization of the simulated annealing algorithm 32
   1.5 Some applications 35
        1.5.1 Benchmark problems of combinatorial optimization 35
        1.5.2 Layout of electronic circuits 36
        1.5.3 Search for an equivalent schema in electronics 40
        1.5.4 Practical applications in various fields 42
   1.6 Advantages and disadvantages of the method 44
   1.7 Simple practical suggestions for the beginners 44
   1.8 Annotated bibliography 45

2 **Tabu Search** 47
   2.1 Introduction 47
   2.2 The quadratic assignment problem 49
   2.3 Basic tabu search 51

2.3.1   Neighborhood.................................... 51
2.3.2   Moves, neighborhood 52
2.3.3   Evaluation of the neighborhood 54
2.4   Candidate list.. 56
2.5   Short-term memory...................................... 57
2.5.1   Hashing tables 57
2.5.2   Tabu list 59
2.5.3   Duration of tabu conditions........................ 60
2.5.4   Aspiration conditions 66
2.6   Convergence of tabu search 66
2.7   Long-term memory 69
2.7.1   Frequency-based memory......................... 69
2.7.2   Obligation to carry out move 71
2.8   Strategic oscillations 72
2.9   Conclusion 72
2.10 Annotated bibliography 72

3   **Evolutionary Algorithms**................................... 75
3.1   From genetics to engineering 75
3.2   The generic evolutionary algorithm 77
3.2.1   Selection operators 77
3.2.2   Variation operators 78
3.2.3   The generational loop............................ 78
3.2.4   Solving a simple problem 79
3.3   Selection operators 81
3.3.1   Selection pressure 81
3.3.2   Genetic drift 82
3.3.3   Proportional selection 83
3.3.4   Tournament selection 88
3.3.5   Truncation selection 90
3.3.6   Replacement selections 90
3.3.7   Fitness function 92
3.4   Variation operators and representation 93
3.4.1   Generalities about the variation operators 93
3.4.2   Binary representation 97
3.4.3   Real representation 101
3.4.4   Some discrete representations for the permutation
        problems 108
3.4.5   Representation of parse trees for the genetic
        programming 113
3.5   Particular case of the genetic algorithms 118
3.6   Some considerations on the convergence of the evolutionary
      algorithms...119
3.7   Conclusion 120
3.8   Glossary 121

3.9   Annotated bibliography 122

4   **Ant Colony Algorithms** 123
    4.1   Introduction 123
    4.2   Collective behavior of social insects 124
        4.2.1   Self-organization and behavior 124
        4.2.2   Natural optimization: pheromonal trails 127
    4.3   Optimization by ant colonies and the traveling salesman
           problem 129
        4.3.1   Basic algorithm 130
        4.3.2   Variants... 131
        4.3.3   Choice of the parameters 134
    4.4   Other combinatorial problems 134
    4.5   Formalization and properties of ant colony optimization 135
        4.5.1   Formalization 135
        4.5.2   Pheromones and memory.......................... 137
        4.5.3   Intensification/diversification 137
        4.5.4   Local search and heuristics....................... 138
        4.5.5   Parallelism 138
        4.5.6   Convergence 139
    4.6   Prospect 139
        4.6.1   Continuous optimization 139
        4.6.2   Dynamic problems................................ 147
        4.6.3   Metaheuristics and ethology 147
        4.6.4   Links with other metaheuristics 148
    4.7   Conclusion 149
    4.8   Annotated bibliography 150

Part II   Variants, Extensions and Methodological Advices

5   **Some Other Metaheuristics** 153
    5.1   Introduction 153
    5.2   Some variants of simulated annealing 154
        5.2.1   Simulated diffusion 154
        5.2.2   Microcanonic annealing 155
        5.2.3   The threshold method 157
        5.2.4   "Great deluge" method 157
        5.2.5   Method of the "record to record travel" 157
    5.3   Noising method 159
    5.4   Method of distributed search 159
    5.5   "Alienor" method 160
    5.6   Particle swarm optimization method 162
    5.7   The estimation of distribution algorithm 166
    5.8   GRASP method.. 169

5.9   "Cross-Entropy" method 170
5.10 Artificial immune systems 172
5.11 Method of differential evolution 173
5.12 Algorithms inspired by the social insects 175
5.13 Annotated bibliography 176

6 **Extensions** 179
6.1   Introduction 179
6.2   Adaptation for the continuous variable problems 179
      6.2.1 General framework of "difficult" continuous optimization179
      6.2.2 Some continuous metaheuristics 185
6.3   Multimodal optimization 196
      6.3.1 The problem 196
      6.3.2 Niching with the sharing method 196
      6.3.3 Niching with the deterministic crowding method ...199
      6.3.4 The clearing procedure 201
      6.3.5 Speciation 203
6.4   Multiobjective optimization 206
      6.4.1 Formalization of the problem 206
      6.4.2 Simulated annealing based methods 208
      6.4.3 Multiobjective evolutionary algorithms 211
6.5   Constrained evolutionary optimization 216
      6.5.1 Penalization methods 217
      6.5.2 Superiority of the feasible individuals 219
      6.5.3 Repair methods 220
      6.5.4 Variation operators satisfying the constraint structures .221
      6.5.5 Other methods dealing with constraints 223
6.6   Conclusion 223
6.7   Annotated bibliography 224

7 **Methodology** 225
7.1   Introduction 225
7.2   Problem modeling 227
7.3   Neighborhood choice 228
      7.3.1 "Simple" neighborhoods 228
      7.3.2 Ejection chains 230
      7.3.3 Decomposition into subproblems: POPMUSIC 231
      7.3.4 Conclusions on modeling and neighborhood 233
7.4   Improving method, simulated annealing, taboo search...? .235
7.5   Adaptive Memory Programming 235
      7.5.1 Ant colonies 236
      7.5.2 Evolutionary or memetic algorithms 236
      7.5.3 Scatter Search 236
      7.5.4 Vocabulary building 238
      7.5.5 Path relinking 239

7.6   Iterative heuristics comparison 240
       7.6.1   Comparing proportion 241
       7.6.2   Comparing iterative optimization methods 243
7.7   Conclusion 244
7.8   Annotated bibliography 247

Part III Case Studies

8   Optimization of UMTS Radio Access Networks with Genetic Algorithms 251

8.1   Introduction 251
8.2   Introduction to mobile radio networks 252
       8.2.1   Cellular network 252
       8.2.2   Characteristic of the radio channel 253
       8.2.3   Radio interface of the UMTS 255
8.3   Definition of the optimization problem 261
       8.3.1   Radio planning of a UMTS network 261
       8.3.2   Definition of the optimization problem 262
8.4   Application of the genetic algorithm to automatic planning . . . 265
       8.4.1   Coding 265
       8.4.2   Genetic operators 266
       8.4.3   Evaluation of the individuals 267
8.5   Results 267
       8.5.1   Optimization of the capacity 269
       8.5.2   Optimization of the loads 270
       8.5.3   Optimization of the intercellular interferences 272
       8.5.4   Optimization of the coverage 272
       8.5.5   Optimization of the probability of access 273
8.6   Conclusion 274

9   Genetic Algorithms Applied to Air Traffic Management . . . 277

9.1   *En route* conflict resolution 277
       9.1.1   Complexity of the conflict resolution problem 280
       9.1.2   Existing resolution methods 280
       9.1.3   Modeling of the problem 281
       9.1.4   Implementation of the genetic algorithm 285
       9.1.5   Theoretical study of a simple example 288
       9.1.6   Numerical application 292
       9.1.7   Remarks 295
9.2   Ground Traffic optimization 296
       9.2.1   Modeling 296
       9.2.2   BB: the 1-against-n resolution method 300
       9.2.3   GA and GA+BB : genetic algorithms 301
       9.2.4   Experimental results 303

    9.3 Conclusion 306

10 **Constraint Programming and Ant Colonies Applied to**
    Vehicle Routing Problems 307
    10.1 Introduction 307
    10.2 Vehicle routing problems and constraint programming........ 308
        10.2.1 Vehicle routing problems 308
        10.2.2 Constraint programming 309
        10.2.3 Constraint programming applied to PDP: ILOG
               Dispatcher....................................... 314
    10.3 Ant colonies 316
        10.3.1 Behavior of the real ants 316
        10.3.2 Ant colonies, vehicle routing problem and constraint
               programming 316
        10.3.3 Ant colony algorithm with backtracking 317
    10.4 Experimental results 323
    10.5 Conclusion 325

Conclusion 327

Appendices

A **Modeling of Simulated Annealing Through the Markov**
   Chain Formalism 331

B **Complete Example of Implementation of Tabu Search for**
   the Quadratic Assignment Problem 339

References 347

Index 365

Introduction

Introduction

Everyday, the engineers and the decision makers are confronted with problems of growing complexity, which emerge in diverse technical sectors, e.g. in operations research, the design of mechanical systems, image processing, and particularly in electronics (C.A.D. of electrical circuits, placement and routing of components, improvement of the performances or the manufacture yield of circuits, characterization of equivalent schemas, training of fuzzy rule bases or neural networks ...). The problem to be solved can be often expressed as an *optimization problem*. Here one can define an (or several) objective function, or cost function, that is sought to be minimized or maximized vis-à-vis all the parameters concerned. The definition of the optimization problem is often supplemented by the information of *constraints*. All the parameters of the adopted solutions must satisfy these constraints, or otherwise these solutions are not realizable. In this book, our interest is focused to a group of methods, called *metaheuristics* or *meta-heuristics*, which include in particular the simulated annealing method, the evolutionary algorithms, the tabu search method, the ant colony algorithms ..., available from the 1980s, with a common ambition: to solve the problems known as of *difficult optimization*, as well as possible.

We will see that the metaheuristics are largely based on a common set of principles, which make it possible to design solution algorithms; the various regroupings of these principles lead thus to a large variety of metaheuristics.

"Difficult" optimization

Two types of optimization problems can be distinguished: the "discrete" problems and problems with continuous variables. To be more precise, let us quote two examples. Among the discrete problems, one can discuss the famous traveling salesman problem: it is a question of minimizing the length of the round of a "traveling salesman", which must visit a certain number of cities, before turning over to the town of departure. A traditional example of continuous problem is that of the search for the values to be assigned to the parameters of a digital model of a process, so that this model reproduces the real behavior observed, as accurately as possible. In practice, one may also encounter " mixed problems", which comprise simultaneously of discrete variables and continuous variables.

This differentiation is necessary to determine the domain of difficult optimization. Indeed, two kinds of problems are referred, in the literature, as difficult optimization problems (this name is not strictly defined, and bound, in fact, with regard to the state of the art for optimization):

- certain discrete optimization problems, for which there is no knowledge of an exact *polynomial* algorithm (i.e. whose computing time is proportional to N^n, where N is the number of unknown parameters of the problem, and n is an integer constant). It is the case, in particular, of the problems known as "NP-difficult", for which one conjectures that there is no constant n for which the solution time is limited by a polynomial of degree n.
- certain optimization problems of continuous variables, for which there is no knowledge of an algorithm enabling to definitely locate a global optimum (i.e. the best possible solution) and in a completed number of computations.

There were many efforts carried out for a long time, separately, to solve these two types of problems. In the field of continuous optimization, there is thus a significant arsenal of traditional methods used for *global optimization* [Horst and Pardolos, 1995], but these techniques are often ineffective if the objective function does not possess a particular structural property, such as convexity. In the field of discrete optimization, a great number of *heuristics*, which produce solutions close to the optimum, were developed; but the majority of them were conceived specifically for a given problem.

The arrival of the *metaheuristics* mark a reconciliation of both domains: indeed, those apply to all kinds of discrete problems and they can also adapt to the continuous problems. Moreover, these methods have in common the following characteristics:

- they are, at least to some extent, *stochastic*: this approach makes it possible to counter the *combinatorial explosion* of the possibilities;
- generally of discrete origin, they have the advantage, decisive in the continuous case, to be direct, i.e. they do not resort to often problematic calculations of the gradients of the objective function;
- they are inspired by *analogies*: with physics (simulated annealing, simulated diffusion...), with biology (evolutionary algorithms, tabu search...) or with ethology (ant colonies, particle swarms...);

- they share also the same disadvantages: difficulties of *adjustment* of the parameters of the method and the *large computation time*.

These methods are not mutually excluded: indeed, in the current state of research, it is generally impossible to envisage with certainty the effectiveness of a given method, when it is applied to a given problem. Moreover, the current tendency is the emergence of *hybrid methods*, which endeavors to benefit from the specific advantages of different approaches by combining them. One can finally underline another richness of the metaheuristics: they lend themselves to all kinds of *extensions*. Let us quote, in particular:

- *multiobjective* optimization [Collette and Siarry, 2003], where it is a question of optimizing several contradictory objectives simultaneously;
- *multimodal* optimization, where one endeavors to locate a whole set of global or local optima;
- *dynamic* optimization, which faces temporal variations of the objective function;
- the recourse to *parallel implementations*.

These in particular require, for the solution methods, the specific properties which are not present in all the metaheuristics. We will reconsider this subject, which offers a means of guiding the user in the choice of a metaheuristic. The adjustment and the comparison of the metaheuristics are often carried out empirically, by exploiting analytical sets of test functions, whose global and local minima are known. We present, as an example, in figure 0.1, the shape of one of these test functions.

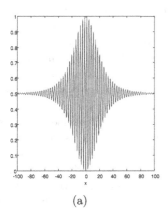
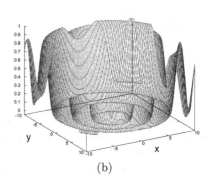

(a) (b)

Fig. 0.1. Shape of the test function F6. (a) one-dimensional representation in the domain $[-100, 100]$, (b) two-dimensional representation in the domain $[-10, 10]$.

Source of the effectiveness of metaheuristics

To facilitate the discussion, let us consider a simple example of optimization problem: that of the placement of the components of an electronic circuit. The objective function to be minimized is the length of connections, and the unknown factors — called "decision variables" — are the sites of the circuit components. The shape of the objective function of this problem can be schematically represented as in the figure 0.2, according to the "configuration": each configuration is a particular placement, associated with a choice of value for each decision variable. Let us note that in the entire book — except otherwise explicitly mentioned — one will seek in the same way to minimize an objective. When the space of the possible configurations has such a tormented structure, it is difficult to locate the global minimum c^*. We explain below the failure of a "classical" iterative algorithm, before commenting on the advantage that we can gain by employing a metaheuristic.

Trapping of a "classical" iterative algorithm in a local minimum

The principle of a traditional "iterative amelioration" algorithm is the following: one starts from an initial configuration c_0, which can be selected at random, or — for example in the case of the placement of an electronic circuit — can be determined by a designer. An elementary modification is then tested, often called a "movement" (for example, two components chosen at random are permuted, or one of them is relocated), and the values of the objective function are compared, before and after this modification. If the change led to a reduction in the objective function, it is accepted, and the configuration c_1 obtained, which is a "neighbor" of the preceding one, is used as the starting point for a new test. In the contrary case, one returns to the preceding configuration, before making another attempt. The process is made iterative until any modification makes the result worse. The figure 0.2 shows that this algorithm of iterative improvement (also indicated as *classical method*, or *descent method*) does not lead, in general, to the global optimum, but only to one local minimum c_n, which constitutes the best accessible solution taking the initial assumption into account.

To improve the effectiveness of the method, one can, of course, apply it several times, with arbitrarily selected different initial conditions, and retain as final solution the best local minima obtained. However, this procedure appreciably increases the computing time of the algorithm, and may not find the optimal configuration for c^*. The repeated application of descent method does not guarantee its determination and it is particularly ineffective when the number of local minima grows exponentially with the size of the problem.

Capability of metaheuristics to be extracted from a local minimum

To overcome the obstacle of the local minima, another idea was demonstrated to be very profitable, so much so that it is the basic core of all metaheuristics

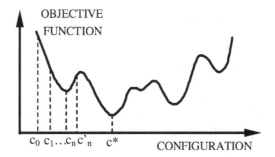

Fig. 0.2. Shape of the objective function of a difficult optimization problem according to the "configuration".

based on a *neighborhood* (simulated annealing, tabu method): it is a question of authorizing, from time to time, the movements *of increase*, in other words to accept a temporary degradation of the situation, during a change in the current configuration. It is the case if one passes from c_n to c'_n (see figure 0.2). A mechanism for controlling the degradations — specific to each metaheuristic — makes it possible to avoid the divergence of the process. It consequently becomes possible to be extracted from the trap which represents a local minimum, to leave to explore another more promising "valley". The "distributed" metaheuristics (such as the evolutionary algorithms) also have mechanisms allowing the departure of a particular solution out of a local "well" of the objective function. These mechanisms (as the *mutation* in the evolutionary algorithms) affect a solution in hand, in this case, to assist the collective mechanism to fight against the local minima, that represents parallel control of a set of "population" of solutions.

Principle of the most widespread metaheuristics

Simulated Annealing

S. Kirkpatrick and his colleagues were specialists in statistical physics (who were precisely interested in the low energy configurations of disordered magnetic materials, gathered under the term of *spin glasses*). The numerical determination of these configurations posed frightening problems of optimization, because the *"energy landscape"* of a spin glass presents several *valleys* of unequal depths; it is similar to the "landscape" in the figure 0.2. S. Kirkpatrick et al. [Kirkpatrick et al., 1983] (and independently V. Cerny [Cerny, 1985]) proposed to deal with these problems by taking as a starting point the experimental technique of the *annealing* used by the metallurgists to obtain a "well ordered" solid state, of minimal energy (by avoiding the "metastable" structures, characteristic of the local minima of energy). This technique consists in carrying material at high temperature, then to lower this temperature slowly. To illustrate the phenomenon, we represent in figure 0.3 the effect of the *annealing* technique, and that of the opposite technique of the *quenching*, on a system of a set of particles.

The *simulated annealing* method transposes the process of the *annealing* to the solution of an optimization problem: the objective function of the problem, similar to the energy of a material, is then minimized, with the help of the introduction of a fictitious *temperature*, which is, in this case, a simple controllable parameter of the algorithm.

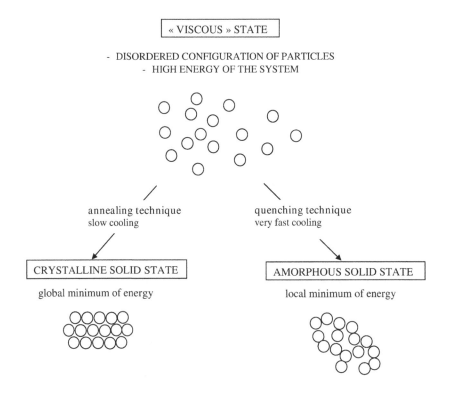

Fig. 0.3. Comparison of the techniques of annealing and quenching.

In practice, the technique exploits the Metropolis algorithm, which enables us to describe the behavior of a thermodynamic system in "equilibrium" at a certain temperature. On the basis of a given configuration (for example, an initial placement of all the components), the system is subjected to an elementary modification (for example, one relocates a component, or one exchanges two components). If this transformation causes to decrease the objective function (or *energy*) of the system, it is accepted . On the other hand, if it causes an increase δE of the objective function, it can also be accepted, but with a probability $e^{-\frac{\delta E}{T}}$. This process is then repeated in an iterative manner, by keeping the constant temperature, until thermodynamic balance is reached, concretely at the end of a "sufficient" number of modifications. Then the temperature is lowered, before implementing a new series of transformations: the law of decrease by stages of the temperature is often empirical, just like the criterion of program termination.

The flow chart of the simulated annealing algorithm is schematically presented in the figure 0.4. When it is applied to the problem of the placement of components, simulated annealing operates a disorder-order transformation, which is represented in a pictorial manner in the figure 0.5. One can also visualize some stages of this ordering by applying the method for the placement of components to the nodes of a grid (see figure 0.6).

The disadvantages of simulated annealing lie on one hand in the "adjustments", like the management of the decrease of the temperature; the user should have the know-how of "good" adjustments. In addition, the computing time can become very significant, which led to parallel implementations of the method. On the other hand, the simulated annealing method has the advantage of being flexible with respect to the evolutions of the problem and easy to implement. It gave excellent results for a number of problems, generally of big size.

The Tabu Search method

The method of search with tabus, or simply *tabu search* or *tabu method*, was formalized in 1986 by F Glover [Glover, 1986]. Its principal characteristic is based on the use of mechanisms inspired by the human *memory*. The tabu method takes, from this point of view, a path opposite to that of simulated annealing, which does not utilize memory at all, and thus is incompetent to learn the lessons from the past. On the other hand, the modeling of the memory introduces multiple degrees of freedom, which opposes — even in the opinion of the author [Glover and Laguna, 1997] — any rigorous mathematical analysis of the tabu method. The guiding principle of the tabu method is simple: like simulated annealing, the tabu method at the same time functions with only one "current configuration" (at the beginning, any solution), which is updated during successive "iterations". In each iteration, the mechanism of passage of a configuration, called s, to the next one, called t, comprises of two stages:

- one builds the set of the *neighbors* of s, i.e. the set of the accessible configurations in only one elementary *movement* of s (if this set is too vast, one applies a technique of reduction of its size: for example, one utilizes a list of candidates, or one extracts at random a subset of neighbors of fixed size); let $V(s)$ be the set (or the subset) of these neighbors;
- one evaluates the objective function f of the problem for each configuration belonging to $V(s)$. The configuration t, which succeeds s in the series of the solutions built by the tabu method, is the configuration of $V(s)$ in which

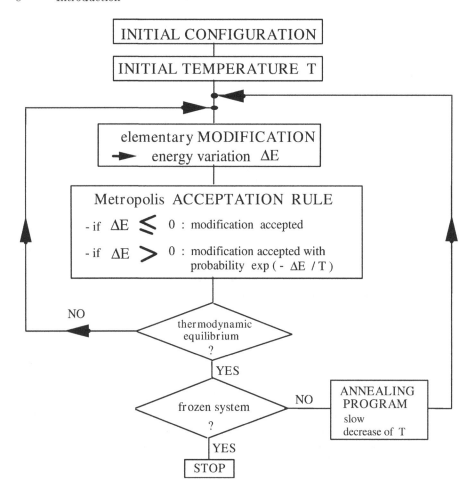

Fig. 0.4. Flow chart of the simulated annealing algorithm.

f takes the minimal value. Let us note that this configuration t is adopted even if it is worse than s, i.e. if $f(t) > f(s)$: due to this characteristic the tabu method facilitates to avoid the trapping of f in the local minima.

Just as mentioned, the preceding procedure is inoperative, because there is a significant risk to return to a configuration already retained at the time of a preceding iteration, which generates a cycle. To avoid this phenomenon, it requires updating and an exploitation, in each iteration, of a list of prohibited movements, the "tabu list": this list — that gave its name to the method — contains m movements $(t \to s)$, which are the opposite of the last m movements $(s \to t)$ carried out. The flow chart of this algorithm known as "simple tabu" is represented in the figure 0.7.

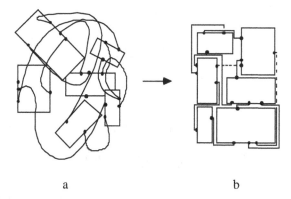

a   b

Fig. 0.5. Disorder-order transformation realized by the simulated annealing applied to the placement of electronic components.

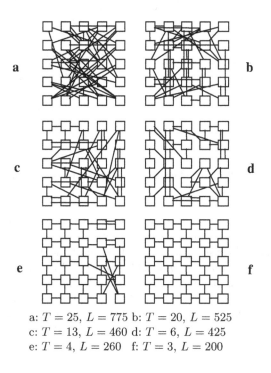

a: $T = 25$, $L = 775$ b: $T = 20$, $L = 525$
c: $T = 13$, $L = 460$ d: $T = 6$, $L = 425$
e: $T = 4$, $L = 260$ f: $T = 3$, $L = 200$

Fig. 0.6. Evolution of the system at various temperatures, on the basis of an arbitrary configuration: L indicates the overall length of connections.

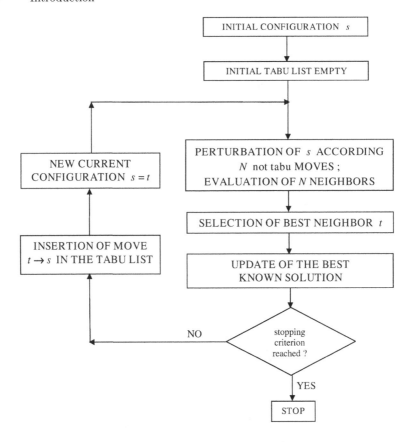

Fig. 0.7. Flow chart of the simple tabu algorithm.

The algorithm thus models a rudimentary form of memory, the *short term memory* of the solutions visited recently. Two additional mechanisms, named *intensification* and *diversification*, are often implemented to also equip the algorithm with a *long term memory*.This process does not exploit more the temporal proximity of particular events, but rather the frequency of their occurrence, over a longer period.   The intensification consists in looking further into the exploration of certain areas of the solution space, identified as particularly promising ones. Diversification is on the contrary the periodic reorientation of the search for an optimum towards areas, seldom visited until now.

For certain optimization problems, the tabu method gave excellent results; moreover, in its basic form, the method comprises less parameters of adjustment than simulated annealing, which makes it easier to use. However, the various additional mechanisms, like the intensification and diversification, bring a notable complexity.

Genetic Algorithms and Evolutionary Algorithms

The evolutionary algorithms (EAs) are the search techniques inspired by the biological evolution of the species and appeared at the end of the 1950s [Fraser, 1957]. Among several approaches [Holland, 1962] [Fogel et al., 1966] [Rechenberg, 1965], the genetic algorithms (GAs) are certainly the most well known example, following the publication of the famous book by D. E. Goldberg [Goldberg, 1989] in 1989: *Genetic Algorithms in Search, Optimization and Machine Learning.* The evolutionary methods initially aroused a limited interest, because of their significant cost of execution. But they have experienced, for the last ten years, a considerable development, that can be attributed to the significant increase in the computing power of the computers, and in particular following the development of massively parallel architectures, which exploit their "intrinsic parallelism" (see for example [Cohoon et al., 1991], for an application to the placement of components). The principle of an evolutionary algorithm can be simply described. A set of N points in a search space, chosen a priori at random, constitutes the *initial population*; each individual x of the population has a certain fitness value, which measures its degree of *adaptation* to the objective aimed. In the case of the minimization of an objective function z, the fitness of x will be higher, if $z(x)$ is smaller. An EA consists in evolving gradually, in successive *generations*, the composition of the population, by maintaining its size constant. During generations, the objective is to overall improve the fitness of the individuals; such a result is obtained by simulating the two principal mechanisms which govern the evolution of the living beings, according to the theory of C. Darwin:

- the *selection*, which supports the reproduction and the survival of the fittest individuals,
- and the *reproduction*, which allows mixing, the recombination and the variations of the hereditary features of the parents, to form offspring with new potentialities.

In practice, a representation must be chosen for the individuals of a population. Classically, an individual could be a list of integers for combinatorial problems, a vector of real numbers for numerical problems in continuous spaces, a string of binary digits for Boolean problems, or will be able to even combine these representations in complex structures, if it is required. The passage from one generation to the next one proceeds in four phases: a phase of selection, a phase of reproduction (or variation), a phase of fitness evaluation and a phase of replacement. The selection phase designates the individuals who take part in the reproduction. They are chosen, possibly several times, a priori all the more often as they have high fitness. The selected individuals are then available for the reproduction phase. This one consists in applying variation operators to copies of the individuals previously selected to generate new individuals; the operators most often used are *crossover* (or *recombination*), which produces one or two offspring from two parents, and *mutation*, which produces a new individual from only one individual (see, for an example in figure 0.8). The structure of the variation operators depends largely on the chosen representation for the individuals. The fitness of the new individuals are then evaluated, during the evaluation phase, from the objectives specified. Lastly, the replacement phase consists in selecting the members of the new generations: one can, for example, replace the lowest fitness individuals of the population by the best produced individuals, in an equal number. The algorithm is terminated after a certain number of generations, according to a termination criterion arbitrarily specified by the user. The principle of an evolutionary algorithm is represented in figure 0.9.

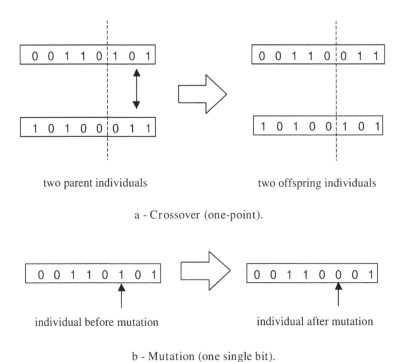

Fig. 0.8. Examples of crossover and mutation operators, in the case of individuals represented by 8-bit strings.

Because they handle a population of solution instances, the evolutionary algorithms are particularly indicated to propose a set of various solutions, when an objective function comprises several global optima. Thus, they can provide a sample of trade-off solutions, when solving problems involving several objectives, possibly contradictory. These possibilities are more specifically discussed in chapter 6.

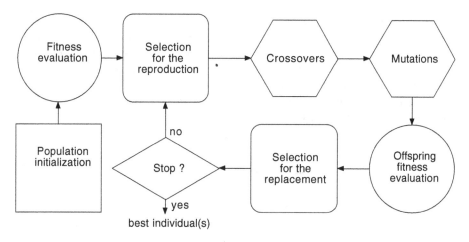

Fig. 0.9. Principle of an evolutionary algorithm.

Ant Colony Algorithms

This approach, proposed by Colorni, Dorigo and Maniezzo [Colorni et al., 1992], endeavors to simulate the collective capability to solve certain problems, as observed in a colony of ants, whose members are individually equipped with very limited faculties. Ants came to existence on earth over 100 million years ago and they are indeed one of the most prosperous species: 10 million billion individuals, living everywhere on the planet. Their total weight is of the same order of magnitude as that of the humans! Their success raises many questions. In particular, the entomologists analyzed the collaboration which is established between the ants in seeking food outside the anthill. It is remarkable that the ants always follow the same path, and this path is the shortest possible one. This control is the result of a mode of indirect communication, via the environment: the "stigmergy". Each ant deposits, along its path, a chemical substance, called "pheromone". All the members of the colony perceive this substance and preferentially direct their walk towards the more "odorous" areas.

It results particularly in a collective faculty to find the shortest path quickly, if this one is blocked fortuitously by an obstacle (see figure 0.10). While this behavior was taken as a starting point to model the algorithm, Dorigo *et al.* proposed a new algorithm for the solution of the traveling salesman problem. Since this research work, the method was extended to many other optimization problems, some combinatorial and some continuous.

The ant colony algorithms have several interesting characteristics; to mention in particular high *intrinsic parallelism*, *flexibility* (a colony of ants is able to adapt to modifications of the environment), *robustness* (a colony is ready to maintain its activity even if some individuals are failing), the *decentralization* (a colony does not obey a centralized authority) and the *self-organization* (a colony finds itself a solution, which is not known in advance). This method seems particularly useful for the problems which are *distributed* in nature, problems of dynamic *evolution*, which require a *strong fault-tolerance*. At this stage of development of these recent algorithms, the transposition with each optimization problem is not however trivial : it must be the subject of a specific treatment, which can be difficult...

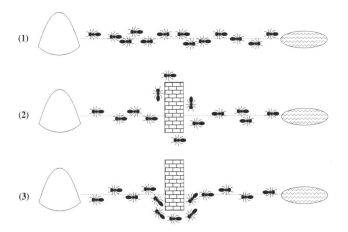

Fig. 0.10. Faculty of an ant colony to find the shortest path, fortuitously blocked by an obstacle.

1. *The real ants follow a path between the nest and a source of food.*
2. *An obstacle appears on the path, the ants choose to turn on the left or right, with equal probabilities; the pheromone is deposited more quickly on the shortest path.*
3. *All the ants chose the shortest path.*

Other metaheuristics

Whether they are the variants of the most famous methods or not, they are legions. The interested reader can refer to the chapter 5 of this book and three other recent books: [Reeves, 1995], [Saït and Youssef, 1999] and [Pham and Karaboga, 2000]; each one of them is devoted to several metaheuristics.

Extensions of the metaheuristics

We review some of the extensions, which were proposed to face characteristics of optimization.

Adaptation for the problems with continuous variables

These problems, by far the most current ones in engineering, evoke less interest among the specialists in informatics ... The majority of metaheuristics, of combinatorial origin, are however adapted to the continuous case, which assumes in particular the recourse to a discretization strategy of the variables. The discretization step must adapt in the course of optimization, to guarantee at the same time the regularity of the progression towards the optimum and the precision of the result. Our proposals relating to simulated annealing, the tabu method and GAs are described in [Siarry and Berthiau, 1997], [Chelouah and Siarry, 2000a] and [Chelouah and Siarry, 2000b].

Multiobjective optimization

More and more problems require the simultaneous consideration of several contradictory objectives. There does not exist, in this case, a single optimum; one seeks, on the other hand, a range of solutions "Pareto optimal", which form the "trade-off surface " for the problem considered. These solutions can be subjected to the final arbitration of the user. The principal methods of multiobjective optimization (either using a metaheuristic, or not) and some applications, in particular in telecommunication, were presented in the book [Collette and Siarry, 2003].

Hybrid methods

Fast success of metaheuristics is due to the difficulties encountered by the traditional optimization methods in the complex engineering problems. After the initial success of using such or such metaheuristic, the time came to make a realistic assessment and to accept the complementary nature of these new methods, between themselves like between other approaches: from where we saw the current emergence of *hybrid methods* (see for example [Renders and Flasse, 1996]).

Multimodal optimization

The purpose is to determine a whole set of optimal solutions, instead of a single optimum. The evolutionary algorithms are particularly well adapted to this task, because they handle a population of solutions. The variants of the "multipopulation" type exploit in parallel several populations, which endeavor to locate different optima.

Parallelization

Multiple modes of parallelization were proposed for the different metaheuristics. Certain techniques were desired to be generalized; others, on the other hand, benefit from characteristics of the problem. Thus, in the problems of placement of components, the tasks can be naturally distributed between several processors: each one of them is responsible to optimize a given geographical area and information is exchanged periodically between nearby processors (see for example [Sechen, 1988] and [Wong et al., 1988]).

Place of metaheuristics in a classification of the optimization methods

In order to recapitulate the preceding considerations, we propose in figure 0.11 a general classification of the mono-objective optimization methods, already published in [Collette and Siarry, 2003]. One finds, in this graph, the principal distinctions made above:

- initially, the combinatorial and the continuous optimizations are differentiated;
- for combinatorial optimization, one can approach different methods, when one is confronted with a difficult problem; in this case, the choice is sometimes possible between "specialized" heuristics, entirely dedicated to the problem considered, and a metaheuristic;
- for continuous optimization, one summarily separates the linear case (which is concerned in particular with the *linear programming*) from the non-linear case, where the framework for difficult optimization can be found. In this case, a pragmatic solution can be to resort to the repeated application of a local method which exploits, or not, the gradients of the objective function. If the number of local minima is very high, the recourse to a global method is essential: those metaheuristics are then found, which offer an alternative to the traditional methods of global optimization, those requiring the restrictive mathematical properties of the objective function;
- among the metaheuristics, one can differentiate the metaheuristics "of neighborhood", which make progress by considering only one solution at a time (simulated annealing, tabu search ...) from the "distributed" metaheuristics, which handle in parallel a complete population of solutions (genetic algorithms ...);
- finally, the hybrid methods often associate a metaheuristic with a local method. This co-operation can take the simple form of a passage of relay between the metaheuristic and the local technique, with the objective to refine the solution. But the two approaches can also be intermingled in a more complex way.

Applications of the metaheuristics

The metaheuristics are from now on regularly employed in all the sectors of engineering, such that it is not possible to draw up an inventory of the applications here. Several examples will be described in the chapters devoted to different metaheuristics. Moreover, the last part of this book is devoted to the detailed presentation of three case studies, in the fields of telecommunications, the air traffic and the vehicle routing.

An open question: the choice of a metaheuristic

This presentation should not elude the principal difficulty with which the engineer is confronted, in the presence of a concrete optimization problem: that of the choice of an "efficient" method, able to produce an "optimal" solution — or of acceptable quality — at the cost of a "reasonable" computing time. Compared to this pragmatic concern of the user, the theory is not yet of a great help, because the convergence theorems are often non-existent, or applicable under very restrictive assumptions. Moreover, the "optimal" adjustment of the various parameters of a metaheuristic which can be recommended theoretically, is often inapplicable in practice, because it induces a prohibitive computing cost. Consequently, the choice of a "good" method, and the adjustment of the parameters of this one, generally calls upon the know-how and the "experience" of the user, rather than the faithful application of well laid down rules. The efforts of research in progress, for example the analysis of the "energy landscape" or the development of a taxonomy of the hybrid methods, aim at rectifying this situation, perilous in the long term for the credibility of the metaheuristics.... Nevertheless, we will try to outline, in the chapter 7 of this book, a technique of assistance for the selection of a metaheuristic.

Plan of the book

The book comprises of three parts.

The first part is devoted to the detailed presentation of the four more widely known metaheuristics:

- the simulated annealing method;
- tabu search;
- the evolutionary algorithms;
- ant colony algorithms.

Each one of these metaheuristics is actually a family of methods, of which we try to discuss the essential elements.

In the second part, we present some other metaheuristics, less widespread or emergent. Then we describe the extensions of metaheuristics (continuous optimization, multiobjective optimization...) and some ways of search.

Lastly, we consider the problem of the choice of a metaheuristic and we describe two unifying methods which tend to attenuate the difficulty of this choice.

The last part concentrates on three case studies:

- the optimization of the 3G mobile networks (UMTS) using the genetic algorithms. This chapter is written by Sana Ben Jamaa, Zwi Altman, Jean-Marc Picard and Benoît Fourestié of France Télécom R&D;
- the application of genetic algorithms to the problems of management of the air traffic. This chapter is written by Nicolas Durand and Jean-Baptiste Gotteland, of the National School of the Civil Aviation (E.N.A.C.);
- constraint programming and ant colony algorithms applied to problems of vehicle routing. This chapter is written by Sana Ghariani and Vincent Furnon, of ILOG.

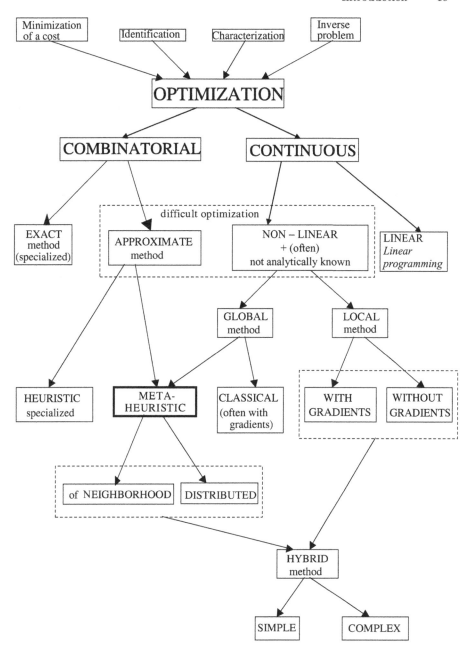

Fig. 0.11. General classification of the mono-objective optimization methods.

Part I

Presentation of the Main Metaheuristics

1

Simulated Annealing

1.1 Introduction

The complex structures of the configuration space of a difficult optimization problem (as shown in the figure 0.2 of the foreword) inspired to draw analogies with physical phenomena, which led three researchers of IBM society — S. Kirkpatrick, C.D. Gelatt and M.P. Vecchi — to propose in 1982, and to publish in 1983, a new iterative method: the simulated annealing technique [Kirkpatrick et al., 1983], which can avoid the local minima. A similar work, developed independently at the same time by V. Cerny [Cerny, 1985], was published in 1985.

Since its discovery, the simulated annealing method has proved its effectiveness in various fields such as the design of the electronic circuits, the image processing, the collection of the household garbage, or the organization of the data-processing network of French Loto... On the other hand it has been found too greedy or incapable of solving certain combinatorial optimization problems, which could be solved better by some specific heuristics.

This chapter starts with initially explaining the principle of the method, with the help of an example of the layout problem of an electronic circuit. This is followed by a simplified description of some theoretical approaches of simulated annealing, which underlines its strong points (conditional guaranteed convergence towards a global optimum) and its weak points (tuning of the parameters, which can be delicate in practice). Then various techniques of parallelization of the method are discussed. This is followed by the presentation of some applications. In conclusion, we recapitulate the advantages and the most significant drawbacks of simulated annealing. To conclude, we put forth some simple practical suggestions, intended for those users who plan to develop their first application based on simulated annealing. In appendix A of this book, we recapitulate the main results of the simulated annealing modeling based on Markov chains.

This chapter partly presents a summary of the synthesis book carried out on the simulated annealing technique [Siarry and Dreyfus, 1989], which we published in the beginning of 1989; this presentation is properly augmented by mentioning the more recent developments [Siarry, 1995, Reeves, 1995]. The references mentioned in the text were selected either because they played a significant role, or because they illustrate a precise point of the discussion. A much more exhaustive bibliography — although old — can be found in the works   [Siarry and Dreyfus, 1989,   Van Laarhoven and Aarts, 1987] [Wong et al., 1988, Sechen, 1988] and in the article [Collins et al., 1988] published on the subject. Interested readers are also requested to go through the elaborate presentations of simulated annealing which appeared in the article [Pirlot, 1992] and in chapter 2 of the book [Reeves, 1995].

1.2 Presentation of the method

1.2.1 Analogy between an optimization problem and some physical phenomena

The idea of simulated annealing can be illustrated by a vision inspired by the layout problem and routing of the electronic circuits: let us assume that a relatively inexperienced electronics specialist randomly spread the components on a plane, and connections were established as indicated in figure 0.4a of the foreword.

It is clear that the solution obtained is an unacceptable one. The purpose of developing a layout-routing program is to transform this disordered situation to lead to an ordered electronic circuit diagram (figure 0.4b of the foreword), where all connections are rectilinear, components are aligned and placed so as to minimize the length of the connections. In other words, this program must carry out a disorder-order transformation which, on the basis of a "liquid of components", leads to an ordered "solid".

However such a transformation occurs spontaneously in nature if the temperature of a system is gradually lowered; there are computer based digital simulation techniques available, which exhibit the behavior of sets of particles in interaction according to the temperature. In order to apply these techniques to the optimization problems, an analogy can be established which is presented in table 1.1.

To lead a physical system to a low energy state, the physicists generally use the annealing technique: we will examine how this method of treatment

Table 1.1. Analogy between an optimization problem and a physical system.

Optimization problem	physical system
objective function	free energy
parameters of the problem	"coordinates" of the particles
find a "good" configuration (even optimal configuration)	find the low energy states

of materials (real annealing) is helpful to deal with an optimization problem (simulated annealing).

1.2.2 Real annealing and simulated annealing

To modify the state of a material, the physicists have an adjustable parameter: the temperature. To be specific, annealing is a strategy where an optimum state can be approached by controlling the temperature. To have a deeper understanding, let us consider the example of the growth of a monocrystal. The annealing technique consists in heating a material beforehand to impart high energy to it. Then the material is cooled slowly, by keeping at each stage a temperature of sufficient duration; if the decrease in temperature is too fast, it may cause defects which can be eliminated by local reheating. This strategy of a controlled decrease of the temperature leads to a crystallized solid state, which is a stable state, corresponding to an absolute minimum of energy. The opposite technique is that of the quenching, which consists in very quickly lowering the temperature of the material: this can lead us to an amorphous structure, a metastable state that corresponds to a local minimum of energy. In the annealing technique the cooling of a material caused a disorder-order transformation, while the quenching technique was responsible in solidifying a disordered state.

The idea to use the annealing technique in order to deal with optimization problems gave rise to the simulated annealing technique. It consists in introducing a control parameter in optimization, which plays the role of the temperature. The "temperature" of the system to be optimized must have the same effect as the temperature of the physical system: it must condition the number of accessible states and lead towards the optimal state, if the temperature is lowered gradually in a slow and well controlled manner (as in the annealing technique) and towards a local minimum if the temperature is lowered abruptly (as in the quenching technique).

To conclude, we have to describe the algorithm in such a way that will enable us to implement the annealing in a computer.

1.2.3 Simulated annealing algorithm

The algorithm is based on two results of statistical physics.

On one hand, when thermodynamic balance is reached at a given temperature, the probability for a physical system to have a given energy E, is proportional to the Boltzmann factor: $e^{\frac{-E}{k_B T}}$, where k_B denotes the Boltzmann constant. Then, the distribution of the energy states is the Boltzmann distribution at the temperature considered.

In addition, to simulate the evolution of a physical system towards its thermodynamic balance at a given temperature, the Metropolis algorithm [Metropolis et al., 1953] can be utilized: on the basis of a given configuration (in our case, an initial layout for all the components), the system is subjected to an elementary modification (for example, a component is relocated, or two components are exchanged); if this transformation causes a decrease in the objective function (or "energy") of the system, it is accepted; on the contrary, if it causes an increase ΔE of the objective function, it is also accepted, but with a probability $e^{\frac{-\Delta E}{T}}$ (in practice, this condition is realized in the following manner: a real number is drawn at random ranging between 0 and 1, and the configuration causing a ΔE degradation in the objective function is accepted, if the random number drawn is lower than or equal to $e^{\frac{-\Delta E}{T}}$). By repeatedly observing this Metropolis rule of acceptance, a sequence of configurations is generated, which constitutes a Markov chain (in a sense that each configuration depends on only that one which immediately precedes it). With this formalism in place, it is possible to show that, when the chain is of infinite length (in practical consideration, of "sufficient"length...), the system can reach (in practical consideration, can approach) thermodynamic balance at the temperature considered: in other words, this leads us to a Boltzmann distribution of the energy states at this temperature.

Hence the role entrusted to the temperature by the Metropolis rule is well understood . At high temperature, $e^{\frac{-\Delta E}{T}}$ is close to 1, therefore the majority of the moves are accepted and the algorithm becomes equivalent to a simple random walk in the configuration space . At low temperature, $e^{\frac{-\Delta E}{T}}$ is close to 0, therefore the majority of the moves increasing energy is refused. Hence the algorithm reminds us of a classical iterative improvement. At an intermediate temperature, the algorithm intermittently authorizes the transformations that degrade the objective function: hence it leaves a scope for the system to be pulled out of a local minimum.

Once the thermodynamic balance is reached at a given temperature, the temperature is lowered " slightly", and a new Markov chain is implemented at this new temperature stage (if the temperature is lowered too quickly, the evolution towards a new thermodynamic balance is slowed down: the theory establishes a narrow correlation between the rate of decrease of the tempera- ture and the minimum duration of the temperature stage). By comparing the successive Boltzmann distributions obtained at the end of the various temper- ature stages, a gradual increase in the weight of the low energy configurations can be noted: when the temperature tends towards zero, the algorithm con- verges towards the absolute minimum of energy. In practice, the process is terminated when the system is "solidified" (it means that either the temper- ature reached the zero value or no more moves causing increase in energy were accepted during the stage). The flow chart of the simulated annealing algorithm has been presented in figure 0.4 of the foreword.

1.3 Theoretical approaches

The simulated annealing algorithm gave rise to many theoretical works because of the two following reasons: on one hand, it was a new algorithm, for which it was necessary to establish the conditions of convergence; in addition, the method comprises of many parameters as well as variations, whose effect or influence on the mechanism should be properly understood, if one wishes to implement this method with maximum effect.

These approaches, specially those which appeared during the initial years of the formulation, are presented in detail in the book [Siarry and Dreyfus, 1989]. Here, we keep ourselves focused to emphasize on the principal aspects treated in the literature. The theoretical convergence of simulated annealing is analyzed first. Then those factors which are influential in the operation of the algorithm are analyzed in detail: the structure of the configuration space, the acceptance rules and the annealing program.

1.3.1 Theoretical convergence of simulated annealing

Many noted mathematicians have invested their research efforts in the convergence of the simulated annealing (see in particular [Aarts and Van Laarhoven, 1985, Hajek, 1988, Hajek and Sasaki, 1989]) or some of them even endeavored to develop a general model for the analysis of the stochastic methods for global optimization (notably [Rinnooy Kan and Timmer, 1987a, Rinnooy Kan and Timmer, 1987b]). The main outcome of these theoretical studies is the following: under certain conditions (discussed later), simulated annealing probably converges towards a global optimum, in a sense that it is made possible to obtain a solution arbitrarily close to this optimum, with a probability arbitrarily close to unity. This result is, in itself, significant because it distinguishes simulated annealing from other metaheuristic competitors, whose convergence is not guaranteed.

However, the establishment of the "conditions of convergence" is not unanimously accepted. Some of these, like those proposed by Aarts et al. [Aarts and Van Laarhoven, 1985], are based on the assumption of decreasing the temperature in stages. This property enables to represent the optimization process in the form of completely connected homogeneous Markov chains, whose asymptotic behavior can be simply described. It has also been shown that the convergence is guaranteed provided that on one hand the reversibility is respected (the opposite change of any change allowed must also be allowed) and on the other hand the connectivity (any state of the system can be reached starting from any other state with the help of a finite completed number of elementary changes) of the configuration space is also maintained. This formalization presents two advantages:

- it enables us to legitimize the lowering of the temperature in stages, which improves the convergence speed of the algorithm;

- it enables us to establish that a "good" quality solution (located significantly close to the global optimum) can be obtained by simulated annealing in a polynomial time, for certain NP-hard problems [Aarts and Van Laarhoven, 1985].

Some of the other authors, Hajek et al. [Hajek, 1988, Hajek and Sasaki, 1989] in particular, were interested in the convergence of the simulated annealing within the more general framework of the theory of the inhomogeneous Markov chains. In this case, the asymptotic behavior was the more sensitive aspect of study. The main result of this work is the following: the algorithm converges towards a global optimum with a probability of unity if, when time t tends towards infinity, the temperature $T(t)$ does not decrease more quickly than the expression $\frac{C}{\log(t)}$, where C is a constant, related to the depth of the "energy wells" of the problem. It should be stressed that the results of this theoretical work, till now, are not sufficiently generalized and unanimous to be used as a guide for the experimental approach, when one is confronted with a new problem. For example, the logarithmic law of decrease of the temperature, recommended by Hajek, is not used in practice for two major reasons: on one hand it is generally impossible to evaluate the depth of the energy wells of the problem, on the other hand this law introduces an unfavorable increase in computing time...

This analysis is now prolonged by careful, individual examination of the various components of the algorithm.

1.3.2 Configuration space

The configuration space plays a fundamental role in the effectiveness of the simulated annealing technique to solve a complex optimization problem. It is equipped with a "topology", originating from the concept of proximity between two configurations: the "distance" between two configurations represents the minimum number of elementary changes required to pass from one configuration to the other. Moreover, there is an energy associated with each configuration, so that the configuration space is characterized by an " energy landscape". All the difficulties of the optimization problem lie in the fact that the energy landscape comprises of a large number of valleys of varying depth, possibly relatively close to each other, which correspond to local minima of energy.

It is clear that the shape of this landscape is not specific to the problem under study, but to a large extent depends on the choice of the cost function and the choice of the elementary changes. On the other hand, the required final solution i.e. the global minimum (or one of the global minima of comparable energy) must depend primarily on the nature of the problem considered, and not (or very little) on the preceding choices. We showed, with the help of an example problem of placement of building blocks, considered specifically for this purpose, that an apparently difficult problem can be largely simplified, either by widening the allowable configuration space, or by choosing a better adapted topology [Siarry and Dreyfus, 1989].

Several authors endeavored to establish general analytical relations between certain properties of the configuration space and the convergence of simulated annealing. In particular, some of these works were directed towards the analysis of the energy landscapes, and they searched to develop any link between the "ultrametricity" and simulated annealing [Kirkpatrick and Toulouse, 1985, Rammal et al., 1986, Solla et al., 1986]: the simulated annealing method would be more effective for those optimization problems whose low local minima (i.e. the required solutions) form an ultrametric set. Thereafter, G.B. Sorkin [Sorkin, 1991] showed that certain fractal properties of the energy landscape induce a polynomial convergence of simulated annealing; such an explanation was provided by the author on the basis of the effectiveness of the method in the field of the electronic circuit layouts. In addition, Azencott et al. [Azencott, 1992] utilized the "theory of the cycles" (originally developed in the context of the dynamic systems) to establish general explicit relations between the geometry of the energy landscape and the expected performances of simulated annealing. This work led them to propose the "method of the distortions" for the objective function, which significantly improved the quality of the solutions for certain difficult problems [Delamarre and Virot, 1998]. However, all these approaches of simulated annealing are still in a nascent stage, and their results are not yet generalized.

Lastly, another aspect of immediate practical interest relates to the adaptation of the simulated annealing for the solution of the continuous optimization problems [Siarry, 1994, Courat et al., 1994]. This subject is examined more in detail in chapter 6 of this book. Here, we put stress only on the transformations necessary to graduate from the "combinatorial simulated annealing" to the " continuous simulated annealing". Indeed, the method was originally developed for application in the domain of the combinatorial optimization problems, where the free parameters can take discrete values only. In the majority of these types of problems encountered in practice, topology is considered almost always as data for the problem: for example, in the traveling salesman problem, the permutation of two cities is a natural way to generate the rounds close to a given round. It is the same in the layout of the components for the exchange of two blocks. On the other hand, when the objective is to optimize a function of continuous variables, the topology has to be updated. This gives rise to the concept of "adaptive topology": here the length of the elementary steps is not imposed any more by the problem. This choice must be dictated by a compromise between two extreme situations: if the step is too small, the program explores only a limited region of the configuration space; the cost function is then improved very often, but in negligible amount. On the contrary, if the step is too large, the tests are accepted only seldom, and they are almost independent of each other. We will examine in the chapter 6 some of the published algorithms , which are generally developed utilizing an empirical step. From the point of mathematical interest, it is necessary to underline the work of L Miclo [Miclo, 1991], which was directed towards the convergence of the simulated annealing in the continuous case.

1.3.3 Rules of acceptance

The principle of simulated annealing requires that one accepts, occasionally and under the control of the "temperature", an increase in the energy of the current state, which enables it to be pulled out of a local minimum. The rule of acceptance generally used is the Metropolis rule described in section 1.2.3. It possesses an advantage that it originates directly from statistical physics. There are however several variations of this rule [Siarry and Dreyfus, 1989], which can be more effective from the point of view of the computing time.

Another aspect can be the examination of the following problem: at low temperature the rate of acceptance of the algorithm becomes very low, hence the method is ineffective. It is a well-known problem encountered in simulated annealing, which can be solved by substituting the traditional Metropolis rule with an accelerated alternative, called "thermostat" [Siarry and Dreyfus, 1989], as soon as the rate of acceptance falls too low. In practice, this methodology is rarely employed.

1.3.4 Annealing scheme

The convergence speed of the simulated annealing methodology depends primarily on two factors: the configuration space and the program of annealing. With regard to the configuration space, the effects on convergence of topology and the shape of the energy landscape were described above. Let us discuss the influence of the "program of annealing": it addresses the problem of controlling the "temperature" as well as the possibility of a system to reach, as quickly as possible, a solution. The program of annealing must specify the following values of the control parameters of the temperature:

• the initial temperature;
• the length of the homogeneous Markov chains, i.e. the criterion for change of temperature stage;
• the law of decrease of the temperature;
• the criterion for program termination.

In the absence of general theoretical results, which can be readily exploited, the user has to take resort to an empirical adjustment of these parameters. For certain problems, the task is even complicated by the great sensitivity of the result (and the computing time) to this adjustment. This aspect — that unites simulated annealing with other metaheuristics — is an indisputable disadvantage of this method.

To elaborate the subject a little more, we deliberate on the characteristic of the program of annealing which drew most attention: the law of decrease

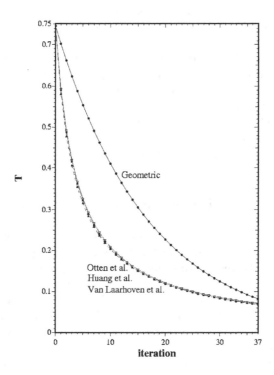

Fig. 1.1. Lowering of the temperature according to the number of stages for the geometrical law and several traditional laws.

of the temperature. The geometrical law of decrease: $T_{k+1} = \alpha \cdot T_k$, $\alpha = constant$, is a widely accepted one, because of its simplicity. An alternative solution, potentially more effective, consists in resorting to an adaptive law of the form: $T_{k+1} = \alpha(T_k) \cdot T_k$, but it is then necessary to exercise a choice among several laws suggested in the literature. One can show, however, that several traditional adaptive laws, having quite different origins and mathematical expressions are, in practice, equivalent (see figure 1.1), and can be expressed in the following generic form:

$$T_{k+1} = \left(1 - T_k \cdot \frac{\Delta(T_k)}{\sigma^2(T_k)}\right) \cdot T_k$$

where:
$$\sigma^2(T_k) = \left\langle f_{T_k}^2 \right\rangle - \left\langle f_{T_k} \right\rangle^2,$$

f denotes the objective function,
$\Delta (T_k)$ depends on the adaptive law selected.

The simplest adjustment, $\Delta (T_k) = constant$, can then be made effective, although it does not correspond to any of the traditional laws.

Due to the inability in synthesizing the results (theoretical and experimental) showing some disparities presented in the literature, the reader is redirected to the paragraph 1.7, where we propose a suitable tuning algorithm for the four parameters of the program of annealing, which can often be useful, at least to start with.

Those readers who are interested in the mathematical modeling of simulated annealing are advised to refer to the appendix A at the end of this book: the principal results produced by the Markov formalism are described there.

1.4 Parallelization of the simulated annealing algorithm

Often, the computing time becomes a critical factor in the economic evaluation of the utility of a simulated annealing technique, for applications in real industrial problems. To reduce this time, a promising research direction is the parallelization of the algorithm, which consists in simultaneously carrying out several calculations necessary for its realization. This step can be considered in the context of the significant activities which have developed around the algorithms and architectures of parallel computation for quite some time now. However, this may appear paradoxical, because of the sequential structure of the algorithm. Nevertheless, several types of parallelization have been considered to date. A book [Azencott, 1992] has been published which was completely devoted to this direction; it has described at once the rigorous mathematical results available and the simulation results, executed on parallel or sequential computers. To have a concrete idea, we shall describe the idea behind two principal modes of parallelization, which are independent of the problem dealt with and were suggested very soon after the invention of simulated annealing. The distinction of these two modes remains relevant to date, as has been shown in the recent status of the state of the art described by Delamarre and Virot in [Delamarre and Virot, 1998].

The first type of parallelization [Aarts et al., 1986] consists in implementing several Markov chain computations in parallel, by using K elementary processors. To implement this, the algorithm is decomposed into K elementary processes, constituting K Markov chains. If L be the length of these Markov chains, assumed constant, each chain is divided into K sub-chains of length $\frac{L}{K}$. The first processor executes the first chain at the initial temperature, and implements the first $\frac{L}{K}$ elements of this chain (i.e. the first sub-chain); then it calculates the temperature of the following Markov chain, starting from the states already obtained. The second elementary processor then begins executing the second Markov chain at this temperature, starting from the final configuration of the first sub-chain of the first chain. During this time, the first processor begins the second sub-chain of the first chain. This process continues for the K elementary processors. It has been shown that this mode of parallelization — described in more detail in the reference [Siarry and Dreyfus, 1989] — allows to divide the total computing time by a factor K, if K is small compared to the total number of Markov chains carried out. However, the procedure presents a major disadvantage: its convergence towards an optimum is not guaranteed. Indeed, the formalism of the Markov chains enables to establish that the convergence of simulated annealing is assured provided that the distribution of the states, at the end of each Markov chain, is close to the stationary distribution. In the case of the algorithm described, this proximity is not established at the end of each sub-chain, and larger the number K of the processors in parallel, larger is the deviation from the proximity.

The second type of parallelization [Kravitz and Rutenbar, 1987] [Roussel-Ragot et al., 1990] consists in carrying out the computation in parallel for several states of the same Markov chain, while the following condition must always be kept in mind: at a low temperature, the number of elementary transformations rejected becomes very important; it is thus possible to consider that these moves are produced by independent elementary processes, which may likely be implemented in parallel. Then the computing time can be divided approximately by the number of processes. A strategy consists in subdividing the algorithm into K elementary processes, each of which is responsible to calculate the energy variations corresponding to one or more elementary moves, and to carry out the corresponding Metropolis tests. Two operating modes are considered:

- at "high temperature", a process corresponds to only one elementary move. Each time K elementary processes were implemented in parallel, one can randomly choose a transition among those which were accepted, and the memory, containing the best solution known, is updated with the new configuration;
- at "low temperature", the accepted moves become very rare: less than one transition is accepted for K moves carried out. Each process then consists in calculating the energy variations corresponding to a succession of disturbances, until one of them is accepted. As soon as any of the processes succeeds, the memory is updated.

These two operating modes can ensure a behavior, and in particular a convergence, which is strictly identical to those of the sequential algorithms. This type of parallelization was tested by experimenting on the optimization problem of the placement of connected blocks [Roussel-Ragot et al., 1990]. We estimated the amount of computing time saved in two cases: the placement of presumed point blocks in predetermined sites and the placement of real blocks on a plane. With 5 elementary processes in parallel, the saving in computing time lies between 60 % and 80 %, according to the program of annealing used. This work was then continued, within the scope of the thesis work of P. Roussel-Ragot [Roussel-Ragot, 1990], by considering a theoretical model, which was validated by programming the simulated annealing using a network of "Transputers".

In addition to these two principal types of parallelization of simulated annealing, which should be applicable for any optimization problem, other methods were proposed to deal with specific problems. Some of these problems are problems of placement of electronic components, problems of image processing and problems of meshing (for the finite element method). In each of these three cases, information is distributed in a plane or in space, and each processor can be entrusted with the task to optimize the data pertaining to a geographical area by simulated annealing; here information is exchanged periodically between the neighboring processors.

Another step was planned to reduce the cost of synchronizations between the processors: the algorithms known as "asynchronous" agree to calculate the energy variations starting from partially out-of-date data. However it seems very complex and sensitive to control the admissible error, except for certain particular problems [Durand and White, 1991].

As an example, let us describe the asynchronous parallelization technique, suggested by Casotto et al. [Casotto et al., 1987] to deal with the problem of the placement of electronic components. The method consists in distributing the components to be placed in K independent groups, respectively assigned to K processors. Each processor applies the simulated annealing technique to seek the optimal site for the components that belong to its group. The processors function in parallel, and in an asynchronous manner to each other. All of them have access to a common memory, which contains the current state of the circuit plan. When a processor plans to exchange the position of a component of its group with that of an affected component in another group pertaining to another processor, it temporarily blocks the activity of this processor. It is clear that the asynchronous working of the processors involves errors, in particular in the calculation of the overlapping between the blocks, and thus in the evaluation of the cost function. In fact, when a given processor needs to evaluate the cost of a move (translation or permutation), it will search, in the memory, the current position of all the components of the circuit . However the information collected is partly erroneous, since certain components are in the course of displacement, because of activities of the other processors. In order to limit these errors, the method is supplemented by the two following provisions. On one hand, the distribution of the components between the processors is in itself an object of optimization by simulated annealing technique, which is performed simultaneously with the optimization process already described: in this manner, the membership of the components geographically close to the same group can be favored. In addition, the maximum amplitude of the moves carried out by the components is reduced as the temperature decreases. Consequently, when the temperature decreases, the moves mainly relate to nearby components, thus generally belonging to the same group . In this process the interactions between the groups can be reduced, thus reducing the frequency of the errors mentioned above. This technique of parallelization of simulated annealing was validated using several examples of real circuits: the algorithm functions approximately six times faster with eight processors than with only one, the results being of comparable quality with those of the sequential algorithm.

1.5 Some applications

The majority of the preceding theoretical approaches are based on asymptotic behaviors which impose several restrictive assumptions, very often causing excessive enhancements in computing times. This is why, to solve real industrial problems under reasonable conditions, it is often essential to adopt an experimental approach, which may frequently result in crossing the barriers recommended by the theory. The simulated annealing method proved to be interesting in solving many optimization problems, NP-hard or not. Some examples of these problems are presented here.

1.5.1 Benchmark problems of combinatorial optimization

The effectiveness of the method was initially tested on the benchmark problem instances of combinatorial optimization. In this type of problem, the practical purpose is secondary: the initial objective is to develop the optimization method, and to compare its performances with those of the other methods. We will detail only one example: that of the traveling salesman problem.

The reason for the choice of this problem is that it is very simple to formulate, and, at the same time, very difficult to solve: the larger problems for which the optimum was found, and proved, comprise of a few thousands of cities. To illustrate the disorder-order transformation, carried out by the simulated annealing technique, as the temperature goes down, we present, in the figure 1.2, three intermediate configurations obtained on 13206 nodes of the Swiss road network..

Bonomi and Lutton considered very high dimensional examples: between 1000 and 10000 cities [Bonomi and Lutton, 1984]. They showed that, to prevent a prohibitive computing time, the domain containing the cities can be deconstructed into areas, and the moves for a round of the traveler can be so forced that they are limited between the cities located in contiguous areas. Bonomi and Lutton compared simulated annealing with the traditional techniques of optimization, for the traveling salesman problem: simulated annealing was slower for small dimensional problems (N lower than 100); on the other hand, it was, by far, more powerful for higher dimensional problems (N higher than 800). The traveling salesman problem has been extensively studied to illustrate and establish several experimental and theoretical developments on the simulated annealing method [Siarry and Dreyfus, 1989].

Many other benchmark problems of combinatorial optimization were also solved using simulated annealing [Siarry and Dreyfus, 1989, Pirlot, 1992]: in particular the problems of the "partitioning of graph", of the "minimal coupling of points", of the "quadratic assignment" ... The comparison with the best known algorithms leads to different results, varying according to the problems and ... the authors. Thus the works by Johnson et al. [Johnson et al., 1989, Johnson et al., 1991, Johnson et al., 1992], which were devoted to a systematic comparison of several benchmark problems, conclude that the only benchmark problem which can find favor with simulated annealing is that of the partitioning of graph. For some problems, promising results with the simulated annealing method could only be observed for high dimensional examples (a few hundreds of variables), and that too at the cost of a high computing time. Therefore, if simulated annealing has the merit to adapt simply to a great diversity of problems, it cannot claim as much to supplement those specific algorithms that already exist for these problems.

We now present the applications of simulated annealing for practical problems. The first significant application of industrial interest was developed in the field of the electronic circuit design; this industrial sector still remains the biggest domain in which maximum number of application works using simulated annealing have been published. Two applications in the area of electronics are discussed in detail in two following paragraphs. This is followed by discussions regarding other applications in some other fields.

1.5.2 Layout of electronic circuits

The first application of the simulated annealing method for practical problems was developed in the field of the layout-routing of the electronic circuits [Kirkpatrick et al., 1983, Vecchi and Kirkpatrick, 1983, Siarry and Dreyfus, 1984]. Till now numerous works have been reported on this subject in several publications and, in particular, two books were completely devoted to this problem [Wong et al., 1988, Sechen, 1988]. A detailed bibliography, concerning the works carried out in the initial period 1982-1988, can be found in the books [Siarry and Dreyfus, 1989, Van Laarhoven and Aarts, 1987, Wong et al., 1988, Sechen, 1988].

The search for an optimal layout is generally carried out in two stages. The first consists in calculating an initial placement quickly, by a constructive method: the components are placed one after another, in order of decreasing connectivity. Then an algorithm for iterative improvement is employed that gradually transforms, by elementary moves (e.g. exchange of components, operations of rotation or symmetry etc.), the initial layout configuration. The algorithms for iterative improvement of the layout differ according to the rule adopted for the succession of elementary moves. Simulated annealing can be used in this second stage.

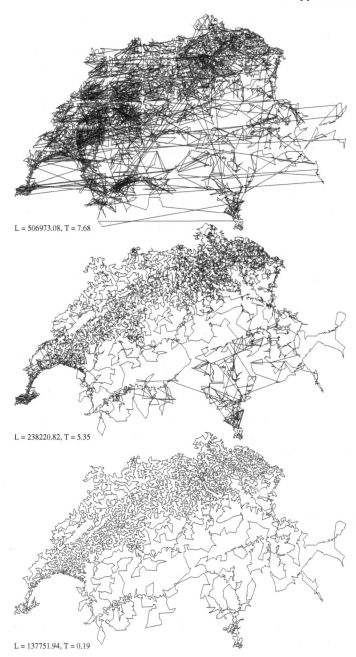

L = 506973.08, T = 7.68

L = 238220.82, T = 5.35

L = 137751.94, T = 0.19

Fig. 1.2. The traveling salesman problem (13206 nodes of the Swiss road network): best known configurations (length: L) at the end of 3 temperature stages (T).

Our interest was concerned with a unit of 25 identical blocks to be placed on predetermined sites, which are the nodes of a planar square network. The list of connections is so that, in the optimal configurations, each block is connected only to its closer neighbors (see figure 0.6 of the foreword): an a priori knowledge about the global minima of the problem then facilitates to study the influence of the principal parameters of the method on its convergence speed. The cost function is the overall Manhattan length (i.e. the L length) of the connections. The only authorized elementary move is the permutation of two blocks. A detailed explanation for this benchmark problem on layout design — which is a form of " quadratic assignment" problem — can be found in the references [Siarry, 1986] and [Siarry et al., 1987]. Here, the discussions will be kept limited to the presentation of two examples of applications. First of all, to appreciate the effectiveness of the method, we start with a completely disordered initial configuration (see figure 0.6 of the foreword), and an initial "elevated" temperature (in the sense that at this temperature 90% of the moves are accepted): the figure 0.6 of the foreword represents the best configurations observed at the end of a few temperature stages. In this example, the temperature profile is that of a geometrical decrease, of ratio 0.9. A global optimum of the problem could be obtained after 12000 moves, whereas the total number of possible configurations is about 10^{25}.

To illustrate the advantages of the simulated annealing technique, we applied the traditional method of iterative improvement (simulated annealing at zero temperature), for the same initial configuration (see figure 1.3b), and by authorizing the same number of permutations as during the preceding test. It was observed that the traditional method got trapped in a local minimum (see figure 1.3c); it is clear that the shifting from this configuration to the optimal configuration as shown in the figure 1.3a would require several stages (at least five), majority of which correspond to an increase in energy, inadmissible by the traditional method. This problem of placement in particular made it possible to empirically develop a program of "adaptive" annealing, which could achieve gain in computing time by a factor of 2; the lowering of the temperature is carried out according to the law $T_{k+1} = D_k \cdot T_k$, with:

$$D_k = \min \left(D_0, \frac{E_k}{\langle E_k \rangle} \right)$$

that includes:

$D_0 = 0.5$ to 0.9

E_k is the minimal energy of the configurations accepted during the stage k

$\langle E_k \rangle$ is the average energy of the configurations accepted during the stage k

(at high temperature, $D_k = \frac{E_k}{\langle E_k \rangle}$ is small: hence the temperature is lowered quickly; at low temperature, $D_k = D_0$, this corresponds to a slow cooling).

Then we considered a more complex problem consisting of positioning components of different sizes, with an objective of simultaneous minimization of the length of the necessary connections and the surface area of the circuit used. In this case, the translation of a block is a new means of iterative transformation of the layout. Here we can observe that the blocks are overlapping with each other, what is authorized temporarily, but must be generally excluded from the final layout. This new constraint can be accommodated within the cost function of the problem, by introducing a new factor called the overlapping surface between the blocks. Calculating this surface area can become very cumbersome when the circuit comprises of many blocks. This is why the circuit was divided into several planar areas, whose size is such that a block can overlap only with those blocks located in the same area, or with one of the immediately close areas. The lists of the blocks belonging to each area are updated after each move, using a chaining method. Moreover, to avoid leading to a circuit congestion such as routing is impossible, a fictitious increase in the dimensions of each block is introduced. The calculation for the length of the connections consists in determining, for each equipotential, the barycentre of the terminations, and then to add the L distances of the barycentre with each termination. Lastly, the topology of the problem is adaptive, which can be described in the following manner: when the temperature decreases, the maximum amplitude of the translations decreases, and the exchanges are considered between the neighboring blocks only.

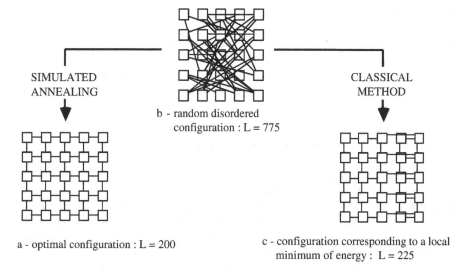

SIMULATED
ANNEALING

b - random disordered
configuration : L = 775

CLASSICAL
METHOD

a - optimal configuration : L = 200

c - configuration corresponding to a local
minimum of energy : L = 225

Fig. 1.3. The traditional method getting trapped in a local minimum of energy.

With the simulated annealing algorithm, it was possible to optimize industrial circuits, in particular in hybrid technology, in collaboration with the

Thomson D.C.H. (Department of the Hybrid Circuits) company. As an example, we present in the figure 1.4, the result of the optimization of a circuit layout comprising of 41 components and 27 equipotentials: the automated layout design procedure causes a gain of 18 % in the connection lengths, compared to the initial manual layout.

This study showed that the flexibility of the method enables it to take into account not only the rules of drawing, which translate the standards of technology, but also the rules of know-how, which are intended to facilitate the routing. Indeed, the rules of drawing impose in particular a minimal distance between two components, whereas the rules of know-how recommend a larger distance, allowing the passage of connections. To balance these two types of constraints, the calculation of the area of overlapping between the blocks, on a two by two basis, is undertaken according to the formula:

$S = S_r + a \cdot S_v$, where the notations indicate:

S_r	the "real" overlapping surface
S_v	the "virtual" overlapping surface
a	a weight factor (typically: 0.1)

Surfaces S_r and S_v are calculated by increasing dimensions of the components fictitiously, with a larger increase in S_v. Here, this induces some kind of an "intelligent" behavior, similar to that of an expert system. We notice, from the figure 1.4, a characteristic of the hybrid technology, which was easily incorporated in the program: the resistances, shown by a conducting link, can be placed under the diodes or the integrated circuits.

The observations noted by the majority of the authors concerning the application of the simulated annealing technique for the layout design problem conform to our observations: the method is very simple to implement, it adapts easily to various and evolving technological standards, and the final result is of good quality, but it is sometimes obtained at the cost of a significant computing time.

1.5.3 Search for an equivalent schema in electronics

We now present an application which mixes the combinatorial and the continuous aspects: automatic identification of the "optimal" structure of a linear circuit pattern. The objective was to automatically determine a model which comprises of the least possible number of elementary components, while ensuring a " faithful" reproduction of experimental data. This activity, in collaboration with the Institute of Fundamental Electronics (IEF, CNRS URA 22, in Orsay), began with the integration, in a single software, of a simulation program of the linear circuits (implemented in the IEF) and of a simulated annealing based optimization program developed by us. We initially validated this tool, by characterizing models of real components having a given structure (described using their distribution parameters S). A comparison with a

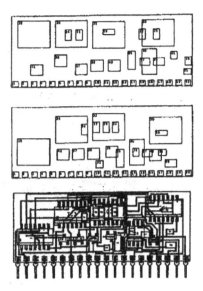

Fig. 1.4. Optimization by simulated annealing of the design of an electronic circuit layout comprising of 41 components.

- *drawing at the top: initial manual layout; length of connections: 9532;*
- *drawing at the middle: final layout, optimized by annealing; length of connections 7861;*
- *drawing at the bottom: manual routing using the optimized layout.*

commercial software (developed using the gradient method), at that moment in use in the IEF, showed that simulated annealing is particularly useful if the orders of magnitude of the parameters of the model are completely unknown: obviously the models under consideration are of this nature, since their structure even is to be determined. We developed an alternative simulated annealing, called logarithmic simulated annealing [Courat et al., 1994], which allows an effective exploration of the space of variations of the parameters, when this space is very wide (more than 10 decades per parameter). Then the problem of structure optimization was approached by the examination — in the case of a passive circuit — of progressive simplification of a general "exhaustive" model: we proposed a method which could be successfully employed to automate all the simplification stages [Courat et al., 1995]. This technique rests on the progressive elimination of the parameters, according to their statistical behavior during the process of optimization by simulated annealing.

We present, with the help of illustrations, the example of the search for an equivalent schema for an MMIC inductance, in the frequency range of 100 MHz to 20 GHz. On the basis of the initial "exhaustive" model with 12 parameters, as shown in the figure 1.5, and allowing each parameter to move over 16 decades, we obtained the equivalent schema shown in the figure 1.6 (the final values of the 6 remaining parameters are beyond the scope of our present interest: they are specified in [Courat et al., 1995]). The layouts in the Nyquist plane of the four S parameters of the quadripole of the figure 1.6 coincide nearly perfectly with the experimental results of MMIC inductance, and this is true for the entire frequency range [Courat et al., 1995].

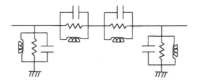

Fig. 1.5. Initial structure with 12 elements.

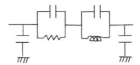

Fig. 1.6. Optimal structure with 6 elements.

1.5.4 Practical applications in various fields

An important field of application for simulated annealing happens to be image processing : here the main problem is to restore the images, by using computer, mainly in three-dimensional forms, starting from incomplete or irregular data. There are numerous practical applications in several domains like robotics, medicine (e.g. tomography), geology (e.g. prospections)... The restoration of an image using an iterative method involves, under normal circumstances, the treatment of a large number of variables. Hence it calls for development of a suitable method, which can limit the computing time of the operation. Based on the local features of the information contained in an image, several authors proposed numerous structures and algorithms specifically addressed to carry out calculations in parallel. Empirically, it appears that the simulated annealing method should be particularly well suited for this task. A rigorous theoretical justification of this property can be obtained starting from the concept of *Markovian field* [Geman and Geman, 1984], which provides a convenient and coherent model of the local structure of information in an image. This concept has been explained in detail in the reference

[Siarry and Dreyfus, 1989]. The "Bayesian approach" for the problem of optimal restoration of an image, starting from a scrambled image, consists in determining the image which presents "the maximum of a posteriori likelihood". It has been shown that this problem is ultimately configured as a well known minimization problem of an objective function, comprising of a very large number of parameters, e.g. light intensities of all the "pixels" of an image, in case of an image in black and white. Consequently, the problem can be considered as a typical problem for simulated annealing. The iterative application of this technique consists in updating the image by modifying the intensity of all the pixels in turn, in a pre-specified order. This procedure leads to a significant consumption of computing time: indeed, the number of complete sweepings of the image necessary to obtain a good restoration is, typically, about 300 to 1000. But as the calculation of the energy variation is purely local in nature, several methods were proposed to update the image by simultaneously treating a large number of pixels, using specialized elementary processors. The formalism of the Markovian fields made it possible to treat, by simulated annealing, several crucial tasks in automated analysis of the images: the restoration of scrambled images, the image segmentation, the image identification ... Apart from this formalism, other problems in the image processing domain were also solved by annealing: for example, the method was utilized to determine the geological structure of the basement, starting from results of seismic experiments.

To finish, we will mention some specific problems, in very diverse fields, where simulated annealing was employed successfully: organization of the data-processing network for the French Loto (it required ten thousand playing machines to be connected to host computers), optimization of the collection of the household garbage in Grenoble, timetable problems (the problem was, for example, to determine the optimal planning of the rest days in a hospital), optimization in architecture (in a project on constructing a 17 floor building for an insurance company, it was necessary to distribute the activities among the various parts, so that the work output from 2000 employees can be maximized)... Several applications of simulated annealing for the scheduling problems can be found (particularly, in the references [Van Laarhoven et al., 1992, Brandimarte, 1992, Musser et al., 1993, Jeffcoat and Bulfin, 1993]). The adequacy of the method for this type of problem has been discussed. For instance, Lenstra et al. [Van Laarhoven et al., 1992] showed that the computing time involved was unsatisfactory. Moreover, in [Fleury, 1995], Fleury underlines several characteristics of the scheduling problems which make them unsuitable for simulated annealing and he recommends a different stochastic method, inspired by simulated annealing and tabu search: the "kangaroo method", for this problem.

1.6 Advantages and disadvantages of the method

From the preceding discussion, the principal characteristics of the method can be established. Firstly, the *advantages*: it is observed that the simulated annealing technique generally achieves a good quality solution (i.e. absolute minimum or good relative minimum for the objective function). Moreover, it is a general method: it is applicable, and easy to implement, for all the problems which can potentially employ the iterative optimization techniques, under the condition that after each transformation the corresponding change in the objective function can be evaluated directly and quickly (often the computing time becomes excessive if complete re-computation of the objective function cannot be avoided, after each transformation). Lastly, it offers great flexibility, as one can add new constraints easily afterwards in the program.

Now, let us discuss the disadvantages. The users are sometimes repelled by the involvement of great many parameters (initial temperature, rate of decrease of the temperature, length of the temperature stages, termination criterion for the program...). Although the standard values published for these parameters generally allow an effective operation of the method, the essential empirical nature of them can never guarantee suitability for a large variety of problems. The second defect of the method — which depends on the preceding one — is the computing time involved, which is excessive in certain applications.

In order to reduce this computing time, it still requires an extensive research effort to determine the best values of the parameters of the method [Siarry, 1994], particularly for the law of decrease of the temperature. Any progress in the effectiveness of the technique and the computing time involved can be obtained by continuing the analysis of the method in three specific directions: utilization of interactive parameter setting, parallelization of the algorithm and incorporation of statistical physics based approaches to analyze and study disordered mediums.

1.7 Simple practical suggestions for the beginners

- *Definition of the objective function* : some constraints are integrated here, others constitute a limitation in allowed disturbances for the problem.
- *Choice of the disturbance mechanisms* for a "current configuration": the calculation of the corresponding ΔE variation of the objective function must be *direct* and rapid.
- *Initial temperature T_0*: it may be calculated as a preliminary step using the following algorithm:
  - initiate 100 disturbances at random; evaluate the average $\langle \Delta E \rangle$ of the corresponding ΔE variations;

- – choose an initial rate of acceptance τ_0 of the "degrading perturbations", according to the assumed "quality" of the initial configuration; for example:
  - · "poor" quality: $\tau_0 = 50\%$ (starting at high temperature),
  - · "good" quality: $\tau_0 = 20\%$ (starting at low temperature),
- – deduce T_0 from the relation: $e^{\frac{-(\Delta E)}{T_0}} = \tau_0$.

- • *Acceptance rule of Metropolis.* it is practically utilized in the following manner: if $\Delta E > 0$, a number r in $[0, 1]$ is drawn randomly, and accept the disturbance if $r < e^{\frac{-\Delta E}{T}}$, where T indicates the current temperature.
- • *Change in temperature stage*: can take place as soon as one of the 2 following conditions is satisfied during the temperature stages:
  - – $12 \cdot N$ perturbations accepted;
  - – $100 \cdot N$ perturbations attempted,
    N indicating the number of degrees of freedom (or parameters) of the problem
- • *Decrease of the temperature*: can be carried out according to the geometrical law: $T_{k+1} = 0.9 \cdot T_k$.
- • *Program termination*: can be activated after 3 successive temperature stages without any acceptance.
- • *Essential verifications during the first executions of the algorithm*:
  - – the generation of the real random numbers (in $[0, 1]$) must be well *uniform;*
  - – the "quality" of the result should not vary significantly when the algorithm is executed *several times*:
    - · with different "seeds" for the generation of the random numbers,
    - · with different initial configurations,
  - – for each initial configuration used, the result of simulated annealing can be favorably compared, theoretically, with that of the *quenching* ("disconnected" Metropolis rule).
- • *An alternative for the algorithm in order to achieve less computation time*: simulated annealing is greedy and not very effective at low temperature; hence the interest may lie in utilizing the simulated annealing technique, prematurely terminated, in cascade with an algorithm of local type, for specific optimization of the problem, of which role is "to refine" the optimum.

1.8 Annotated bibliography

[Siarry and Dreyfus, 1989]: This book describes the principal theoretical approaches and the applications of the simulated annealing in the early years of formation of the method (1982-1988), when the majority of the theoretical bases were established.

[Reeves, 1995]: The principal metaheuristics are explained in great detail in this work. An elaborate presentation of simulated annealing is proposed in the chapter 2. Some applications are presented: in particular, design of electronic circuits and treatment of the scheduling problems.

[Saït and Youssef, 1999]: In this book several metaheuristics have been extensively explained, which includes simulated annealing (in chapter 2). The theoretical elements relating to the convergence of the method are clearly put in detail. The book comprises also the study of an application in an industrial context (that of the TimberWolf software, which is a reference tool for the layout-routing problem). This should be cited as an invaluable contribution for the teachers: each chapter is supplemented by suitable exercises.

[Pham and Karaboga, 2000]: The principal metaheuristics are also explained in this book. Here, chapter 4 is completely devoted to simulated annealing which concludes with an application in the field of the industrial production.

[Teghem and Pirlot, 2002]: This recent book is a collection of the contributions of a dozen authors. Simulated annealing is however not treated in detail.

2

Tabu Search

2.1 Introduction

Tabu search was first proposed by Fred Glover in an article published in 1986 [Glover, 1986], although it borrowed many ideas suggested before during the Sixties. The two articles simply entitled *Tabu Search* [Glover, 1989, Glover, 1990] proposed the majority of tabu search principles which are currently known. Some of these principles did not gain prominence for a long time among the scientific community. Indeed, in first half of the Nineties, the majority of the research works in tabu search did use a very restricted domain of the principles of the technique. They were often limited to a *tabu list* and an elementary *aspiration condition*.

The popularity of tabu search is certainly due to the pioneering works by the team of D. de Werra at the Swiss Federal Institute of Technology, Lausanne, during the late Eighties. In fact, the articles by Glover, the founder of the method, were not well understood at the time when there was not yet a "metaheuristic culture". Hence a significant credit for the popularization of the basic technique must go to [Hertz and de Werra, 1987, Hertz and de Werra, 1991, de Werra and Hertz, 1989] which definitely played a significant role in the dissemination of the technique.

At the same time, a competition developed between simulated annealing (which then had an established convergence theorem as its theoretical advantage) and tabu search. For many applications, tabu search based heuristics definitely showed more effective results [Taillard, 1990, Taillard, 1991, Taillard, 1993, Taillard, 1994], which increased the interest for the method in the research community.

In the beginning of the Nineties, the technique was exported to Canada, more precisely in the Center for Research on Transportation in Montreal, where post-doctoral researchers from the team of D. de Werra worked in this domain. In the process, this created another center of interest in the domain of tabu search. The technique was then quickly disseminated among several research communities and this culminated in the publishing of the first book which was solely dedicated to tabu search [Glover et al., 1993]. In this chapter, we shall not concentrate on the more advanced principles of tabu search, such as those presented in the book of Fred Glover and Manuel Laguna [Glover and Laguna, 1997], but we shall focus on the most significant and most general principles. Here, it should be mentioned that sometimes another author is credited with the original ideas of tabu search. However in our opinion the claim for such an attribution is abusive: no written document presenting those ideas could ever be obtained, they were only presented in front of a very small audience during a congress in 1986 and, according to what we have heard about the presentation, it comprised of a very limited subset of the basic ideas published earlier by Glover.

What unquestionably distinguishes it from the local search technique presented in the preceding chapter, is that tabu search incorporates intelligence. Indeed, there is a huge temptation to direct an iterative search in a prospective, good direction, so that it is not only directed by the chance and the value of an objective function to be optimized. The development of tabu search gives rise to a couple of challenges: firstly, as in any iterative search, it is necessary that the search engine, i.e. the mechanism of evaluation of neighboring solutions, is effective; secondly, pieces of knowledge regarding the problem under consideration should be transmitted to the search procedure, so that it should not get trapped in bad regions of the solution space. On the contrary, it should be directed intelligently in the solution space, if such a term is permitted to be used.

One of the guiding principles can be to constitute a history of iterative search or, equivalently, to provide the search with memory.

The first two sections of this chapter will present a few techniques that will enable to guide an iterative search. Among the whole range of the principles suggested in literature, we chose those that appeared most effective to direct a search. In particular, we shall put stress on a simple and effective manner to elaborate the concepts of short-term and long-term memory. These concepts of memory will be analyzed by studying the effects of varying the parameters which are associated with them.

The third section of this chapter will present some theoretical results on tabu search. It will also show that a mechanism of short-term memory, based on prohibition to visit certain solutions, and very extensively used because of its simplicity and its great practical effectiveness, proves to be insufficient, as the search does visit a subset of the solution space without passing by the global optimum.

Some of these principles of tabu search will be illustrated with the help of a particular problem, namely the quadratic assignment problem, so that these principles do not float "in air". We chose this problem for several reasons. First of all, it finds applications in multiple fields. For example, the problem of the placement of the electronic modules, about which we discussed in the chapter 1 devoted to simulated annealing, is a quadratic assignment problem. Then, its formulation is very simple, because it deals with finding a permutation. Here, it should be noted that many combinatorial optimization problems could be expressed in the form of a search for a specific permutation. Finally, the implementation of a basic tabu search for this problem is very concise, which enables us to present *in extended form*, in appendix B, the code for this algorithm, which is among the most powerful ones presently available.

2.2 The quadratic assignment problem

Given n objects and the flows f_{ij} between the object i and the object j $(i, j = 1 \ldots n)$, and given n sites with distance d_{rs} between the sites r and s $(r, s = 1 \ldots n)$, the problem deals with placing the n objects on the n sites so as to minimize the sum of the products, flows × distances. Mathematically, this is equivalent to find a permutation $\mathbf{p}$, whose i^{th} component p_i denotes the place of the object i, which minimizes $\sum_{i=1}^{n} \sum_{j=1}^{n} f_{ij} \cdot d_{p_i p_j}$.

This problem has multiple practical applications; among them undoubtedly the most popular ones are the assignment of offices or services in buildings (e.g. university campus, hospital, etc.), the assignment of the departure gates to airplanes in an airport, the placement of logical modules in electronic circuits, the distribution of the files in a database and the placement of the keys of keyboards of typewriters. In these examples, the matrix of the flows represents, respectively, the frequency with which the people may move from a building to another, the number of people that may transit from one airplane to another, the number of electrical connections to be established between two modules, the probability of asking the access to a second file if one accesses the first one and, finally, the frequency with which two particular characters appear consecutively in a given language. The matrix of the distances has an obvious meaning in the first three examples; in the fourth, it represents the transmission time between the databases and, in the fifth, the time separating striking from two keys.

The quadratic assignment problem is NP-hard. One can easily show it by noting that the traveling salesman problem can be formulated as a quadratic assignment problem. Unless $P = NP$, there is no polynomial approximation scheme for this problem. This can be shown simply by considering two problem instances which differ only in the flows matrix. If one withdraws an appropriate constant from all the components of the first problem to obtain the second, this one will have an optimum solution value of zero. Consequently, all ϵ-approximations of this second problem would give an optimal solution, which is possible to implement in polynomial time only if $P = NP$. However, the problems generated at random (flows and distances drawn uniformly) satisfy the following property: when $n \to \infty$, the value of any solution (even the worst one) tends towards the value of an optimal solution [Burkard and Fincke, 1985].

Example.

Let us consider the placement of 12 electronic modules $(1, \ldots, 12)$ on 12 sites $(a, b, \ldots, l)$. The number of connections required between the modules is known and given in table 2.1. This problem instance is referred as SCR12 in literature.

Table 2.1. Number of connections between the modules.

Module	1	2	3	4	5	6	7	8	9	10	11	12
1	—	180	120	—	—	—	—	—	—	104	112	—
2	180	—	96	2445	78	—	1395	—	120	135	—	—
3	120	96	—	—	—	221	—	—	315	390	—	—
4	—	2445	—	—	108	570	750	—	234	—	—	140
5	—	78	—	108	—	—	225	135	—	156	—	—
6	—	—	221	570	—	—	615	—	—	—	—	45
7	—	1395	—	750	225	615	—	2400	—	187	—	—
8	—	—	—	—	135	—	2400	—	—	—	—	—
9	—	120	315	234	—	—	—	—	—	—	—	—
10	104	135	390	—	156	—	187	—	—	—	36	1200
11	112	—	—	—	—	—	—	—	—	36	—	225
12	—	—	—	140	—	45	—	—	—	1200	225	—

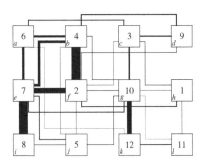

Fig. 2.1. Optimum solution of a problem of connection between electronic modules. The thickness of the lines is proportional to the number of connections.

The sites are distributed on a 3×4 rectangle. Connections can be implemented only horizontally or vertically, implying wiring lengths measured with Manhattan distances. In the solution of the problem presented in figure 2.1, which is optimal, module 6 was placed on the site a, module 4 on the site b, etc.

2.3 Basic tabu search

Thereafter and without being restrictive, one can make the assumption that the problem to be solved can be formulated in the following way:

$$\min_{s \in S} f(s)$$

In this formulation, f denotes the objective function, s a feasible solution of the problem and S the entire set of feasible solutions.

2.3.1 Neighborhood

Tabu search is primarily centered on a non-trivial exploration of all the solutions in a neighborhood. Formally, one can define, for any solution s of S, a set $N(s) \subset S$ that is a set of the neighboring solutions of s. For example, for the quadratic assignment problem, s can be a permutation of n objects and the set $N(s)$ can be the possible solutions obtained by exchanging two objects in a permutation. The figure 2.2 illustrates one of the moves of the set $N(s)$, where objects 3 and 7 are exchanged.

1 2 3 4 5 6 9 7 8 $\longrightarrow$ 1 2 7 4 5 6 9 3 8

Fig. 2.2. A possibility of creating a neighboring solution, for a problem where a permutation is sought.

Local search methods are almost as old as the world itself. As a matter of fact, how will a person behave when someone supplies him with a solution of a problem for which he himself could not find a solution or did not have enough patience to find an optimal solution? The person may try to modify the proposed solution slightly and may check to find out whether it is possible to find better solutions while carrying out these local changes. In other words, it will stop as soon as it meets a *local optimum* related to the modifications authorized to make on a solution. In this process, it never indicates whether the solution thus obtained is a *global optimum* — and, in practice, this is seldom the case. In order to be able to find solutions better than the first local optimum met, one can try to continue the process of local modifications. However, if precautions are not taken, one may be exposed to visit a restricted number of solutions, in a cyclic manner . Simulated annealing and tabu search are two local search techniques which try to eliminate this disadvantage.

Some of these methods, like simulated annealing, were classified as artificial intelligence techniques. However, this classification is certainly abusive as they are guided almost exclusively by a chance — some authors even compare simulated annealing to the peregrination of a person, suffering from amnesia, moving in the fog. Perhaps others describe these methods as intelligent because, often after a large number of iterations during which they have enumerated several poor quality solutions, they produce a good quality solution which would have otherwise required a remarkably extensive human effort.

In essence, tabu search is not centered on chance, although one can introduce random components for primarily technical reasons. The basic idea of tabu search is to make use of memories during the exploration of a part of the solutions of the problem, which consists in moving repeatedly from one solution to a neighboring solution. It is thus primarily a local search, if we look beyond the limited meaning of this term and dig for a broader meaning. However, some principles enabling to carry out jumps in the solution space were proposed; in this direction, tabu search, contrary to simulated annealing, is not a pure local search.

Origin of tabu search *name.*

The origin of the name of this method can be traced back to the idea that a local search can continue beyond a local optimum, but to ensure that it will not return periodically to the same local optimum, it is necessary to prohibit the search to come back to solutions already visited, in other words, some solutions must have a tabu status.

2.3.2 Moves, neighborhood

Local searches are based on the definition of a set $N(s)$ of solutions in the neighborhood of any solution s. But, from a practical point of view, one may find it beneficial to consider the set of the modifications which one can make to s, rather than the set $N(s)$. A modification made to a solution is called a *move*. Thus, the modification of a solution of the quadratic assignment problem (see the figure 2.2) can be considered as a move, characterized by the two elements to be transposed in the permutation. The figure 2.3 gives the structure of the neighborhood based on the exchanges for the set of the permutations of 4 elements, and this is presented under the form of a graph whose vertices represent the solutions and the edges are the neighbors relative to the transpositions.

The set $N(s)$ of the solutions in the neighborhood of the solution s can be expressed as the set of the feasible solutions which one can obtain while applying a move m to the solution s, m pertaining to a set of moves M. The application of m to s can be noted as $s \oplus m$ and one can determine the equivalent $N(s) = \{s'| = s \oplus m, m \in M\}$. Expressing the neighborhood in terms of move facilitates, when it is possible, to characterize the set M more easily. Thus, in the above example of modification of a permutation, M will be characterized by all pairs (place 1, place 2), of which one transposes

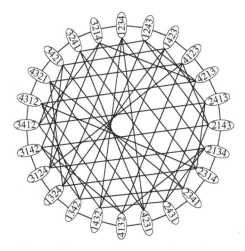

Fig. 2.3. Set of the permutations of 4 elements (represented by vertices) with relations of neighborhood relative to the transpositions (represented by edges).

the elements, independent of the current solution. It should be noted that in the case of permutations with transpositions neighborhood, $|S| = n!$ and $|M| = \frac{n \cdot (n-1)}{2}$. Thus, the entire set of the solutions is much larger than that of the moves, which varies as the square of the number of elements.

However, in some applications this simplification can lead to that definition of moves which would produce non-feasible solutions and, in general, we have $|N(s)| \leq |M|$, without $|M|$ being much larger than $|N(s)|$. For a given problem with few constraints, it can typically be $|N(s)| = |M|$.

Examples of neighborhoods for problems on permutations.

Many combinatorial optimization problems can be naturally formulated as a search for a permutation of n elements. Assignment problems (which include that of the quadratic assignment), the traveling salesman problem or task scheduling problems are representative examples of such problems. For these problems, several definitions of neighboring solutions are possible; some examples are illustrated in figure 2.4. Among the simplest neighborhoods, one can find the inversion of two elements placed successively in the permutation, the transposition of two distinct elements and finally, the move of an element to another place in the permutation. Depending on the problem considered, more elaborate neighborhoods that suit to the structure of good solutions should be considered. This is typically the case for the traveling salesman problem where there are innumerable neighborhoods suggested that do not represent simple operations, if a solution is regarded as a permutation.

The first type of neighborhood, shown in the example above, is the most limited one as it is of size $n-1$. The second type defines a neighborhood taking

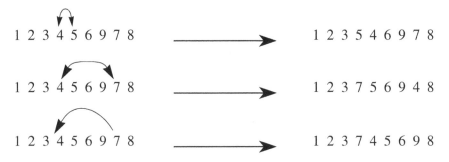

1 2 3 4 5 6 9 7 8   ⟶   1 2 3 5 4 6 9 7 8

1 2 3 4 5 6 9 7 8   ⟶   1 2 3 7 5 6 9 4 8

1 2 3 4 5 6 9 7 8   ⟶   1 2 3 7 4 5 6 9 8

Fig. 2.4. Three possible neighborhoods on permutations (inversion, transposition, displacement).

$\frac{n \cdot (n-1)}{2}$ moves into consideration and the third one is of size $n(n-2)+1$. The capabilities of these various types of neighborhoods to direct a search in few iterations towards good solutions are very different; generally, the first type shows the worst behavior for many problems. The second one can be better than the third one for some problems (like the quadratic assignment problem), whereas, for scheduling applications, the third type often shows better performance [Taillard, 1990].

2.3.3 Evaluation of the neighborhood

In order to implement an effective local search engine, it is necessary that the relationship between the quality or the suitability of the type of moves and the computing time necessary for their evaluation is as healthy as possible. If the quality of a type of move can be justified only by the intuition and in an empirical manner, the evaluation of the neighborhood can, on the other hand, often be accelerated considerably by algebraic considerations. Let us define $\Delta(s, m) = f(s \oplus m) - f(s)$. In many cases it is possible to simplify the expression $f(s \oplus m) - f(s)$ and thus to evaluate $\Delta(s, m)$ quickly. An analogy can be drawn with continuous optimization: the numerical evaluation of $f(s \oplus m) - f(s)$ would be the equivalent to a numerical evaluation of the gradient, whereas the calculation of the simplified function $\Delta(s, m)$ would be the equivalent of the evaluation of the gradient by means of a function implemented with the algebraic expressions of the partial derivatives.

Moreover, if the move m' was applied to the solution s in the preceding iteration, it is often possible to evaluate $\Delta(s \oplus m', m)$ for the current iteration as a function of $\Delta(s, m)$ (which was evaluated in the preceding iteration) and to examine very rapidly the entire neighborhood, simply by memorizing the values of $\Delta(s, m)$.

It may appear that the evaluation of $\Delta(s, m)$ is very difficult and expensive to undertake. For example, for vehicle routing problems (VRP, see section 7.1), a solution s can consist of partitioning goods into subsets whose weights are not more than the capacities of the vehicles. To calculate $f(s)$, we have to find an optimal order in which one will deliver the goods for each subset, which is a difficult problem in itself. This is the well known traveling salesman problem. Therefore, the calculation of $f(s)$, and consequently that of $\Delta(s,m)$, cannot be reasonably performed for any eligible move (i.e. pertaining to M); possibly $\Delta(s,m)$ should be calculated for each move elected (that is in fact carried out) but, in practice, one will have to be satisfied by calculating the true value of $f(s)$ for a limited number of solutions. Hence the computational complexity is limited by evaluating $\Delta(s,m)$ in an approximate manner.

Algebraic example of simplification for the quadratic assignment problem.

As any permutation is a feasible solution for the quadratic assignment problem, its modeling is also trivial. For the choice of the neighborhood, it should be realized that to move the element in the ith position in the permutation to put it in the jth position implies a very significant modification of the solution. This is because all the elements between the ith and the j^{ith} position are moved. The inversion of the objects in the ith and the $i + 1^{\text{th}}$ position in the permutation generates a too restricted neighborhood. Actually, if the objective is to limit ourselves to the neighborhoods modifying the sites assigned to two elements only, it is only reasonable to transpose the elements i and j occupying the sites p_i and p_j. Each of these moves can be evaluated in $O(n)$ (where n is the size of the problem). With flows matrix $\mathcal{F} = (f_{ij})$ and distances matrix $\mathcal{D} = (d_{rs})$, the value of a move $m = (i, j)$ for a solution $\mathbf{p}$ is given by:

$$\Delta(\mathbf{p}, (i,j)) = (f_{ii} - f_{jj})(d_{p_j p_j} - d_{p_i p_i}) + (f_{ij} - f_{ji})(d_{p_j p_i} - d_{p_i p_j})$$
$$+ \sum_{k \neq i,j} (f_{jk} - f_{ik})(d_{p_i p_k} - d_{p_j p_k}) + (f_{kj} - f_{ki})(d_{p_k p_i} - d_{p_k p_j}) \tag{2.1}$$

By memorizing the value of $\Delta(\mathbf{p}, (i,j))$, one can calculate the value of $\Delta(\mathbf{q}, (i,j))$ in $O(1)$, for all $i \neq r, s$ and $j \neq r, s$ by using equation 2.2, where $\mathbf{q}$ is the permutation obtained by exchanging the elements r and s in $\mathbf{p}$, i.e. $q_k = p_k$, $(k \neq r, k \neq s)$, $q_r = p_s$, $q_s = p_r$.

$$\Delta(\mathbf{q}, (i,j)) = \Delta(\mathbf{p}, (i,j))$$
$$+(f_{ri} - f_{rj} + f_{sj} - f_{si})(d_{q_s q_i} - d_{q_s q_j} + d_{q_r q_j} - d_{q_r q_i}) \tag{2.2}$$
$$+(f_{ir} - f_{jr} + f_{js} - f_{is})(d_{q_i q_s} - d_{q_j q_s} + d_{q_j q_r} - d_{q_i q_r})$$

The figure 2.5 illustrates the modifications that are necessary for $\Delta(\mathbf{p}, (i,j))$ to obtain $\Delta(\mathbf{q}, (i,j))$, if the move selected for going from $\mathbf{p}$ to $\mathbf{q}$ is (r, s). It should be noted here that the quadratic assignment problem can be regarded as that of the permutation of the lines and the columns of the distances matrix, so that the scalar product of the two matrices is as small as possible.

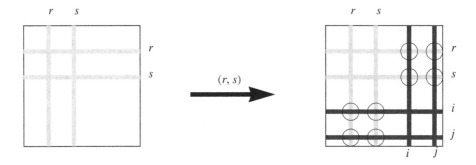

Fig. 2.5. On the left: in gray, the elements for which it is necessary to recalculate the scalar product of the matrices to evaluate the move (r, s) applied to **p** (leading to the solution **q**); on the right: those elements are surrounded for which it is necessary to recalculate the product to evaluate the move (i, j) applied to **q**, compared to those which were calculated if the move (i, j) had been applied to **p**.

Consequently, while memorizing the $\Delta(\mathbf{p}, (i, j))$ values for all i and j, the entire neighborhood can be calculated in $O(n^2)$: by using equation 2.2 one can evaluate the $O(n^2)$ moves that do not utilize the indices r and s and by using equation 2.1, one can evaluate the $O(n)$ moves which precisely utilize these indices.

2.4 Candidate list

Generally, a local search does not necessarily evaluate all the solutions of $N(s)$ in each iteration, but only a subset of it. In fact, simulated annealing is even considering only one neighbor. On the contrary, tabu search is supposed to make an "intelligent" choice of a solution from $N(s)$. One possible way of accelerating the evaluation of the neighborhood can be to reduce its size; this reduction may also have the goal of directing the search.

To reduce the number of eligible solutions of $N(s)$, some authors adopt the policy of randomly choosing a small number of solutions from $N(s)$. When a neighborhood is given by a static set M of moves, one can also consider partitioning M into subsets; in each iteration, only one of these subsets will be examined. In this manner, one can make a partial but cyclic examination of the neighborhood, which will allow to elect a move more quickly, with an associated deterioration in quality since all the moves are not taken into consideration in each iteration. However, on a global level, this limitation might not have a too bad influence on the quality of the solutions produced, because a partial examination can generate a certain diversity in the visited solutions, precisely because moves which were elected were not those which would have been, if a complete examination of the neighborhood were carried out.

Finally, in accordance with the intuition of F. Glover when he proposed the concept of candidate list, one can make the assumption that a good quality move for a solution will remain good for solutions not too different. Practically, this can be implemented by ordering, in a given iteration, the entire set of all feasible moves by decreasing quality. During the later iterations, only those moves that were classified among the best will be considered. This is implemented in form of a data structure called *candidate list*. Naturally, the order of the moves will get degraded during the search, since the solutions become increasingly different from the solution used to build the list, and it is necessary to periodically evaluate the entire neighborhood to preserve a suitable candidate list.

2.5 Short-term memory

When one wishes to make use of memory in an iterative process, the first idea that comes to mind is to check if a solution in the neighborhood was already visited. However, the practical implementation of this idea can be difficult and even worse, it may prove to be not very effective. Indeed, it requires to memorize each solution visited and to test in each iteration for each eligible solution, if the later were already enumerated. This can possibly be done efficiently by using hashing tables, but it is not possible to prevent a significant growth in memory space requirement as it increases linearly with the number of iterations.

Moreover, the pure and simple prohibition of solutions can lead to absurd situations: let us assume that the entire set of the feasible solutions can be represented by points whose co-ordinates are given on a surface in the plane and that one can move from any feasible solution to any other by a number of displacements of unit length. In this case, one can easily find trajectories which disconnect the current solution from an optimal solution or which block an iterative search due to lack of feasible neighboring solutions, if it is tabu to pass by an already visited solution. This situation is schematically illustrated in the figure 2.6.

2.5.1 Hashing tables

A first idea, which is very easy to implement to direct an iterative search, is to prohibit the return to a solution whose value was already obtained during t last iterations. Thus one can prevent a cycle of length t or less. This type of tabu condition can be implemented in an effective manner: Let L be a whole number, relatively large, so that it is possible to memorize an array of L entries in the computer used.

If $f(s_k)$, the value of solution s_k in iteration k, is assumed to be an integer, that is not restrictive (computationaly speaking), one can memorize in

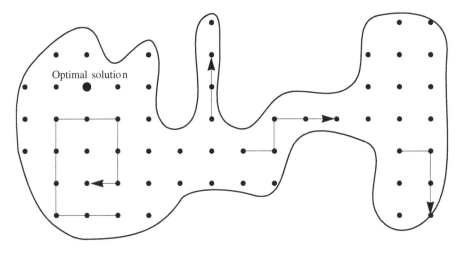

Fig. 2.6. Trajectories blocking search or disconnecting it from the optimal solution in a strict tabu search.

$T[f(s_k)$ modulo $L]$ the value $k + t$. If a solution s' of the potential neighborhood of the solution in the iteration k' is such that $T[f(s')$ modulo $L)] > k'$, s' will not be considered any more as an eligible solution. This effective manner of storing the tabu solutions only approximates the initial intention which was to prevent to return to a solution of a given value, as not only any solution of a given value is prohibited during t iterations, but also all those which have this value modulo L. Nevertheless, only a very weak modification of the search behavior can be noticed in practice, if L is selected sufficiently large.

This type of tabu condition works only if the objective function has a vast span of values. However, there are many problems where the objective function takes a limited span of values. One can circumvent the difficulty by associating, in place of the objective, another function taking a large span of values. In the case of a problem on permutations, one can associate, for example, the hashing function:$\sum_{i=1}^{n} i^2 \cdot p_i$ which takes a number of different values in $O(n^4)$.

More generally, if a solution of a problem can be expressed under the form of a vector $\mathbf{x}$ of binary variables, one can associate the hashing function $\sum_{i=1}^{n} z_i \cdot x_i$ with z_i, a set of n numbers randomly generated at the beginning of the search [Woodruff and Zemel, 1993].

When hashing functions are used for implementing tabu conditions, one needs to focus on three points. Firstly, as already mentioned, it is necessary that the function used take a vast span of possible values. Secondly, the evaluation of the hashing function for a neighboring solution should not impose significantly higher computational burden than the evaluation of the objective. In the case of problems on permutations with neighborhood based on transpositions, the functions mentioned above can be evaluated in constant time for each neighboring solution, if the value of the hashing function for the starting solution is known. Thirdly, it should be noticed that even with a very large hashing table, the collisions (different solutions with identical hashing value) are frequent. Thus, for a problem on permutations of size $n = 100$, with the transpositions neighborhood, approximately 5 solutions in the neighborhood of the solution in the second iteration will enter in collision with the starting solution, if a table of 10^6 elements is used. To reduce the risk of collision effectively, one may find it beneficial to use several hashing functions and several tables simultaneously [Taillard, 1995].

2.5.2 Tabu list

As it can be ineffective to restrict the neighborhood $N(s)$ to those solutions which are not yet visited, tabu conditions are rather based on M, the set of moves applicable for a solution. This set is often of relatively modest size (typically of size $O(n)$ or $O(n^2)$ if n is the size of the problem) and must have the characteristic of *connectivity*, i.e. an optimal solution can be reached from any feasible solution. Initially, to simplify, we assume that M also has the property of *reversibility*: for any move m applicable to a solution s, there must exist a move m^{-1} such that $(s \oplus m) \oplus m^{-1} = s$. As it does not make sense to apply m^{-1} immediately after applying m, it is possible, in all cases, to limit all the moves applicable to $s \oplus m$ to those different from m^{-1}. Moreover, in this process one could avoid visiting s and $s \oplus m$ repeatedly, if s would be a local optimum relative to the neighborhood selected and where the best neighbor of $s \oplus m$ would be precisely s.

By generalizing this technique of limiting the neighborhood, i.e. by prohibiting during several iterations to apply the reverse of a move which has just been made, one can prevent other cycles composed of a number of intermediate solutions. Once it is again possible to carry out the reverse of a move, one hopes that the solution was sufficiently modified, so that it is improbable — but not impossible — to return to an already visited solution. Nevertheless, if such a situation arises, it is hoped that the list of tabu moves (or *tabu list*) would have changed, therefore the future trajectory of the search would change. The number of tabu moves must remain limited enough. Let us assume that M does not depend on the current solution. In this situation, it is reasonable to prohibit only a fraction of the M moves. Thus, the tabu list implements a short-term memory, relating typically to a few units or a few tens of iterations.

For easier understanding, we had assumed that the reverse moves to those that were carried out are stored. However, it is not always possible or obvious to define what a reverse move is. Let us take the example of a problem where the objective is to find an optimal permutation of n elements. A reasonable move can be to transpose the elements i and j of the permutation ($1 \leq i < j \leq n$). In this case, all the M moves applicable to an unspecified solution are given by the entire set of the pairs (i, j). But, thereafter if the move (i, j)

is carried out, the prohibition of (i, j) (which is its own reverse) will prevent visiting certain solutions without preventing the phenomena of cycling: indeed, the moves $(i, j)(k, p)(i, p)(k, j)(k, i)(j, p)$ applied successively do not modify the solution. One thus should not inevitably avoid making a reverse move too quickly, but attribute tabu conditions to certain attributes of moves or solutions. In the preceding example, if p_i is the assignment of the element i in an iteration, it is not the move (i, j) which should be prohibited when one has just carried it out, but it is, for example, simultaneously assigning the element i at the place p_i and the element j at the place p_j that should be prohibited. One can thus at least preserve those cycles which are of lengths smaller or equal to the number of tabu moves, i.e. the tabu list length.

2.5.3 Duration of tabu conditions

Generally speaking, the short-term memory will prohibit to perform some moves, either directly by storing tabu moves or tabu solutions, or indirectly by storing attributes of moves or attributes of solutions that are prohibited. If the minimization problem can be represented as a landscape limited by a territory which may define the feasible solutions and where altitude may correspond to the value of the objective function, the effect of this memory is to visit valleys, (without always being at the bottom of the valley because of tabu moves) and, sometimes, to cross a pass issuing into another valley.

The higher the number of tabu moves is, the more likely one is to cross these mountains, but the less thoroughly will the valleys be visited. Conversely, if the moves are prohibited during few iterations only, there will be less chances of crossing the passes surrounding the valleys because, almost surely, there will be an allowed move which will lead to a solution close to the bottom of the valley; but, on the other hand, the bottom of the first visited valley will most probably be found.

More formally, for a very small number of tabu moves, the iterative search will tend to visit the same solutions over and over again. If this number is increased, the probability of remaining confined to a very limited number of solutions decreases and, consequently, the probability of visiting several good solutions increases. However, the number of tabu moves should not be very large, because, it then becomes less probable to find good local optima, for lack of available moves. To some extent, the search is directed by the few allowed moves rather than by the objective function.

The figure 2.7 illustrates these phenomena in the case of the quadratic assignment problem: for each of 3000 instances of size 12, drawn at random, 50 iterations of a tabu search were performed. According to the number of iterations during which a reverse move is prohibited, this figure gives the two following statistics: firstly, the average value of all the solutions visited during search and, secondly, the average value of the best solutions found by each search. It should be noted that the first statistics grows with the number of prohibited moves, which means that the average quality of the visited solutions degrades. On the other hand, the quality of the best found solutions improves with the increase in the number of tabu moves, which establishes the fact that the search succeeds to escape from more or less poor local optima; then, their quality worsens, but this tendency is very limited here. Thus, it can be concluded that the size of the tabu list must be chosen carefully, in accordance with the problem under consideration, the size of the neighborhood, the problem instance, the total number of iterations performed, etc. It is relatively easy to determine the order of magnitude that should be assigned to the number of tabu iterations, but the optimal value cannot be obtained without testing all the possible values.

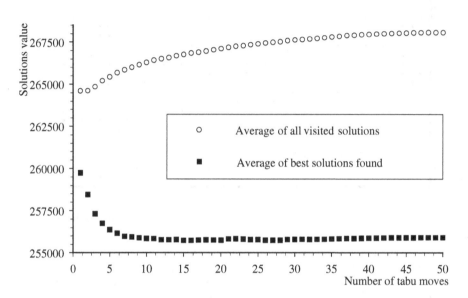

Fig. 2.7. The influence of number of iterations during which the moves are tabu.

Random tabu duration

To obtain simultaneous benefits from the advantages of a small — for several deepened visits of a valley — and a large number of tabu moves — for the ability to escape valleys —, one may find it useful to modify this number during the search process. Several methodologies can be considered for this choice: for example, it can be decided at random between a lower and an upper limit, in each iteration or after a certain number of iterations. These limits can often be easily identified; it can also be increased or decreased on the basis of certain characteristics observed during the search, etc. These were the various strategies employed till the end of the eighties [Taillard, 1990, Taillard, 1991, Taillard, 1995]. These strategies were shown to be much more efficient than tabu lists of fixed size (often implemented under the form of a circular list, although this may not be the best alternative, as it can be seen in the paragraph 2.7.2).

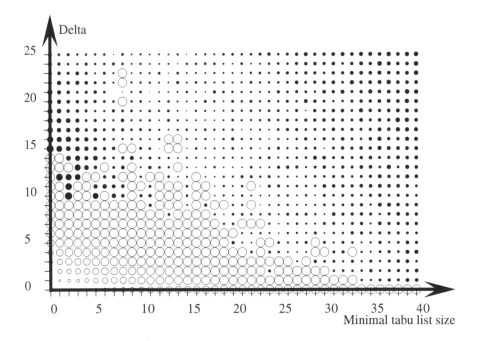

Fig. 2.8. The effect of random selection of the number of iterations during which the moves are prohibited (for quadratic assignment problem instances of size 15 drawn at random). The number of iterations during which the reverse of a move is prohibited is drawn at random, uniformly between a minimum value and this value increased by *Delta*. The size of the discs grows with the average number of iterations necessary for the resolution of the problems until the optimum is found. A circle indicates that a cycling phenomenon has appeared. The size of the circles is proportional to the number of problem instances solved optimally.

Still for the quadratic assignment problem, the figure 2.8 gives the average number of iterations necessary for the resolution of 500 examples of problems of size 15, generated at random, when the technique is to choose the number of tabu moves at random between a minimal value and this value increased by *Delta*. The surface of the black discs depends on the average number of iterations necessary to obtain the optimal solutions for the 500 problems. An empty circle indicates that at least one of the problems was not solved optimally. The surface of these circles is proportional to the number of problems for which it was possible to find the optimum. For *Delta* = 0, i.e. when the number of tabu moves is constant, cycles appear even for relatively large sizes of tabu lists. On the other hand, the introduction of a positive *Delta*, even a very small one, can ensure much more protection against cycling. As it can be noted in figure 2.7, the lower the tabu list size is, the smaller is the average number of iterations required to obtain the optimum. However, below a certain threshold, cycles appear, without passing through the optimum. From the robustness point of view, one is thus constrained to choose sizes of tabu lists slightly larger than the optimal value (for this size of instances it seems that it should be [7, 28] (minimum size = 7, *Delta* = 21), but it is noticed that for [8, 28] a cycle appeared).

This technique of randomly selecting the number of tabu moves can thus direct the search automatically towards good solutions. Let us note that such a mechanism could be described as myopic because it is directed mainly by the value of the objective function. Although it provides very encouraging results considering its simplicity, it cannot be considered as an intelligent way of directing the search, but must rather be viewed as a basic tool for implementing the search process.

Type of tabu list for the QAP

A solution for the quadratic assignment problem can be represented under the form of a permutation $\mathbf{p}$ of n elements. A type of move very frequently used for this problem is to transpose the positions of two objects i and j. Indeed, it is possible to evaluate effectively, in $O(n^2)$, the entire set of moves applicable for a solution.

As it was discussed earlier, a technique for directing search at short-term is to prohibit, during t iterations, the application of the reverse of the moves which have just been carried out. If the move (i, j) is applied to the permutation $\mathbf{p}$, the reverse of a move can be defined as a move which simultaneously places the object i on the site p_i and the object j on the site p_j. There are other possible definitions of the reverse of a move, but this is one of the most effective ones to prevent cycles and appears to be the least sensitive one to the value of the parameter t, the number of iterations during which one avoids applying the reversal of a move. A fixed value of t does not produce a robust search, because cycles may appear (see figure 2.8) even for large values of t. To overcome this problem, it is proposed in [Taillard, 1991] that t should be uniformly drawn at random, between $\lfloor 0, 9 \cdot n \rfloor$ and $\lceil 1, 1 \cdot n + 4 \rceil$. Indeed, experiments have shown that a tabu duration equal to the size of the problem, or slightly larger for small examples, seems rather effective. This paved the way for the idea to select the value of t in a dynamic manner during the course of search, by choosing an average value slightly higher than the value which would have been ideal in the static case. The program described in appendix B was one of the earliest ones to propose this technique.

To implement this tabu mechanism in practice, a matrix $\mathcal{T}$ will be used whose t_{ir} entry will give the iteration number during which the element i was moved the last time from the site r (to go to the site p_i), number to which one adds the tabu duration t. Thus, the move (i, j) will be prohibited if both entries t_{ip_j} and t_{jp_i} contain values higher than the current iteration number (see appendix B).

Let us consider the small 5×5 instance of quadratic assignment problem, known in literature as NUG5, with flows matrix $\mathcal{F}$ and distances matrix $\mathcal{D}$:

$$\mathcal{F} = \begin{pmatrix} 0 & 5 & 2 & 4 & 1 \\ 5 & 0 & 3 & 0 & 2 \\ 2 & 3 & 0 & 0 & 0 \\ 4 & 0 & 0 & 0 & 5 \\ 1 & 2 & 0 & 5 & 0 \end{pmatrix}, \quad \mathcal{D} = \begin{pmatrix} 0 & 1 & 1 & 2 & 3 \\ 1 & 0 & 2 & 1 & 2 \\ 1 & 2 & 0 & 1 & 2 \\ 2 & 1 & 1 & 0 & 1 \\ 3 & 2 & 2 & 1 & 0 \end{pmatrix}$$

Iteration 0. On the basis of the initial solution $\mathbf{p} = (2, 4, 1, 5, 3)$ of cost 72, the following evolution of tabu search will take place. The search can be started by initializing the matrix $\mathcal{T} = \mathbf{0}$.

Iteration 1. Then, the value $\Delta(\mathbf{p}, (i, j))$ is calculated for each transposition (i, j):

move	$(1, 2)$	$(1, 3)$	$(1, 4)$	$(1, 5)$	$(2, 3)$	$(2, 4)$	$(2, 5)$	$(3, 4)$	$(3, 5)$	$(4, 5)$
cost	2	-12	-12	2	0	-10	-12	4	8	6

It can be realized that three moves produce a maximum profit of 12: to exchange the objects on the sites $(1, 3)$, $(1, 4)$ and $(2, 5)$. One can assume that it is the first of these moves, $(1, 3)$, which is retained. If it is prohibited during $t = 9$ iterations (i.e. until the iteration 10) to put element 2 on site 1 and element 1 on site 3 simultaneously, the following matrix is obtained:

$$\mathcal{T} = \begin{pmatrix} 0 & 0 & 10 & 0 & 0 \\ 10 & 0 & 0 & 0 & 0 \\ 0 & 0 & 0 & 0 & 0 \\ 0 & 0 & 0 & 0 & 0 \\ 0 & 0 & 0 & 0 & 0 \end{pmatrix}$$

Iteration 2. The move chosen during iteration 1 leads to the solution $\mathbf{p} = (1, 4, 2, 5, 3)$ of cost 60. Calculating the value of each transposition one can obtain:

move	$(1, 2)$	$(1, 3)$	$(1, 4)$	$(1, 5)$	$(2, 3)$	$(2, 4)$	$(2, 5)$	$(3, 4)$	$(3, 5)$	$(4, 5)$
cost	14	12	-8	10	0	10	8	12	12	6
tabu		yes								

For this iteration, it should be noticed that the reverse of the preceding move is now prohibited. The authorized move $(1, 4)$, giving minimum cost, is retained, for a profit of 8. If the randomly selected tabu duration of the reverse move is $t = 6$, the matrix $\mathcal{T}$ becomes:

$$T = \begin{pmatrix} 8\,0\ 10\,0\ 0 \\ 10\,0\ \ 0\,0\,0 \\ 0\,0\ \ 0\,0\,0 \\ 0\,0\ \ 0\,0\,0 \\ 0\,0\ \ 0\,8\,0 \end{pmatrix}$$

Iteration 3. The solution $\mathbf{p} = (5, 4, 2, 1, 3)$ of cost 52 is reached which is a local optimum. Indeed, at the beginning of iteration 3, no move has a negative cost:

move	$(1,2)$	$(1,3)$	$(1,4)$	$(1,5)$	$(2,3)$	$(2,4)$	$(2,5)$	$(3,4)$	$(3,5)$	$(4,5)$
cost	10	24	8	10	0	22	20	8	8	14
tabu		*yes*								

The selected move $(2, 3)$ in this iteration has a zero cost. It should be noticed here that the move $(1, 3)$, which was prohibited during iteration 2 is again authorized, since the element 5 was never in third position. If the random selection of the tabu duration results in $t = 8$, the following matrix is obtained:

$$T = \begin{pmatrix} 8\ \ 0\ 10\,0\,0 \\ 10\ \ 0\ 11\,0\,0 \\ 0\ \ 0\ \ 0\,0\,0 \\ 0\ 11\ \ 0\,0\,0 \\ 0\ \ 0\ \ 0\,8\,0 \end{pmatrix}$$

Iteration 4. One can then obtain a solution $\mathbf{p} = (5, 2, 4, 1, 3)$ of cost 52 and the data structures situation will be as follows:

move	$(1,2)$	$(1,3)$	$(1,4)$	$(1,5)$	$(2,3)$	$(2,4)$	$(2,5)$	$(3,4)$	$(3,5)$	$(4,5)$
cost	24	10	8	10	0	8	8	22	20	14
tabu		*yes*		*yes*						

However, it is not possible any more to choose the move $(2, 3)$ corresponding to the minimum cost, which could bring us back to the preceding solution, because this move is prohibited. Similar situation arises for the move $(1, 4)$ which would replace the element 5 in position 1 and the element 1 in fourth position, although this exchange leads to a solution not yet visited. Hence we are forced to choose an unfavorable move $(2, 4)$, that increases the cost of the solution by 8. With a selected tabu duration of $t = 5$, one can obtain:

$$T = \begin{pmatrix} 8\ \ 0\ 10\ 9\ 0 \\ 10\ \ 9\ 11\,0\,0 \\ 0\ \ 0\ \ 0\,0\,0 \\ 0\ 11\ \ 0\,0\,0 \\ 0\ \ 0\ \ 0\,8\,0 \end{pmatrix}$$

Iteration 5. The solution at the beginning of this iteration is: $\mathbf{p} = (5, 1, 4, 2, 3)$. The calculation of the cost of the moves gives:

move	$(1,2)$	$(1,3)$	$(1,4)$	$(1,5)$	$(2,3)$	$(2,4)$	$(2,5)$	$(3,4)$	$(3,5)$	$(4,5)$
cost	12	−10	12	10	0	−8	4	14	20	10
tabu			*yes*		*yes*	*yes*				

It is noticed that the move degrading the quality of the solution at the preceding iteration was beneficial, because it now facilitates to arrive at an optimal solution $\mathbf{p} = (4, 1, 5, 2, 3)$ of cost 50, by choosing the move $(1, 3)$.

2.5.4 Aspiration conditions

Sometimes, some tabu conditions are absurd. For example, a move which leads to a solution better than all those visited by the search in the preceding iterations does not have any reason to be prohibited. In order to not to miss this solution, it is required to disregard the possible tabu status of such moves. According to the tabu search terminology, this move is called *aspired*. Naturally, it is possible to assume other aspiration criteria, less directly related to the value of the objective to be optimized.

It should be noted here that the first presentations on tabu search insisted heavily on the aspiration conditions, but, in practice, these were finally limited to authorize a tabu move which helped to improve the best solution found so far during the search. As this last criterion became implicit, little research were carried out later in defining more elaborate aspiration conditions. On the other hand, aspiration can also be sometimes described as a form of long-term memory, consisting in forcing a move never carried out over several iterations, irrespective of its influence on the objective function.

2.6 Convergence of tabu search

Formally, one cannot speak about "convergence" for a tabu search, since in each iteration the solution is modified. On the other hand, it is definitely interesting to pass at least once through a global optimum. This was the focus of discussion in [Hanafi, 2001], on the theoretical level, using an elementary tabu search. It was shown that the search could be blocked if one prohibits passing through the same solution twice. Consequently, it is necessary to enable one to revisit the same solution. By considering a search which memorizes all the solutions visited and which chooses, if all the neighboring solutions were already visited, the oldest that was visited, it can be shown that all the solutions of the problem will be enumerated. This is valid if the set of solutions is finite, if the neighborhood is either reversible (or symmetric: any neighboring solution to s has s in its neighborhood) or strongly connected (there is a succession of moves enabling to reach any solution s' starting from any solution s). Here, all the solutions visited must be memorized (possibly in an implicit form) and it should be understood that this result remains theoretical.

There is another theoretical result on the convergence of tabu search presented in [Faigle and Kern, 1992]. These authors have considered probabilistic tabu conditions. It is then possible to choose probabilities such that the search process is similar to that of a simulated annealing. Starting from this observation, it can be well assumed that the convergence theorems for simulated annealing can be easily adapted for a process called *probabilistic tabu search*. Again, it should be understood that the interest of this result remains of purely theoretical nature.

Conversely, it can be shown that tabu search in its simplest form can simultaneously appear to be effective in practice but unable to pass through a global optimum for some badly conditioned instances. To show this property, it is important to work with a powerful method in practice. It is obviously uninteresting to show that a method that does not work for a given problem could never pass through a global optimum.

To prove that, one can define an elementary tabu search for the vehicle routing problem (VRP) which enables it to obtain, if its single parameter is well chosen, the optimal solution for a problem instance that is at the limits of possibilities of exact methods. Then, one can show that this iterative search, irrespective of the value of its single parameter, never passes through the global optimum of a small badly conditioned instance.

To define an iterative search, firstly one has to model the problem and the moves applicable to a solution. In the case of the VRP, one can authorize those solutions where all the goods are not delivered, but attributing a penalty of $2d$ for a good not delivered which is at a distance d from the depot (as if one urged for a vehicle only to deliver it). The maximum number of vehicles engaged is limited by a value m (which can be easily determined). A move consists in removing a good i ($i = 1, \ldots, n$) currently served by the tour T_j ($j = 0, \ldots, m$, the tour T_0 represents the non-served goods) and to deliver it during the tour T_k ($k = 0, \ldots, m, k \neq j$). When a good i is withdrawn from T_j, it is directly passed from the place of delivery of the good preceding i in T_j to that succeeding i in the T_j. In tour T_k, a good i is added at the position which causes the shortest detour. Then, it is necessary to define how the search is directed. One possibility may consist in prohibiting the reverse of those moves which were just carried out during t iterations. If the good i of the tour T_j was removed to be inserted in the tour T_k, then it will be prohibited to put the good i back in the tour T_j during the next t iterations (unless that leads to a solution better than the best one found until now). The search is initialized with a (very bad) solution consisting of no delivery at all (all the goods are placed in the dummy tour T_0).

In spite of its simplicity, this procedure can quickly find good quality solutions for classical instances from the literature when the parameter t is appropriately selected. A small poorly conditioned instance is as follows: let us consider a maximum of 2 vehicles of capacity 24 and 11 goods to deliver with the respective volumes of 3, 8, 8, 8, 3, 3, 3, 3, 3, 3, 3 in only one place, at a distance d from the depot. The optimal solution for this problem consists in delivering all the goods of volume 3 by a vehicle and those of volume 8 by the other. It is observed that the tabu search described earlier, for any value of the parameter t, either returns periodically at the same state (same current solution and same tabu moves, prohibited for an identical number of iterations), or arrives at a solution where no authorized move exists, without ever passing through an optimal solution.

The figure 2.9 represents, according to the parameter t, the iteration from which one enters a cycle and the length of this cycle, i.e. the number of intermediate solutions that this iterative search visits before returning to a given state. The number of different solutions is 258 for this problem (removing some equivalences). It is noticed that sometimes one can move through a number of solutions larger than the total number of solutions of the problem, before entering a cycle. Perhaps that explains why this search behaves relatively well for better conditioned instances. This figure also shows that the length of the cycles is always slightly larger than the value of the parameter t. If the value of the later exceeds 32, the search gets blocked in a solution such that all its neighbors are tabu.

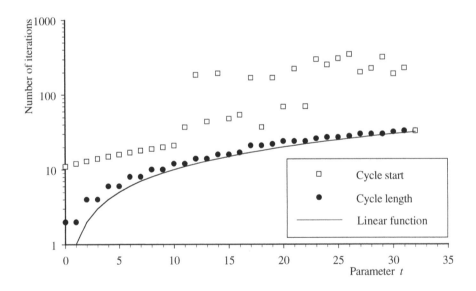

Fig. 2.9. Appearance of cycles as a function of the parameter t, for an ill-conditioned instance.

Thus, it is shown that the most elementary tabu search is not capable of avoiding cycles. Hence, it is necessary to resort to other principles to implement a more effective search. These principles were presented in the first articles concerning tabu search, but were not used during several years before they were established as indispensable principles of this technique. In facts, some of the articles published in the mid Nineties still presented heuristics based exclusively on a short term memory presented above. The following section discusses those techniques which facilitate to direct a longer-term search.

2.7 Long-term memory

In the case of a neighborhood defined by a static set of moves, i.e. when it does not depend on the solution found in the process, statistics of the moves chosen during the search can be of great utility. If some moves are elected much more frequently than others, it is necessary to assume that the search faces difficulties in exploring solutions of varied composition and that it may remain confined in a "valley". In practice, it is frequently observed problem instances that comprise extended valleys. Thus, these can be visited using moves of small amplitude, considering the absolute difference of the objective function. If only the mechanism of prohibiting the moves which are reverse of those recently carried out is employed to direct search, then the number of prohibited moves is so low that it becomes almost impossible to escape from some valleys. It was also seen that an attempt to increase this number of tabu moves may force the search procedure to often reside on the hillside and even if the search can change the valley, it can not succeed in finding good solutions in the new valley because of the moves which will be prohibited after the visit of the preceding valley. It is thus necessary to introduce other mechanisms to direct effectively a search at long-term.

2.7.1 Frequency-based memory

In order to ensure certain diversity throughout the search without prohibiting too many moves, a technique consists in penalizing the moves frequently used. Several penalization methods can be imagined, for instance the prohibition to carry out those moves whose frequency of occurrence during the search may exceed a given threshold, or the addition of a value proportional to their frequency of usage at the time of the evaluation of the moves. Moreover, the addition of a penalty proportional to the frequency will have a beneficial effect for the problems where the objective function takes only a small number of values, which can generate awkward equivalence to direct search, when several neighboring solutions have same evaluation. In these situations, the search will then tend to choose those moves which are least employed rather than to select a move more or less at random.

The figure 2.10 illustrates the effect of a penalization of the moves which adds a factor proportional to their frequency of usage at the time of their evaluation. For this purpose, the experiment carried out to show the influence of tabu duration was repeated (see figure 2.7), but this time by varying the coefficient of penalization; the moves are thus penalized, but never tabu. This experiment relates again to the 3000 quadratic assignment instances of size 12 generated at random. In figure 2.10, the average of the best solutions found after 50 iterations and the average value of all the solutions visited are given as functions of the coefficient of penalization. It should be noticed that the behavior of these two statistics is almost the same as that shown in figure 2.7, but overall, the solutions generated are worse than those obtained by the use of a short-term memory.

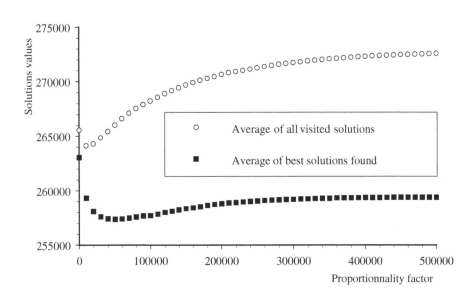

Fig. 2.10. Effect of the coefficient of penalization on the frequencies.

As implemented for the short-term memory, this can be generalized for a long-term memory of non-static set of moves, i.e. where M depends on s: then the frequency with which one employed certain characteristics of move is recorded rather than the moves themselves. Here the similarities in implementing these two forms of memories should be noticed: one stores the iteration in which one can again resort to a characteristic of move, whereas the other memorizes the number of times that this characteristic was used in the chosen moves.

Value of the penalization.

As implemented for the tabu duration in the short-term mechanism, it is necessary to tune the importance that one associates with a penalization based on the frequencies. This tuning can be carried out on the basis of the following considerations: Firstly, if $freq(m)$ denotes the frequency of usage of the move m, it seems reasonable to penalize this move by a factor proportional to $freq(m)$, though another possible function can be, for example, $freq^2(m)$.

Secondly, if the objective is a linear function and if a new problem instance is considered where all the data were multiplied by a constant, it is not desired that this mechanism of penalization based on the frequencies depends on the value of the constant. In the same way, the mechanism of penalization should not work in a different manner if one adds a constant to the objective. Consequently, it also seems legitimate to use a penalization which is proportional to the average amplitude of two neighboring solutions.

Thirdly, the larger the neighborhood is, the more the distribution of the frequencies concentrates on small values. The penalty should be multiplied by a function strictly increasing with the size of the neighborhood, so that the penalty does not become zero when the size of the problem increases. If the identity function proves to be too large in practice (cf. [Taillard, 1993, Taillard, 1994]), one can consider, for example, a factor proportional to $\sqrt{|M|}$.

Naturally, the concept of using a penalization based on the frequencies also requires taking the mechanism of aspiration into consideration. If not, then it is highly likely that we may pass beside excellent solutions.

2.7.2 Obligation to carry out move

Another long-term mechanism consists in performing a move which was never used during a large number of iterations, irrespective of its influence on the quality of the solution. Such a mechanism can be useful in destroying the structure of a local optimum, therefore escaping from the valley in which it was confined. This is also valid for high dimensional problems, as well as instances having more modest size but very structured (i.e. for which the local optima are separated by very bad solutions).

In the earlier example of the quadratic assignment problem, it is not even necessary to introduce a new data structure to implement this mechanism. In fact, it is enough to implement the tabu list under the form of a matrix of two dimensions (element, position), whose entries indicate in which iteration each element is authorized to occupy a given position, either to decide if a move is prohibited (the entries in the matrix corresponding to the move contain values larger than the number of the current iteration) or, on the contrary, if a given element did not occupy a given position during the last v iterations. If the matrix contains an entry whose value is lower than the number of the current iteration decreased by the parameter v, the corresponding move is elected, independent of its evaluation. It may happen that several moves could be simultaneously elected because of this rule. This problem can be solved by considering that, before the search was started, one had carried out all the $|M|$ moves (a static, definite neighborhood of all the M moves is assumed) during hypothetical iterations $-|M|, -|M|+1, \ldots, -1$. Of course, it is necessary that the parameter v be (sufficiently) larger than $|M|$, so that these moves are only imposed after v iterations. This is illustrated in the program given in appendix B.

2.8 Strategic oscillations

The effect of the penalization on the frequencies produces fruitful results for long searches and can thus be considered like a long-term memory. To obtain effective implementations of tabu search, this long-term memory must be exploited in collaboration with a short-term memory. This collaboration can be either constant, where the number of tabu moves and the coefficient of penalization are chosen once and for all, or varied, by alternating search phases where the long-term memory can have a dominating role or a limited one. The purpose of these phases will be to *intensify* the search (when the number of prohibited moves is reduced and/or the long-term memory is evoked to choose solutions with characteristics close to the best solutions enumerated by the search), or to *diversify* the search (when the number of tabu moves is large and/or the long-term memory is used to support solutions or moves with characteristics seldom met).

2.9 Conclusion

Only some basic concepts of tabu search were presented in this chapter. Other principles may lead to a more effective and intelligent method. When possible, the graphical representations of the solutions visited successively during the search will actively stimulate the spirit of the designer and will suggest, often in an obvious way, how to direct the search more intelligently. The development of a tabu search is an iterative process: it is almost improbable to propose an excellent method at first attempt; adaptations, depending on the type as well as on the problem instance dealt with, must certainly be required. This chapter described only those principles which should enable a designer to proceed towards an effective algorithm more quickly. Finally, let us mention that other principles, often presented within the framework of tabu search as suggested by F. Glover — such as scatter search, vocabulary building or path relinking — will be presented in the chapter 7 devoted to the methodology.

2.10 Annotated bibliography

[Glover and Laguna, 1997]: This book is undoubtedly the most important reference on tabu search. It describes the technique extensively, including certain extensions which will be discussed in this book, in the chapter 7.

[Glover, 1989, Glover, 1990]: These two articles can be considered as the founders of the discipline, even if the denomination of tabu search and certain ideas already existed previously. They are not easily accessible; hence certain concepts presented in these articles, like path relinking or scatter search, were studied by the research community only several years after their publication.

3

Evolutionary Algorithms

3.1 From genetics to engineering

The biological evolution generated extremely complex autonomous living beings which can solve extraordinarily difficult problems, such as the continuous adaptation to complex, uncertain environments and in perpetual transformation. For that, the superior living beings, like the mammals, are equipped with excellent capabilities of pattern recognition, training and intelligence. The large variety of the situations to which the life adapted shows that the process of evolution is robust and is capable of solving many classes of problems. This allows a spectator of the living world to conceive that there are other ways than establishing precise processes, patiently derived from quality knowledge of the natural laws, to satisfactorily build up complex and efficient systems.

According to C. Darwin [Darwin, 1859], the original mechanisms of evolution of the living beings rest on the competition which selects the most well adapted individuals to their environment while ensuring a descent, as in the transmission of the useful characteristics to the children which allowed the survival of the parents. This inheritance mechanism is based, in particular, on a form of cooperation implemented by the sexual reproduction.

The assumption that the Darwin theory, enriched by the current knowledge of the genetics, accounts for the mechanisms of evolution is still not justified. Nobody can confirm till today that these mechanisms are well understood, and that there is no essential phenomenon that remains unexplored. In the same manner that it was necessary to wait for a long time to understand that if the birds fly, it is not so much because of the beating of their wings, which gives a visible and misleading demonstration, but rather because of the profile of their wings, which creates the desired aerodynamic phenomenon.

However, the Neo-Darwinism is the only theory of evolution available that has never failed until now. The development of the electronic calculators facilitated the study of this theory in simulation and some researchers desired to test it on engineering problems, long back in the 1950s. But this work was not convincing because of insufficient knowledge, available at that time, of the natural genetics and also because of the weak performances of the calculators available. In addition, the extreme slowness of the evolution crippled the thinking that such a process can be usefully exploited.

During the 1960s and 1970s, as soon as calculators of more credible power came in existence, many attempts to model the process of evolution were undertaken. Among those, three approaches emerged independently, being mutually unaware of presence of others, until the beginning of the 1990s:

- the *evolution strategies (ES's)* of H.P. Schwefel and I Rechenberg [Rechenberg, 1965, Beyer, 2001] which were designed in the middle of the 1960s like an optimization method for problems using continuously varying parameters;
- the *evolutionary programming (EP)* of L.J. Fogel *et al.* [Fogel et al., 1966] which aimed, during the middle of the 1960s, to make evolve the structure of finite-state automata with iterated selections and mutations; it was desired to be an alternative for the artificial intelligence of the epoch;
- the *genetic algorithms* (GA's) were presented in 1975 by J.H. Holland [Holland, 1992], with the objective to understand the subjacent mechanisms of self-adaptive systems.

Thereafter, these approaches underwent many modifications according to the variety of the problems faced by their founders and their pupils. The genetic algorithms became extremely popular after the publication of the book *"Genetic Algorithms in Search, Optimization and Machine Learning"* by D. E. Goldberg [Goldberg, 1989] in 1989. This book, published world wide, resulted in an exponential growth in interest in this field. While there were about a few hundreds of publications in this area over 20 years duration before this book appeared, there are several tens of thousands of them available today. Researchers in this field have organized common international conferences presenting and combining their different approaches.

Genetic algorithms or evolutionary algorithms?

The widespread term *Evolutionary Computation*, appeared in 1993 as the title of a new journal published by the MIT Press, and then it was widely used to designate all the techniques based on the metaphor of the biological evolution theory. However, some specialists use the term "Genetic Algorithms" to designate any evolutionary technique even though they have few common points with the original propositions of Holland and Goldberg.

The various evolutionary approaches are based on a common model presented in section 3.2. The sections 3.3 and 3.4 describe various alternatives of selection and variation operators, basic building blocks of any evolutionary algorithm. The genetic algorithms are the most "popular" evolutionary algorithms. This is why the section 3.5 is especially devoted to them. It shows how it is possible to build a simple genetic algorithm from an adequate combination of specific selection and variation operators. Finally the section 3.6 briefly presents some questions related to the convergence of the evolutionary algorithms. This chapter concludes with a mini-glossary of terminologies usually used in the field and a bibliography with accompanying notes.

3.2 The generic evolutionary algorithm

In the world of the evolutionary algorithms, the *individuals* subjected to evolution are the solutions, more or less efficient, for a given problem. These solutions belong to the search space of the optimization problem. The set of the individuals treated simultaneously by the evolutionary algorithm constitutes a *population*. It evolves during a succession of iterations called *generations* until a termination criterion, which takes into account a priori the quality of the solutions obtained, is satisfied.

During each generation, a succession of operators is applied to the individuals of a population to generate the new population for the next generation. When one or more individuals are used by an operator, they are called the *parents*. The individuals originating from the application of the operator are its *offspring*. Thus, when two operators are applied successively, the offspring generated by one can become parents for the other.

3.2.1 Selection operators

In each generation, the individuals reproduce, survive or disappear from the population under the action of two *selection operators*:

- the selection for the reproduction, or simply: *selection*, that determines how many times an individual will be reproduced in a generation;
- the selection for the replacement, or simply: the *replacement*, that determines which individuals will have to disappear from the population in each generation so that, from generation to generation, the population size remains constant, or more rarely, is controlled according to a definite policy.

In accordance with the Darwinist creed, the better an individual, the more often it is selected to reproduce or survive. It may be, according to the alternative of the algorithm, that one of the two operators does not favor the good individuals compared to the others, but it is necessary that the application of the whole of the two operators during a generation introduces a bias in favor of the best. To make a selection possible, a fitness value, which depends obviously on the objective function, must be attached to each individual. This implies that, in each generation, the fitnesses of the offspring are evaluated, which can be computation intensive. The construction of a good *fitness function* from an objective function is rarely easy.

3.2.2 Variation operators

In order that the algorithm can find solutions better than those represented in the current population, it is required that they are transformed by the application of *variation operators* or *search operators*. A large variety of them can be imagined. They are classified into two categories:

- the *mutation* operators, which modify an individual to form another;
- the *crossover* operators, which generate one or more offspring from combinations of two parents. The designations of these operators are based on the real life concept of the sexual reproduction of the living beings, with the difference that the evolutionary computation, not knowing the biological constraints, can be generalized to implement the combination of more than two parents, possibly the combination of the entire population.

The way of modifying an individual depends closely on the structure of the solution that it represents. Thus, if it is desired to solve an optimization problem in a continuous space, e.g. a domain of $\mathbb{R}^n$, then it will be a priori adequate to choose a vector of $\mathbb{R}^n$ to represent a solution, and the crossover operator must implement a means so that two vectors of $\mathbb{R}^n$ for the parents correspond to one (or several) vector of $\mathbb{R}^n$ for the offspring. On the other hand, if one wishes to use an evolutionary algorithm to solve instances of the traveling salesman problem, it is common that an individual corresponds to a round trip. It is possible to represent it as a vector where each component is the number designated to a city. The variation operators should then generate only legal round trips, i.e. rounds for which each city of the circuit is present only once. These examples show that it is impossible to design universal variation operators, independent of the problem under consideration. They are necessarily related to the *representation* of the solutions in the search space. As a general rule, for a representation chosen, it is necessary to define the variation operators used, because they closely depend on it.

3.2.3 The generational loop

In each generation, an evolutionary algorithm implements a "loop iteration" that incorporates the application of these operators on the population:

1. for the reproduction, selection of the parents among a population of μ individuals to generate λ offspring;
2. crossover and mutation of the λ selected individuals to generate λ offspring;
3. fitness evaluation for the offspring;
4. selection for the survival of μ individuals among the λ offspring and μ parents, or only among the λ offspring, according to the choice made by the user, in order to build the population for the next generation.

The figure 3.1 graphically represents this loop with the insertion of the stopping test and addition of the initialization phase of the population. It will be noted that the hexagonal forms refer to the variation operators dependent on the representation chosen, while the "rounded squares" represent the selection operators that are independent of the solution representation.

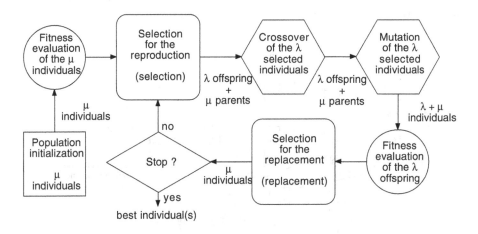

Fig. 3.1. The generic evolutionary algorithm.

3.2.4 Solving a simple problem

Following our own way of illustrating the operation of an evolutionary algorithm, let us consider the maximization of the function $C(x) = 400 - x^2$ for x in the interval $[-20, 20]$. There is obviously no practical interest for using this type of algorithm to solve such a simple problem, the objectives here are exclusively didactic. This example will be considered again and commented on throughout the part presenting the basics of the evolutionary algorithms. The figure 3.2 shows the succession of the operations from the initialization phase of the algorithm until the end of the first generation. In this figure, an individual is represented by a rectangle partitioned into two zones. The top zone represents the value of the individual x ranging between -20 and +20. The bottom zone contains the corresponding value of the objective function $C(x)$ after it was calculated during the evaluation phase. When this value is not known, the zone is shown in gray. As we are confronted with a problem of maximization and that the problem is very simple, the objective function is also the fitness function. The ten individuals of the population are represented in a row while the vertical axis describes the temporal sequence of the operations.

The choice of using ten individuals to constitute a population should not be a misleading one. This can be useful in practice when the computation of the objective function takes much time suggesting to reduce the computational burden by choosing a small population size. One prefers to use populations of the order of at least one hundred individuals to increase the chances to discover an acceptable solution. According to the problems under considerations, the population size can exceed ten thousands of individuals, which then requires a treatment on a multiprocessor computer (up to several thousands processing units) so that the execution times are not crippling.

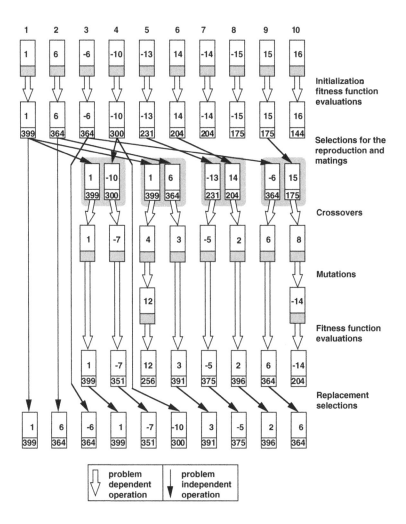

Fig. 3.2. Application of an evolutionary algorithm on a population of $\mu = 10$ parents and $\lambda = 8$ offspring.

Our evolutionary algorithm works here with the integer representation. This means that one individual is represented by an integer and that the variation operators must generate integers from the parents. For searching the optimum of $C(x) = 400 - x^2$, it is decided that the crossover will generate two offspring from two parents, each offspring being an integer number drawn randomly in the interval defined by the values x of the parents. The mutation is only the random generation of an integer in the interval $[-20, +20]$. The result of the mutation does not depend on the value of the individual before mutation, which could appear destructive. However, one can notice in the figure 3.2 that the mutation is applied seldom in our model of evolution, which makes this policy acceptable.

3.3 Selection operators

In general, the capability of an individual of being selected, for reproduction or replacement, depends on its fitness. The selection operator thus determines a number of selections for each individual according to its fitness.

In our "guide" example (see figure 3.2), the ten parents generate eight offspring. This number is a parameter of the algorithm. According to the figure, the selection operator thus copied twice the best parent, and once six other parents to produce the population of the offspring. Those are generated by the variation operators from the copies. Then the replacement operator is activated and selects the ten best individuals among the parents and the offspring to constitute the population of the parents for the next generation. It is noticed that four parents survived, while two offspring, who were of very bad quality, disappeared from the new population.

3.3.1 Selection pressure

The individuals having the best fitnesses are reproduced more often than the others and replace the worst ones. If the variation operators are inhibited, the best individual should reproduce more quickly than the others, until its copies completely take over the population. This observation leads to a first definition of the selection pressure suggested by Goldberg and Deb in 1991 [Goldberg and Deb, 1991]. The *takeover time* τ^* is defined as the number of generations necessary to fill the population with the copies of the best individual under the action of the selection operators only. The selection pressure will be higher as τ^* will be low.

The *selection intensity* S is another method, borrowed from the population genetics, to define the selection pressure. Let $\bar{f}$ be the average fitness of the μ individuals of the population before the selection. Let $\bar{g}$ be the average fitness of the λ offspring of the population after the selection. Then S measures the increase in the average fitness of the individuals of a population determined before and after selection with the standard deviation σ_f of the individual fitnesses before selection taken as a unit of measure:

$$S = \frac{\bar{g} - \bar{f}}{\sigma_f}$$

If the selection intensity is computed for the reproduction, then $\bar{f} = \sum_{i=1}^{\mu} f_i/\mu$, with f_i be the fitness of the individual i, and $\bar{g} = \sum_{i=1}^{\lambda} g_i/\lambda$, with g_i be the fitness of the individual i. The definitions presented above are general and are applicable to any selection technique. It is possible to present other definitions, whose validity is possibly limited to certain techniques, as we will see later with regard to the *proportional selection*.

With a high selection pressure, there is a great risk of *premature convergence*. This current situation occurs when the copies of a super-individual, nonoptimal but who reproduces much more quickly than the others, take over the population. Then, the exploration of the search space becomes local, since it is limited to a search randomly centered on the super-individual, and there will be huge risks that the global optimum is not approached in the event of existence of local optima.

3.3.2 Genetic drift

The *genetic drift* is also a concept originating from the population genetics. It is concerned about a random fluctuation of the frequency of the alleles in a population of small size, where an *allele* is a variant of an element of a sequence of DNA[1] having a specific function. For this reason, hereditary features can disappear or be fixed at random in a small population even without any selection pressure.

This phenomenon also occurs within the framework of the evolutionary algorithms. At the limit, even for a population formed by different individuals but of the same fitness, in the absence of variation generated by mutation and cross-over operators, the population converges towards a state where all the individuals are identical. It is the consequence of the stochastic nature of the selection operators. The genetic drift can be evaluated by the time required to obtain a homogeneous population using a Markovian analysis. But these results are approximations and are difficult to generalize on the basis of the case studies in the literature. However, it is verified that the time of convergence towards an absorption state becomes longer as the population size increases.

Another technique to study the genetic drift measures the reduction of the variance of the fitness in the population for each generation, under the action of the selection operators only, each parent having a number of offspring independent of its fitness (neutral selection). This last hypothesis must be satisfied to ensure that the reduction of variance is not due to the selection pressure. Let r be the ratio of the expectation of the fitness variance in a given generation to the variance in the previous generation. In this case, A. Rogers and

---

[1]desoxyribonucleic acid: a giant molecule that supports part of the hereditary features of the living beings

A. Prügel-Bennett [Rogers and Prügel-Bennett, 1999] showed that r depends only on the variance V_s of the number of offspring for each individual and on the population size, assumed constant:

$$r = \frac{E(V_f(g+1))}{V_f(g)} = 1 - \frac{V_s}{P-1}$$

where $V_f(g)$ is the variance of the population fitness distribution at generation g. V_s is a characteristic of the selection operator. It is seen that increasing the population size or reducing the variance V_s of the selection operator decreases the genetic drift.

The effect of the genetic drift is prevalent when the selection pressure is low and this situation leads to a loss of diversity. This involves a premature convergence a priori far away from the optimum since it does not depend on the fitness of the individuals.

In short, in order that an evolutionary algorithm can work adequately, it is necessary that the selection pressure is neither too strong, nor too weak, for a population of sufficient size, with the choice of a selection operator characterized by a low variance.

3.3.3 Proportional selection

This type of selection was originally proposed by J Holland for the genetic algorithms. It is used only for the reproduction. The expected number of selections λ_i of an individual i is proportional to its fitness f_i. This implies that the fitness function is positive in the search domain and that it must be maximized, which can already impose some simple transformations of the objective function to satisfy these constraints. Let μ be the population size and let λ be the total number individuals generated by the selection operator, λ_i can be expressed as:

$$\lambda_i = \frac{\lambda}{\sum_{j=1}^{\mu} f_j} f_i$$

Table 3.1 gives the expected number of selections λ_i of each individual i for a total of $\lambda = 8$ offspring in the population of 10 individuals from our "guide" example.

However, the effective number of offspring can be only integers. For example, the situation in the figure 3.2 was obtained with a proportional selection technique. It shows that the individuals 7, 8 and 10, whose respective fitnesses 204, 175 and 144 are among the worst ones, do not have offspring. Except the best individual that is selected twice, the others take part only once in the process of crossover. To obtain this, a stochastic sampling procedure constitutes the core of the proportional selection operator. Two techniques are widespread and are described below: the *roulette wheel selection method, RWS* because it is the operator originally proposed for the genetic algorithms, but it suffers from high variance, and the *stochastic universal sampling method, SUS* because it guarantees a low variance of the sampling process [Baker, 1987].

Table 3.1. Expected number of offspring in the population of 10 individuals.

i	1	2	3	4	5
f_i	399	364	364	300	231
λ_i	1.247	1.138	1.138	0.938	0.722

i	6	7	8	9	10
f_i	204	204	175	175	144
λ_i	0.638	0.638	0.547	0.547	0.450

Proportional selection algorithms

The RWS method exploits the metaphor of a biased roulette game, which comprises of as many compartments as individuals in the population and where the size of these compartments would be proportional to the fitness of each individual. Once the game is started, the selection of an individual is indicated by the stopping of the ball on its compartment. If the compartments are unrolled on a straight line segment, the selection of an individual corresponds to choosing, at random, a point of the segment with a uniform probability distribution (figure 3.3). The variance of this process is high. It is possible that an individual having a good fitness value is never selected. In extreme cases, it is also possible, by sheer misfortune, that bad quality individuals are selected as many as there are offspring. This phenomenon creates a genetic drift that facilitates some poor individuals to have offspring with the detriment of better individuals. To reduce this risk, the population size must be sufficiently large.

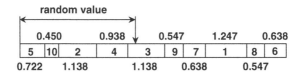

Fig. 3.3. RWS method: individual 3 is selected after drawing a random number.

It is the SUS method which was used in our "guide" example. One always considers a straight line segment partitioned in as many zones as there are individuals in the population, each zone having its size proportional to the fitness. But this time the selected individuals are designated by a set of equidistant points, their number being equal to the number of offspring (figure 3.4). This method is different from the RWS method as here only one random drawing is required to place the origin of the series of equidistant points and thus to generate all the offspring in the population. According to the figure, individuals 7, 8 and 10 are not selected, the best individual is selected twice, while the others are selected only once. For an expected number of selections λ_i of the individual i, the effective number of selections will be either the integer part of λ_i, or its immediate higher integer number. The variance of the process being weaker than in the RWS method, the genetic drift appears much less and, if $\lambda \geq \mu$, the best individuals are certain to have at least an offspring each.

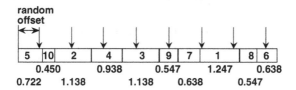

Fig. 3.4. SUS method: the selected individuals are designated by equidistant points.

Proportional selection and selection pressure

In the case of proportional selection, the expected number of selections of the best individual with fitness $\hat{f}$ among μ selections for a population of μ parents is appropriate to define the selection pressure:

$$p_s = \frac{\mu}{\sum_{j=1}^{\mu} f_j} \hat{f} = \frac{\hat{f}}{\bar{f}}$$

where $\bar{f}$ is the average of the fitnesses of the population. If $p_s = 1$, then all the individuals have equal chances to be selected, indicating an absence of selection pressure.

Let us consider the search for the maximum of a continuous function, e.g. $f(x) = \exp(-x^2)$. The individuals of the initial population are assumed to be uniformly distributed in the domain $[-2, +2]$. Some of them have a value close to 0, which is also the position of the optimum, and thus their fitness $\hat{f}$ will be close to 1. The average fitness of the population $\bar{f}$ will be

$$\bar{f} \approx \int_{-\infty}^{+\infty} f(x)p(x)dx$$

where $p(x)$ is the probability density of presence of an individual at x. An uniform density is chosen in the interval $[-2, +2]$, thus $p(x)$ is $1/4$ in this interval, and is 0 elsewhere. Thus

$$\bar{f} \approx \frac{1}{4} \int_{-2}^{+2} e^{-x^2} dx$$

that is $\bar{f} \approx 0.441$, which gives a selection pressure of the order of $p_s = \hat{f}/\bar{f} \approx$ 2.27. The best individual will thus have an expected number of offspring close to two (figure 3.5a).

Now let us consider that the majority of the individuals of the population are in a much smaller interval around the optimum, for example $[-0.2; +0.2]$. This situation spontaneously occurs after some generations, because of the selection pressure, which favors the reproduction of the bests, these ones being also closest to the optimum. In this case, assuming an uniform distribution again $\bar{f} \approx 0.986$, and $p_s \approx 1.01$ (see figure 3.5b). The selection pressure becomes almost non-existent: the best individual has practically as many expected offspring as any other individual, and it is the genetic drift which will prevent the population from converging towards the optimum as precisely as desired.

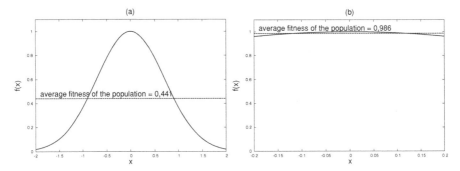

Fig. 3.5. The selection pressure decreases when the population concentrates in the neighborhood of the optimum.

This undesirable behavior of the proportional selection, where the selection pressure strongly decreases when the population approaches the optimum in the case of continuous functions, is overcome by techniques of fitness function scaling.

Linear scaling of the fitness function

With a technique of proportional selection, the expected number of selections of an individual is proportional to its fitness. In this case, the effects of misadjusted selection pressure can be overcome by a linear transformation of the fitness function f. The adjusted fitness value f_i' for an individual i is equal to $f_i - a$, where a is a positive value if it is desired to increase the pressure, or otherwise negative, identical for all the individuals. a should be chosen so that the selection pressure is maintained at a moderate value, neither too large, nor too small, typically about two. With such a technique, one should pay attention to the fact that the values of f' are never negative. They can be possibly lower bounded to 0, or to a small positive value, so that any individual, even of bad quality, has a small chance to be selected. This disposition contributes in the maintenance of the diversity in the population. Assuming that no individual becomes negative, the value of a can be calculated in each generation from the value of the desired selection pressure p_s:

$$a = \frac{p_s \bar{f} - \hat{f}}{p_s - 1} \text{ with } p_s > 1$$

In the context of the above example, if the individuals are uniformly distributed in the interval $[-0.2; +0.2]$, then $a = 0.972$ for a desired selection pressure $p_s = 2$. The figure 3.6 illustrates the effect of the transformation $f' = f - 0.972$. It can be noticed that there are values of x for which the function f' is negative, whereas this situation is forbidden for a proportional selection. To correct this drawback, the fitnesses of the individuals concerned can be kept clamped at zero, or a small constant positive value, which has the side effect of decreasing the selection pressure.

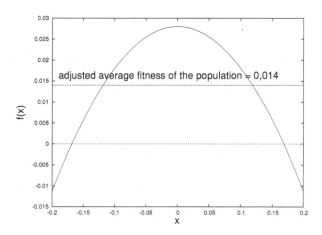

Fig. 3.6. Adjustment of the selection pressure by subtraction of a positive constant from $f(x)$.

Exponential scaling of the fitness function

Rather than operating a linear transformation to adjust the selection pressure, currently another common alternative exercised is to elevate the fitness function to an adequate k power to obtain the desired selection pressure:

$$f'_i = f_i^k$$

the parameter k depends on the problem. The Boltzmann selection (De La Maza and Tidor 1993 [De La Maza and Tidor, 1993]) is another alternative, where the scaled fitness can be expressed as:

$$f'_i = \exp(f_i/T)$$

The value of the parameter T, known as the "temperature", determines the selection pressure. T is usually a decreasing function of the number of generations, thus enabling the selection pressure to grow with time.

Rank based selection

These techniques for adjusting the selection pressure proceed by ranking the individuals i according to the values of the raw fitnesses f_i. The individuals are ranked from the best (first) to the worst (last). The fitness value f'_i actually assigned to each individual depends only on its rank by decreasing value (see figure 3.7) according to, for example, the formula given below which is usual:

$$f'_r = \left(1 - \frac{r}{\mu}\right)^p$$

where μ is the number of parents, r is the rank of the individual considered in the population of the parents after ranking. p is an exponent which depends on the desired selection pressure. After ranking, a proportional selection is applied according to f'. With our definition of the pressure p_s, the relation is: $p_s = 1 + p$. Thus, p must be greater than 0. This fitness scaling technique is not affected by a constraint of sign: f_i can either be positive or negative. It is appropriate for a maximization problem as well as for a minimization problem, without the necessity to operate a transformation. However, it does not consider the importance of the differences between the fitnesses of the individuals, so that the individuals of very bad quality, but who are not at the last row of the ranking will be able to persist in the population. It is not inevitably a bad situation because it contributes to a better diversity. Moreover, this method does not require the exact knowledge of the objective function, but of simply being able to rank the individuals by comparing each with the others. These good properties make that, overall, it is preferred by the users of evolutionary algorithms compared to the linear scaling technique.

3.3.4 Tournament selection

The tournament selection is an alternative to the proportional selection techniques which, as explained before, presents difficulties to control the selection pressure during the evolution, while being relatively expensive in the computation power involved.

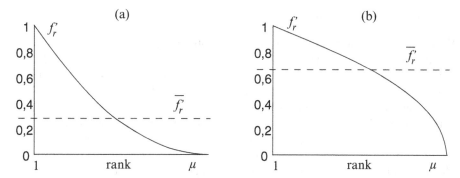

Fig. 3.7. Performance obtained after ranking. (a) : $f_r' = (1 - r/\mu)^2$ – strong selection pressure, (b): $f_r' = \sqrt{(1 - r/\mu)}$ – weak selection pressure.

Deterministic tournament

The simplest tournament consists in choosing at random a number k of individuals in the population, and selecting for the reproduction the one that has the best fitness. During a selection step, there are as many tournaments as selected individuals. The individuals who take part in a tournament can be replaced in the population, or they can be withdrawn from it, according to the choice of the user. A drawing without replacement makes it possible to conduct $\lfloor N/k \rfloor$ tournaments with a population of N individuals. A copy of the population is re-generated when it is exhausted, and this is implemented as many times as necessary, until the desired number of selections is reached. The variance of the tournament process is high, which favors the genetic drift. It is however weaker in the case of drawing without replacement. This method of selection is very much used, because it is much simpler to implement than a proportional reproduction with a behavior and properties similar to the ranking selection.

The selection pressure is adjusted by the number of participants k in a tournament. Indeed, let us consider the case where the participants in a tournament are replaced in the population. Then the probability that the best individual of the population is not selected in k drawings is $((N-1)/N)^k$. By making the assumption that N is very large compared to k, this probability is approximately $1 - k/N$ by a binomial expansion limited to first order. Thus, the probability that the best individual is drawn at least once in a tournament is close to k/N. If there are M tournaments in a generation, the best individual will have kM/N expected selections, that involve a selection pressure of k, by considering again the definition that was proposed for the proportional reproduction (with $M = N$). This pressure will be necessarily greater than or equal to 2.

Stochastic tournament

With the stochastic binary tournament, involving two individuals in competition, the best wins with a probability p ranging between 0.5 and 1. It is still easy to calculate the selection pressure generated by this process. The best individual takes part in a tournament with a probability of $2/N$ (see the preceding paragraph). The best individual of the tournament will be selected with a probability p. The two events being independent, the probability that the best individual of the population is selected after a tournament is thus $p \cdot 2/N$. If there are N tournaments, the best will thus have $2p$ expected offspring. The selection pressure thus will range between 1 and 2.

Another alternative, the Boltzmann tournament, ensures that the distribution of the fitness values in a population is close to a Boltzmann distribution. This method allows to build a bridge between evolutionary algorithms and simulated annealing.

3.3.5 Truncation selection

This selection is very simple to implement, as it does nothing but to choose the n best individuals from a population, n being a parameter chosen by the user. If the operator is used for the reproduction to generate λ offspring from n selected parents, each parent will have λ/n offspring. If the operator is used for the replacement and thus generates the population of μ individuals for the next generation, then $n = \mu$.

3.3.6 Replacement selections

Generational replacement

This type of replacement is the simplest, since the population of the parents for the generation $g + 1$ will be composed of all the offspring, and only them, generated in generation g. Thus: $\mu = \lambda$. The canonical genetic algorithm uses a generational replacement.

Replacement for the Evolution Strategies "(μ, λ)– ES"

A truncation selection of the best μ individuals among λ offspring forms the population for the next generation. Usually, λ is larger than μ.

Steady state replacement

In each generation, a small number of offspring (one or two) are generated and they replace a lower or equal number of parents, to form the population for the next generation. This strategy is useful especially when the representation of a solution is distributed on several individuals, possibly the entire population.

In this way, the loss of a small number of individuals in each generation: those that are replaced by the offspring, does not disturb the solutions excessively and thus they evolve gradually.

The choice of the replaced parents obeys various criteria. With the uniform replacement, the replaced parents are designated at random. The choice can also depend on the fitness: the worst parent is replaced, or it is selected stochastically, according to a probability distribution that depends on the fitness or other criteria.

The steady state replacement generates a population where the individuals are subject to large variations of lifespan measured in number of generations and thus in number of offspring. The high variance of these values augments the genetic drift, which appears more especially as the population is small [De Jong and Sarma, 1993].

Elitism

An elitist strategy consists in preserving in the population, from one generation to the next one, at least the individual having the best fitness. The example shown in the figure 3.2 implements an elitist strategy since the best individuals of the population composed by the parents and the offspring are selected to form the population of the parents for the next generation. The fitness of the best individual from the current population is thus monotonically nondecreasing from one generation to the next generation. It is noticed, for this example, that four parents of generation 0 find themselves in generation 1.

There are various elitist strategies. The strategy employed in our "guide" example originates from the Evolution Strategies known as "$(\mu + \lambda)$–ES". In other current alternatives, the best parents in generation g are copied systematically in $\mathbf{P}(g+1)$, the population for the generation $g+1$. Or, if the best individual of $\mathbf{P}(g)$ is better than that of $\mathbf{P}(g+1)$, because of the action of the variation or selection operators, then the best of $\mathbf{P}(g)$ will be copied in $\mathbf{P}(g+1)$, by usually replacing the lowest fitness individual.

It appears that such strategies improve considerably the performance of evolutionary algorithms for some classes of functions, but prove to be disappointing for other classes, by increasing the rate of premature convergences. For example, an elitist strategy is harmful to seek the global maximum of the F5 function of De Jong (figure 3.8). In fact, such a strategy increases the exploitation of the best solutions, resulting in an accentuated local search, with the detriment of the exploration of the search space.

objective

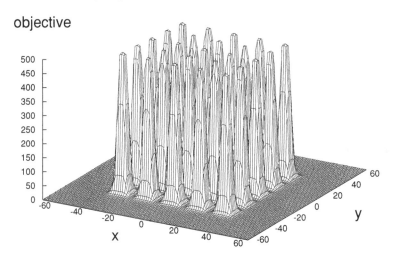

Fig. 3.8. F5 Function of De Jong.

Choosing a non-elitist strategy can be advantageous, but there is then no guarantee that the fitness function of the best individual is increasing during the evolution. That obviously implies to keep a copy of the best solution found by the algorithm since the initiation of the evolution, without however this copy taking part in the evolutionary process. It is an indispensable precaution for any stochastic optimization algorithm anyhow.

3.3.7 Fitness function

Fitness function associates a fitness value to each individual in order to determine the number of times it will be selected to be reproduced, or whether it will be replaced or not. In the case of the function $C(x)$ chosen for example, the fitness function is also the objective function for our maximization problem. This kind of situation is exceptional, and it is often necessary to carefully construct the fitness function for a given problem. The quality of this function greatly influences the efficiency of a genetic algorithm.

Construction

If a proportional selection is chosen, it is possibly necessary to transform the problem under consideration so that it becomes a maximization problem of a numerical function with positive values on its domain of definition. For example, the solution of a system of equations $\mathbf{S}(\mathbf{x}) = 0$ could be obtained by searching the maxima of $1/(a + |\mathbf{S}(\mathbf{x})|)$, where the notation $|\mathbf{V}|$ represents the modulus of the vector $\mathbf{V}$. a is a non-null positive constant.

The construction of a good fitness function should consider the chosen representation and the nature of the variation operators so that it can give non deceptive indications on the progress towards the optimum. For example, it would be necessary to try to reduce the presence of local optima on the top of broad peaks as long as a priori knowledge available on the problem allows it. This relates to the study of the fitness landscapes that will be introduced in the section 3.4.1 referring to the variation operators.

Moreover, a good fitness function must satisfy several criteria which refer to its complexity, to the satisfaction of the constraints of the problem, and also to the adjustment of the selection pressure during the evolution. When the fitness function appears excessively complex, consuming a considerable computing power, the search for an approximation is desirable, sometimes indispensable.

Reduction of the required computing power

In general, in the case of industrial problems, the evaluation of the fitness function consumes by far the greatest amount of the computing power during an evolutionary optimization. Let us assume that the calculation of a fitness value takes 30 seconds, that there are 100 individuals in the population, and that acceptable solutions are discovered after a thousand of generations, each one implying each time the evaluation of all the individuals, then it will require 35 days of computation. Now in the case of industrial problems, the fitness evaluations usually involve computation intensive numerical methods, for example: finite element methods. Strategies must be used to reduce these computation times. Parallel computing can be considered. This kind of approach is efficient but hardware expensive. One can also consider fitness approximation calculations, which will be refined gradually, as the generations pass. Thus, for finite element methods for example, it is natural to start by using a coarse mesh, in the beginning of the evolution. The difficulty is then to determine when the fitness function should be refined so that the optimizer does not converge prematurely to false solutions generated by the approximations. Another solution to simplify calculation can exploit a tournament selection or also, a ranking selection (section 3.3.3). Indeed, in these cases, it is not necessary to know the precise values of the objective function, because only the ranking of the individuals is significant.

3.4 Variation operators and representation

3.4.1 Generalities about the variation operators

The variation operators belong to two categories:

- the crossover operators, that use several parents (often two) to create one or more offspring;
- the mutation operators, that transform one individual.

They make it possible to create diversity in a population by building "off-spring" individuals, who inherit, partly, the characteristics of "parent" individuals. They must be able to serve two mandatory functions during the search for an optimum:

- the exploration of the search space, in order to discover the interesting areas, which are most likely to contain the global optima;
- the exploitation of these interesting areas, in order to concentrate search there and to discover the optima with the required precision, for those which contain them.

For example, a purely random variation operator, where solutions are drawn at random independent of each other, will have excellent qualities of exploration, but will not be able to discover an optimum in a reasonable time. A local search operator that performs "hill climbing" will be able to effectively discover an optimum in an area of space, but there will be a great risk that it will be a local solution, and the global solution will not be obtained. A good search algorithm for the optimum will have to thus carry out an adequate balance between the exploration capabilities and exploitation of the variation operators it uses. It is not easy to conceive and it strongly depends on the properties of the problem under consideration.

The study of the *fitness landscape* helps to understand why a variation operator will be more effective than any other operator for the same problem and the same choice of representation. The notion was introduced within the framework of the theoretical genetics in the 1930s by S. Wright [Wright, 1932]. A fitness landscape is defined by:

- a search space Ω whose elements are called "configurations";
- a fitness function $f: \Omega \to \mathbf{R}$;
- a relation of neighborhood or accessibility χ.

It can be noticed that the relation of accessibility is not a part of the optimization problem. This relation depends on the characteristics of the variation operators chosen. Utilizing a configuration in the search space, the application of these stochastic operators potentially gives access to a set of accessible configurations with various probabilities. The relation of accessibility can be formalized within the framework of a discrete space Ω by a directed hypergraph whose hyperarcs have values given by the probabilities of access to an "offspring" configuration from a set of "parent" configurations.

For the mutation operator, the hypergraph of the relation of accessibility becomes a directed graph which, from an individual, or configuration X, represented by a node of the graph, gives a new configuration X', with a probability given by the value of the arc (X, X'). For a crossover operation between two individuals X and Y that produce an offspring Z, the probability of generating Z knowing that X and Y are crossed is given by the value of the hyperarc $(\{X, Y\}, \{Z\})$.

The definition of the fitness landscape given above shows that it depends at the same time on the optimization problem under consideration, on the chosen representation defined by space Ω and on the relation of accessibility defined by the variation operators. What is obviously expected is that the application of the latter offers a sufficiently high probability to improve the fitness of the individuals from one generation to another. This point of view will be wise to adopt when designing relevant variation operator for a given representation and a problem, while benefiting from all knowledge, formalized or not, that is available for this problem.

After some general considerations regarding the crossover and mutation operators, the following paragraphs present examples of traditional operators applicable in various popularly used search spaces:

- the space of the binary strings;
- the real representation in domains of $\mathbb{R}^n$;
- the representations of permutations usable for various combinatorial problems, like the traveling salesman problem, and the problems of scheduling;
- the representation of parse trees, for the resolution of problems by automatic programming.

Crossover

The crossover operator uses two parents to generate one or two offspring. The operator is generally stochastic, hence the repeated crossover of the same couple of distinct parents gives different offspring. As the crossovers of the evolutionary algorithms are not subject to biological constraints, more than two parents, in the extreme case the complete population, can participate in mating for crossover [Eiben et al., 1995].

The operator generally respects the following properties:

- the crossover of two identical parents will produce offspring, identical to the parents.
- By extension, on the basis of an index of proximity depending on the chosen representation defined in the search space, two parents which are close in the search space will generate offspring, close to them.

These properties are satisfied by the "classical" crossover operators like most of those described in this chapter. These are not absolute, as in the current state of knowledge of the evolutionary algorithms, the construction of a crossover operator does not follow a precise rule.

The *crossover rate* determines the proportion of the individuals crossed among the offspring. For the example in the figure 3.2, this rate was fixed at 1, i.e. all offspring are obtained by crossover. In the simplest version of an evolutionary algorithm, the individuals are mated at random among the offspring generated by the selection without taking account of their characteristics. This strategy can prove to be harmful when the fitness function has several optima. Indeed, it is generally not likely that the crossover of high quality individuals located on different peaks will give good quality individuals (refer to figure 3.9). A crossover is known as *lethal* if it produces one or two offspring having too low a fitness to reproduce from good parents.

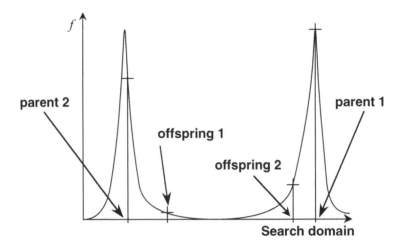

Fig. 3.9. Crossover of two individuals placed on different peaks of a fitness function f.

A solution to avoid a too strong proportion of lethal crossovers consists in preferentially mating the individuals who resemble each other. A distance being defined in the search space, the simplest way to proceed consists in selecting two individuals according to the probability distribution of the selection operator and then to cross the individuals only if their distance is lower than a threshold r_c called *restriction radius*. If the latter is small, this will lower the rate of effective crossover significantly, which can be prejudicial. It is preferable then to select a first parent by the selection operator, then, if there are individuals in its neighborhood, one of them is selected to become the second parent. In all situations, if r_c is selected too small, it significantly reduces the exploration of the search space by accentuating local search, and it can lead to premature convergences. This is especially sensitive to the initialization of the evolution when the crossover of two individuals distant from each other makes it possible to explore new areas of the search space that potentially contain peaks of the fitness function. Thus, to make the technique efficient, the major problem consists in choosing a good value for r_c; however it depends largely on the fitness landscape, which is in general not known. It is also possible to consider a decreasing radius r_c during the evolution.

Mutation

Classically, the mutation operator modifies an individual at random to generate an offspring who will replace it. The proportion of the mutated individuals in the offspring population is equal to the *mutation rate*. Its order of magnitude can vary substantially according to the chosen model of evolution. In the example of the figure 3.2, two individuals are mutated among the eight offspring obtained by the selection process. For the genetic algorithms, the mutation is considered as a minor operator, aimed at maintaining a minimum diversity in the population, which the crossover cannot ensure. With this model, the mutation rate is typically low, about 0.01 to 0.1, whereas the crossover rate is high. On the contrary, the mutation in the original model of the Evolution Strategies is essential since there is no crossover. The mutation rate is then 100%.

Most of the mutation strategies modify an individual in such a way that the result of the transformation is close to it. In this way, the operator performs a random local search around each individual to mutate. The mutation can considerably improve the quality of the solutions discovered compared to the crossover which loses its importance when most of the population is located in the neighborhood of the maxima of the fitness function. Indeed, the individuals located on the same peak are often identical because of the process of the selection for the reproduction and do not undergo any modification by the crossover operator. If they belong to different peaks, the offspring generally have low fitnesses. On the other hand, the local random search due to the mutations gives a chance to each individual to approach the exact positions of the maxima, as much as the characteristics of the chosen operator allow it.

The mutation with a sufficiently high rate plays an important part in the preservation of the diversity, useful for an efficient exploration of the search space. This operator can fight the negative effects of a strong selection pressure or a strong genetic drift, those phenomena which tend to reduce the variance of the distribution of the individuals in the search space.

If the mutation rate is high and, moreover, the mutation is so strong that the individual produced is almost independent of that which generated it, the evolution of the individuals in the population is equivalent to a random walk in the search space, and the evolutionary algorithm will require an excessive time to converge.

The utilization of the mutation, as a local search operator, suggests combining it with other more effective local techniques, although more problem-dependent, such as a gradient technique for example. This kind of approach led to the design of *hybrid* evolutionary algorithms.

3.4.2 Binary representation

The idea to make evolve a population in a space of binary vectors originated mainly from the genetic algorithms, which was inspired by the transcription *genotype-phenotype* existing in the living world. Within the framework of the genetic algorithms, the genotype is constituted by a string of binary symbols, or more generally, a string of symbols belonging to a low-cardinality alphabet. The phenotype is a solution of the problem in a "natural" representation. The genotype undergoes the action of the genetic operators: selections and variations, while the phenotype is used only for the fitness evaluation.

For example, if a solution is expressed naturally as a vector of real numbers, the phenotype will be this vector. The genotype will thus be a binary string which codes this vector. To code the set of the real variables of a numerical problem as a binary string, the simplest way is to convert each variable in binary format, then these binary numbers are concatenated to produce the genotype. Lastly, the most immediate technique to code a real number in binary format consists in representing it in fixed point format with a number of bits corresponding to the desired precision.

Crossover

For a binary representation, there exists three classical variants of crossovers:

- the "single point" crossover;
- the "two point" crossover;
- the uniform crossover.

A pair of individuals being chosen by random drawing from the population, the "single point" crossover [Holland, 1992] is applied in two stages:

1. random choice of an identical cut point on the two bit strings (see figure 3.10a);
2. cut of the two strings (figure 3.10b) and exchanges of the two fragments located on the right (figure 3.10c).

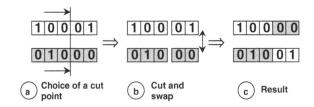

Fig. 3.10. "Single point" crossover of two genotypes of 5 bits.

This process produces two offspring from two parents. If only one offspring is used by the evolutionary algorithm employed, it is chosen at random from the pair and the other one is discarded. The "single point" crossover is the simplest and the most traditional for codings using an alphabet with low cardinality, like the binary coding. An immediate generalization of this operator consists in multiplying the cut points on each string. The "single point" and "two point" crossovers are usually employed in practice for their simplicity and their good effectiveness.

The uniform crossover [Ackley, 1987] can be viewed as a multipoint crossover where the number of cuts is unspecified a priori. Practically, one uses a "template string", which is a binary string of the same length as the individuals. A "0" at the nth position of the template leaves the symbols in the nth position of the two strings unchanged and a "1" activates an exchange of the corresponding symbols (in figure 3.11). The template is generated at random for each pair of individuals. The values "0" or "1" of the elements of the template are generally drawn with a probability of 0.5.

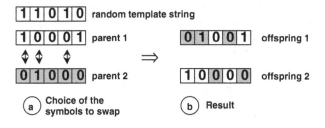

Fig. 3.11. Uniform crossover.

Mutation

Classically, the mutation operator on bit strings modifies at random the symbols of a genotype, with a low probability within the framework of genetic algorithms, typically from 0.01 to 0.1 per individual. This probability is equal to the mutation rate. The most common variants are the *deterministic mutation* and the *bit-flip mutation*. With the "deterministic" mutation, a fixed number of bits chosen at random are reversed for each mutated individual, i.e. a "1" becomes "0" and *vice versa*, while with the "bit-flip" mutation, each bit can be reversed independent of the others with a low probability. If the mutation rate is too high with a great number of mutated bits per individual, the evolution of the individuals of the population is equivalent to a random walk in the search space, and the genetic algorithm loses of its effectiveness.

When a bit string represents a vector of integer or real numbers, the positive effects of the mutation are countered by the difficulty of crossing the *Hamming cliffs*, which appear because of the conversion of the bit strings towards real number vectors. For example, let us consider the function $D(x)$:

$$D(x) = \begin{cases} 256 - x^2 & \text{if } x \leq 0 \\ 0 & \text{otherwise} \end{cases}$$

Let us use a string $b(x) = \{b_1(x), \ldots, b_5(x)\}$ of five bits to represent an individual x that ranges from -16 to +15, and thus has 32 possible different values. $b(x)$ can be simply defined as the number $x + 16$ with a base of 2. The optimum of $D(x)$ is obtained for $x = 0$, which thus corresponds to $b(0) = \{1, 0, 0, 0, 0\}$. The value $x = -1$, obtained from the string $\{0, 1, 1, 1, 1\}$, gives the highest fitness apart from the maximum: this value will thus be favored by the selection operators. However, it is noticed that there is no common bit between $\{1, 0, 0, 0, 0\}$ and $\{0, 1, 1, 1, 1\}$. This means that there is no other individual with whom $\{0, 1, 1, 1, 1\}$ can be mated to give $\{1, 0, 0, 0, 0\}$. As for the mutation operator, it will have to change the 5 bits of the genotype $\{0, 1, 1, 1, 1\}$ simultaneously to give the optimum because the Hamming distance [2] between the optimum and the individual which has the nearest fitness is equal to the size of the strings. Hence, we encounter a *Hamming cliff* here. It is not very likely to cross it with a "bit-flip" mutation, and it is impossible with the "deterministic" mutation unless it flips all the bits of a bit string, which is never used. But the mutation will be able to easily produce the optimum if there are individuals in the population that differ in only one bit of the optimal string, here these individuals are:

string $b(x)$	x	$D(x)$
$\langle 0, 0, 0, 0, 0 \rangle$	-16	0
$\langle 1, 1, 0, 0, 0 \rangle$	8	0
$\langle 1, 0, 1, 0, 0 \rangle$	4	0
$\langle 1, 0, 0, 1, 0 \rangle$	2	0
$\langle 1, 0, 0, 0, 1 \rangle$	1	0

Unfortunately they have completely null fitness and thus have very few chances to "survive" from one generation to the next one.

This tedious phenomenon, which hinders the progress towards the optimum, can be eliminated by choosing a *Gray code* which ensures that two successive integers will have binary representations that differ only in one bit. Starting from strings $b(x)$ that represent integer numbers in base two, it is easy to obtain a Gray code $g(x) = \{g_1(x), \ldots, g_l(x)\}$ by performing, for each bit i, the operation:

$$g_i(x) = b_i(x) \oplus b_{i-1}(x)$$

where the operator $\oplus$ implements the "exclusive or" operation and $b_0(x) = 0$. Conversely, the string of l bits $b(x) = \{b_1(x), \ldots, b_l(x)\}$ can be obtained from the string $g(x) = \{g_1(x), \ldots, g_l(x)\}$:

$$b_i(x) = \bigoplus_{j=1}^{i} g_j(x)$$

---

[2] Hamming distance: number of different bits between two bit strings of the same length.

The Gray codes of $\{0, 1, 1, 1, 1\}$ and $\{1, 0, 0, 0, 0\}$ are respectively $\{0, 1, 0, 0, 0\}$ and $\{1, 1, 0, 0, 0\}$. The mutation of the bit g_1 is then enough to reach the optimum. A Gray code is thus desirable from this point of view. Moreover, it modifies the landscape of the fitness function by reducing the number of local optima created by transcribing a "real or integer vector" towards a "binary string". It will be noted however that the Hamming cliffs are generally not responsible for dramatic fall in the performance of the algorithm.

3.4.3 Real representation

The real representation allows an evolutionary algorithm to operate on a population of vectors in a bounded search domain Ω included in $\mathbf{R}^n$. Let us assume that, in a given generation, the individuals X of a population are drawn in the search domain according to a probability distribution characterized by a density $p(x)$, where x is a point in Ω. Then this distribution has an expectation:

$$E(X) = \int_\Omega xp(x)dx$$

and a variance:

$$V(X) = \int_\Omega x^2 p(x)dx - E^2(X)$$

If λ, the size of the population of the offspring, is large enough, these values are approached by the empirical expectation:

$$\hat{E} = \frac{\sum_{i=1}^\lambda x_i}{\lambda}$$

and the empirical variance:

$$\hat{V} = \frac{\sum_{i=1}^\lambda x_i^2}{\lambda} - \hat{E}^2$$

The empirical variance can be regarded as a measurement of diversity in the population. If it is null, then all the individuals are at the same point in Ω. By adopting a mechanical analogy, $\hat{E}$ is the centroid of the population, while $\hat{V}$ is its moment of inertia from the centroid, by allotting to each individual a mass unit. It is interesting to evaluate these values after application of the variation operators.

Crossover

Let us consider two points x and y in space $\mathbb{R}^n$ corresponding to two individuals selected to generate offspring. After application of the crossover operator, one or two offspring x' and y' are drawn randomly, according to a probability distribution which depends on x and y.

Crossover by exchange of components.

It is about an immediate generalization of the binary crossovers, which consists in exchanging some real components of two parents. One can thus find all the variants of the binary crossover, in particular the "single point", "two point" and "uniform" crossovers (see figure 3.12). The last variant is also called "discrete recombination" according to the terminologies of the *Evolution Strategies*. This type of crossover modifies neither $E(X)$ nor $V(X)$.

Voluminal BLX-α crossover.

The *voluminal BLX-α* operator generates two offspring chosen uniformly inside a hyper-rectangle with sides parallel to the coordinate axes, such that the two parents and the coefficient α define one of its longest diagonals (see figure 3.13). Let x_i and y_i be the components of the two parents x and y respectively, for $1 \leq i \leq n$; an offspring z will have its components:

$$z_i = x_i - \alpha(y_i - x_i) + (1 + 2\alpha)(y_i - x_i) \cdot \mathcal{U}(0, 1)$$

where $\mathcal{U}(0, 1)$ indicates a random number drawn uniformly in the interval $[0, 1]$.

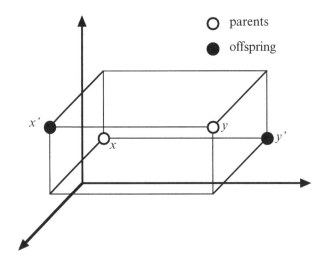

Fig. 3.12. Uniform crossover; an individual resulting from the crossover of x and y is located on a vertex of a hyper-rectangle with sides parallel to the coordinate axes such that a longest diagonal is the segment (x, y).

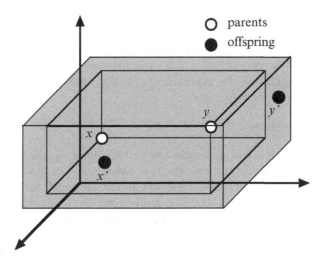

Fig. 3.13. Voluminal BLX-α crossover; an individual resulting from the crossover of x and y is located inside a hyper-rectangle with sides parallel to the coordinate axes such that a longest diagonal passes through x and y.

The voluminal BLX-α crossover does not modify $E(X)$, but changes the value of $V(X)$. Let $V_c(X)$ be the variance of distribution of the population after crossover:

$$V_c(X) = \frac{(1 + 2\alpha)^2 + 3}{6} V(X)$$

The variance after crossover decreases if:

$$\alpha < \frac{\sqrt{3} - 1}{2} \approx 0.366$$

In this case, it is said that the crossover is *contracting*, and the iterative application of the operator alone leads the population to collapse on its centroid. In particular, if $\alpha = 0$, z is located in the hyper-rectangle such that a longest diagonal is the line segment (x, y). In this case, $V_c(X) = \frac{2}{3}V(X)$. After iterative application of this operator alone for g generations, and for an initial population variance $V_0(X)$, the variance becomes:

$$V_{cg}(X) = \left(\frac{2}{3}\right)^g V_0(X)$$

The variance tends quickly towards 0! It is thus seen that the risk of premature convergence is increased with a BLX-0 operator.

If $\alpha > \frac{\sqrt{3}-1}{2}$, the variance is increasing if the domain is $\mathbf{R}^n$. In practice, for a bounded search domain Ω, the variance is stabilized with a non-null value. The "borders" of the search domain can be explored. The possible optima which are there will be more easily found and retained. A usual value is $\alpha = 0.5$.

It is also possible to show [Nomura and Shimohara, 2001] that the operator reduces the possible correlations which exist between the components of the vectors of the population. Its repeated application makes the coefficients of correlation converge towards zero.

Linear BLX-α crossover .

The *linear BLX-α* operator generates one or two offspring, randomly chosen on a line segment passing through the two parents, α being a parameter of the evolutionary algorithm. This crossover is known under several denominations, according to the authors who studied it, like *arithmetic crossover*, or the *intermediary recombination* for the "Evolution Strategies", which are equivalent to BLX-0.

x and y are the points corresponding to two individuals in the search space. An individual z resulting from the crossover of x and y is chosen according to a uniform distribution on a line segment passing through x and y:

$$z = x - \alpha(y - x) + (1 + 2\alpha)(y - x) \cdot \mathcal{U}(0, 1)$$

where $\mathcal{U}(0, 1)$ indicates a random number uniformly drawn in the interval $[0, 1]$. If I is the length of the line segment $[x, y]$, z could be on the segment of length $I \cdot (1 + 2\alpha)$ centered on the segment $[x, y]$ (figure 3.14).

The linear BLX-α crossover does not modify $E(X)$, but changes the value of $V(X)$ in a way similar to the voluminal BLX-α crossover. On the other hand, it is noted that the possible correlations existing between the components of the individuals of a population are preserved by the linear operator, which shows a behavior basically different from that observed for the voluminal operator.

Mutation

The mutation generally consists in the addition of a "small" random value to each component of an individual, according to a zero average distribution, with a variance possibly decreasing with time. In this way, it is assured that the mutation leaves the centroid of the population unchanged.

Uniform mutation.

The simplest mutation technique adds to an individual x, belonging to a domain Ω in $\mathbb{R}^n$, a random variable of uniform distribution in a hyper-cube $[-a, +a]^n$. However, such a mutation does not allow an individual trapped in a

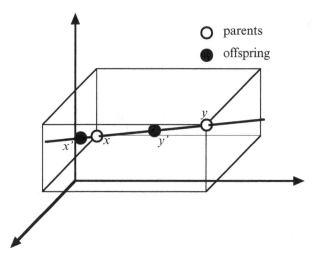

Fig. 3.14. BLX-α crossover; an individual resulting from the crossover of x and y is located on the line defined by x and y, possibly outside the segment $[x, y]$.

local optimum located on a peak broader than the hypercube to escape from it. To avoid this disadvantage, it is preferable to use an unlimited support distribution.

Gaussian Mutation

The Gaussian mutation is one of the most widely used for the real representation. It adds to an individual x a Gaussian random variable $\mathcal{N}(0, \sigma)$, of zero average and standard deviation σ which has a probability density of

$$f(y) = \frac{1}{\sigma\sqrt{2\pi}}e^{-\frac{1}{2}\left(\frac{y}{\sigma}\right)^2}$$

The problem is then an adequate choice of σ, presumably identical for the n components of the vector x in the simplest versions of the operator. In theory, it is possible to escape from a local optimum irrespective of the width of the peak where it is, since the support of a Gaussian distribution is unlimited, but if σ is too small that could happen after far too many attempts. A solution would be to use distributions with thicker tails, such as the Cauchy or Laplace distributions which have proven their advantages [Yao and Liu, 1996] [Montana and Davis, 1989].

However, the Gaussian mutation is often preferred, by adapting the value of σ during the evolution, according to various strategies.

Gaussian mutation and the 1/5 rule.

From a study on two very different test functions with an elitist Evolution Strategy $(1 + 1)-\text{ES}$[3], Rechenberg [Rechenberg, 1973] [Beyer, 2001] calculated optimal standard deviations for each test function that maximize the convergence speed. He observed that for these optimal values, approximately one fifth of the mutations allow to reduce the distance between the individual and the optimum. It deduced the following rule, termed as "one fifth" to adapt σ: *if the rate of the successful mutations is larger than 1/5, increase σ, if it is smaller, reduce σ*. The "Rate of the successful mutations" is the proportion of mutations which make it possible to improve the value of fitness of an individual. Schwefel [Schwefel, 1981] proposed the following rule in practice:

> estimate the rates of beneficial mutations p_s on $10n$ mutations
> **IF** $p_s < 0.2$ **THEN**
> $\quad \sigma(g) \leftarrow \sigma(g - n) \cdot 0.85$
> **ELSE IF** $p_s > 0.2$ **THEN**
> $\quad \sigma(g) \leftarrow \sigma(g - n)/0.85$
> **ELSE**
> $\quad \sigma(g) \leftarrow \sigma(g - n)$

where n is the dimension of the search space and g the index of the current generation.

Self-adaptive Gaussian mutation .

The rule of the "one fifth" requires that σ should have the same value for all the components of a vector x. In this manner, the step of progression towards the optimum is the same in all directions: the mutation is isotropic. However, the isotropy does not make it possible to approach the optimum as quickly as expected when, for example, the isovalues of the fitness function locally take the shape of "flattened" ellipsoids in the neighborhood of the optimum (see figure 3.15). If the step is well adapted in a particular direction, it will not be in the other directions. It is then preferable to consider for each individual an n-dimensional vector of standard deviations σ whose each component σ_i refers to a component of x. The σ_i evolves in a similar way as the variables of the problem under the action of the evolutionary algorithm [Schwefel, 1981]. σ is thus likely to undergo mutations. Let us assume that an individual is represented by a couple of vectors (x, σ). Schwefel proposed that the couple (x', σ'), obtained after mutation, is such that:

$$x'_i = x_i + \mathcal{N}(0, \sigma'_i)$$

$$\sigma'_i = \sigma_i \exp(\tau_0 N + \tau \mathcal{N}(0, 1))$$

---

[3]$(1 + 1)$-*ES*: the population is composed of only one parent individual that generates only one offspring, the best of both is preserved for the next generation.

$$\tau_0 \approx \frac{1}{\sqrt{2n}} \qquad \tau \approx \frac{1}{\sqrt{2\sqrt{n}}}$$

where N indicates a Gaussian random value of average 0 and variance 1, computed for the entire set of n components of $\boldsymbol{\sigma}$, and $\mathcal{N}(0, s)$ represents a Gaussian random variable of average 0, and standard deviation s. σ_i' is thus updated by application of a lognormal perturbation.

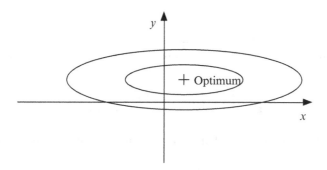

Fig. 3.15. The isotropic mutation is not appropriate if, in the neighborhood of the optimum, the isovalues of the fitness function take the shape of "flattened" ellipsoids.

This self-adaptive mutation was generalized so that it can take into account possible correlations between variables, as in the case of the fitness function whose isovalues are represented in figure 3.16.

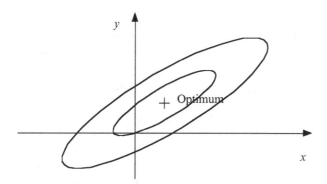

Fig. 3.16. Isovalues of a fitness function for which there is a correlation between variables.

An individual is then considered as a triplet (x, σ, α) where σ is a vector of n standard deviations, as in the previous case, and α is a vector composed of $n(n-1)/2$ elementary rotation angles, which evolve under the action of the evolutionary algorithm. The components α_{kl} undergo mutations according to the following formula:

$$\alpha'_{kl} = \alpha_{kl} + \beta \mathcal{N}(0, 1)$$

where α_{kl} expresses the rotation angle in the plane generated by the basic vectors k and l. The mutated vector x' is obtained from x by the addition of a zero average Gaussian random vector with a covariance matrix $\mathbf{C}$:

$$x' = x + \mathcal{N}(0, \mathbf{C})$$

where $\mathbf{C}$ is obtained from a diagonal matrix $\mathbf{S}$ whose diagonal coefficients are the components of σ' and from a product of elementary rotation matrices $\mathbf{R}(\alpha'_{kl})$:

$$\mathbf{C} = \left(\mathbf{S} \prod_{k=1}^{n-1} \prod_{l=k+1}^{n} \mathbf{R}(\alpha'_{kl}) \right)^T \left(\mathbf{S} \prod_{k=1}^{n-1} \prod_{l=k+1}^{n} \mathbf{R}(\alpha'_{kl}) \right)$$

In practice, the vector x' is calculated according to the following expression:

$$x' = x + \left(\prod_{k=1}^{n-1} \prod_{l=k+1}^{n} \mathbf{R}(\alpha'_{kl}) \right) \mathcal{N}(0, \sigma')$$

where $\mathcal{N}(0, \sigma')$ is a zero average Gaussian random vector with a standard deviation σ'_i for each component i.

Schwefel suggests fixing β at a value close to 0.087 radian, that is approximately 5 degrees. This technique of mutation, although it is powerful, is seldom used because of the amount of memory used by an individual, and its algorithmic complexity of the order of n^2 matrix products for a problem of n variables.

The procedures of self-adapting mutations presented above were specifically studied in detail by the supporters of the *Evolution Strategies*.

3.4.4 Some discrete representations for the permutation problems

There exist many types of combinatorial optimization problems and it is not possible to describe all of them within a restricted space. We will consider here only the *permutation problem* which consist in discovering an order in a list of elements, maximizing or minimizing a given criterion. The traveling salesman problem can be considered as an example. Knowing a set of "cities", as well as the distances between these cities, the traveling salesman must discover the shortest possible path passing by each city once and only once. This NP-complete problem is classically used as a benchmark making it possible to evaluate the effectiveness of an algorithm. Typically, the problems considered comprise several hundreds of cities.

A solution can be represented like a list of integers, each one associated with a city. The list comprises of as many elements as the cities, and each city associated with an element must satisfy the constraint of uniqueness. It is choosen to build individuals satisfying the structure of the problem and to possibly specialize the genetic operators.

Ordinal representation

It is tempting to consider an individual representing an order like an integer vector, and to apply crossovers to the individuals by exchanging components similar to those described in the parts dedicated to the binary or real representations (see sections 3.4.2 and 3.4.3). The ordinal representation makes it possible to satisfy the constraint of uniqueness with the use of these standard crossovers. It is based on an order of reference, for example the natural order of the integers. First the list of the cities O satisfying this order of reference is built, for example: $O = (123456789)$ for 9 cities numbered from 1 to 9. Then an individual is read from left to right. The nth integer read gives the order number in O of the nth visited city. When a city is visited, it is withdrawn from O. For example, let us consider now the individual $\langle 437253311 \rangle$:

- the first integer read in the individual is 4. The first visited city is thus the fourth element in the list of reference O, i.e. the city 4. This city is withdrawn from O. One then obtains $O_1 = (12356789)$;
- the second integer read is 3. According to O_1, the second visited city is 3. This city is withdrawn from O_1 to give $O_2 = (1256789)$;
- the third integer read is 7. The third visited city is thus 9 and one obtains $O_3 = (125678)$ which will be used as list of reference for the next step.

One thus continues until the individual is entirely interpreted. Hence for this example the path is given as: $4 \rightarrow 3 \rightarrow 9 \rightarrow 2 \rightarrow 8 \rightarrow 6 \rightarrow 7 \rightarrow 1 \rightarrow 5$.

But, experimentally, this representation associated with the standard variation operators does not give good results. This shows that it is not well adapted to the problem under consideration, and that the simple satisfaction of the uniqueness constraint is not sufficient. Other ways were explored, which enable the offspring to inherit partially of the order of the cities, or then of the relations of adjacency, which exist in their parents.

Path or sequence representation

In this representation, two successive integers of a list account for two nodes adjacent in the path represented by an individual. Each number in a list must be present once and only once. Useful information lies in the order of these numbers compared to the others. Many variation operators were proposed for this representation. A crossover preserving the order and another preserving the adjacencies, chosen from the most common alternatives in the literature, are presented below.

Uniform order-based crossover.

With the uniform order-based crossover, an offspring inherits a combination of the orders existing in two "parent" sequences. The operator has the advantages of simplicity and, according to L. Davis, one of its proposers [Davis, 1991], it shows a good effectiveness. The crossover is held in three stages (figure 3.17):

- A binary template is generated at random (figure 3.17a).
- Two parents are mated. The "0" (respectively "1"), of the binary mask defines the position preserved in the sequence of the parent "1 " (respectively "2") (see figure 3.17b).
- To generate the offspring "1" (respectively "2"), the non-preserved elements of the parent "1" (respectively "2") are permuted in order to satisfy the order they have in the parent "2" (respectively "1") (figure 3.17c).

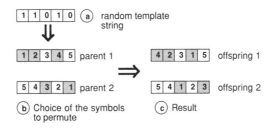

Fig. 3.17. Uniform order-based crossover.

Crossover by edge recombination.

With this class of crossover operators, an offspring inherits a combination of the adjacencies existing in the two parents. This is useful for the non-oriented traveling salesman problem, because the cost does not depend on the route direction in a cycle, but depends directly on the weights between the adjacent nodes of a Hamiltonian cycle.

The edge recombination operator was improved by several authors over several years. The "edge-3" version of Mathias and Whitley [Mathias and Whitley, 1992] is presented now. Let two individuals be selected for mating. They are, for example: ⟨ b, g, j, k, i, e, a, c, l, h, f, d⟩ and ⟨ f, c, b, e, k, a, h, i, l, j, g, d⟩. The first action builds an "edge table" of the adjacencies (see table 3.2) such that to each node corresponds a list of adjacent nodes in both parents: their numbers are from two to four. The adjacencies common to both parents are marked by a * in the edge table.

At the time of action 2, an initial active node is selected at random and all the references to this node are removed from the table.

Action 3 consists in choosing the edge which, from the active node, led to an adjacent node marked by a * or, failing this, having the shortest list of adjacencies. If there are several equivalent options, the choice of the next node is carried out at random. The adjacent node chosen becomes the new active node added in the "offspring" tour. All the references to this node are removed from the adjacency lists of the edge table.

Table 3.2. A table of adjacencies.

nodes	edge lists	nodes	edge lists
a	c, e, h, k	g	*j, b, d
b	d, g, e, c	h	f, l, i, a
c	l, a, b, f	i	e, k, l, h
d	b, *f, g	j	k, *g, l
e	a, i, k, b	k	i, j, a, e
f	*d, h, c	l	h, c, j, i

Action 4 builds a string or possibly a complete tour. It consists of the repetition of action 3 as long as the adjacency list of an active node is non-empty. If it is empty, then the initial node is reactivated to start again from the beginning of the string, but in the reverse direction, until the adjacency list of the active node is empty again. Then action 4 is concluded. It should be noted that the initial node could not be reactivated because its adjacency list is empty due to previous removing of the edges.

As long as a complete tour is not generated, another active node is chosen at random, among those which do not belong to any partial tour already built by previous action 4 executions. Then action 4 is initiated again. The application of the operator is thus summarized with the sequence of actions 1, 2 and as many actions 4 as necessary.

It is hoped that the operator will create few partial tours, and thus few foreign edges which do not belong to the two parents. The "edge-3" operator is powerful from this point of view.

Let us assume that the node a is selected at random as being initially active in the example of table 3.2. Table 3.3 shows an example of execution of the algorithm. The progress in the construction of the Hamiltonian cycle is presented in the last row. The active nodes are underlined. When an active node is marked (1), this means that the next one has to be chosen at random because of the existence of several equivalent possibilities. When it is marked (2), it is about an end of the string: there is no more possible adjacency, which implies to move again in reverse direction by reactivating the initial node a. It was necessary to apply the action 4 only once, which thus generated a complete tour $\langle$l,i,h,a,c,f,d,g,j,k,e,b$\rangle$. Thus, except for the edge (bl), all the other edges originate from one of the two parents.

Table 3.3. Example of bearing of the problem.

stages:	1	2	3,4	5,6	7,8,9	10	11
a	c,e,h,k	e,h,k	e,h,k	e,h,k	h		
b	d,g,e,c	d,g,e	g,e	e			
c	l,b,f	l,b,f	l,b	l,b	l	l	l
d	b,*f,g	b,*f,g	b,g	b			
e	i,k,b	i,k,b	i,k,b	i,k,b	i	i	
f	*d,h,c	*d,h	h	h	h		
g	*j,b,d	*j,b,d	*j,b	b			
h	f,l,i	f,l,i	l,i	l,i	l,i	l,i	l
i	e,k,l,h	e,k,l,h	e,k,l,h	e,k,l,h	l,h	l	l
j	k,*g,l	k,*g,l	k,*g,l	k,l	l	l	l
k	i,j,e	i,j,e	i,j,e	i,e	i	i	
l	h,c,j,i	h,j,i	h,j,i	h,i	h,i	i	
tour:	$\underline{a}^{(1)}$	a,$\underline{c}$	a,c,f,$\underline{d}^{(1)}$	a,c,f,d, $\underline{g}$,$\underline{j}^{(1)}$	a,c,f,d, g,j,k,e, $\underline{b}^{(2)}$	a,c,f,d, $\underline{h}^{(1)}$,a,c, f,d,g,j, k,e,b	$\underline{i}$,h,a,c, f,d,g,j, k,e,b

Mutations of adjacencies.

The "2-opt" mutation is common for the path representation. It is usually used for the Euclidean traveling salesman problem because of its geometrical properties. It consists in randomly choosing two positions in a sequence, then to reverse the sub-sequence delimited by the two positions. Let the sequence be $\langle 987654321 \rangle$, where two positions drawn at random are 3 and 8. Then the sub-sequence located between positions 3 and 8 is reversed which gives the new sequence: $\langle 98\underline{4567}321 \rangle$. The figure 3.18 shows the effect of the operator applied to this sequence for the path representation. The operator can be generalized by choosing more than two inversion positions of sub-sequences.

Mutations of permutations.

If an individual represents a solution for a scheduling problem, the "2-opt" operator modifies the order of a large number of elements, on an average of $l/2$ if l is the size of a sequence. However, the route direction of a sub-sequence, which was irrelevant for the traveling salesman problem, is essential in this new context. Thus, the modifications which the adjacency mutation applies to a sequence are important. However, a mutation should be able to apply often small perturbations to a solution in order to explore its close neighborhood. This is why other types of mutations were also proposed. The simplest one consists in withdrawing an element chosen at random within a sequence to

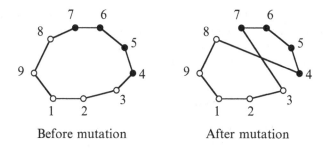

Before mutation   After mutation

Fig. 3.18. An example of 2-opt mutation.

insert it into another position. Several operators are described in the literature, like the mutation by exchange, where two positions in a sequence are chosen at random and the elements in these positions are exchanged. The performances offered by the variants of mutations depend closely on the properties of the problem dealt with.

3.4.5 Representation of parse trees for the genetic programming

Till now we have seen that, for a problem under consideration, an evolutionary algorithm has to discover the optimal values (or almost) of a well defined set of parameters of a representation built a priori. Still it is necessary that an adequate representation be proposed. When this is not the case, the problem can be considered at another level: to discover the data-processing program that will be able to solve it. Admittedly, the human approach to this option results in building a good model so that it is possible to write the program, but as this model is not available by hypothesis, this approach seems to be fruitless. However, the modeling stage is not essential if the program is built *automatically*, with the help of the definition of a quality criterion, which will have to be maximized by the building process. This last problem can involve an evolutionary technique and in this context, we can talk about genetic *programming*, a term popularized by John Koza [Koza, 1992], who strongly contributed to the field by his elegant approach, the quantity of his work and by the organization of the first series of international conferences in the field.

A program modifiable by the action of evolutionary operators, such as the crossover and the mutation, is not represented as a sequence of instructions with rigid and complex syntax, such as those encountered in the current programming languages. These are indeed designed so that the least clerical error is sanctioned by a rejection on behalf of the compiler or the interpreter in order to reduce the risks to generate a wrong program by multiplying the consistency checks. For example, if a program written in language C was viewed like a vector of characters, variation operators that would act on the level of the character would be almost unlikely to generate any valid program. It would be necessary that the chosen representation for a valid program be such that a random modification of some of its elements often leads, if not always, to another valid program. This representation exists, it is generated by a compiler in the form of a *parse tree*, after the parsing stage.

A parse tree is a tree containing two types of nodes:

- the "internal nodes" or "non-terminal symbols";
- the leaves or "terminal symbols".

It is such that the children $(c_{i1}, \ldots, c_{in})$, read from left to right of an internal node i, defines a rule $i \rightarrow (c_{i1}, \ldots, c_{in})$. The root of the tree is a non-terminal symbol. A parse tree associated with the arithmetic expression sqrt(b * b - 4 * a * c) is given in figure 3.19a by using the rules of current priority of the arithmetic operators with only one non-terminal symbol Expr.

It is noted that the set of the non-terminal symbols contains only the element Expr and that it thus does not bring any information. The representation of the tree can be reduced by considering the associativity of multiplication and by placing the symbols of the operators "sqrt ", "*" and "-" in the internal nodes (see figure 3.19b).

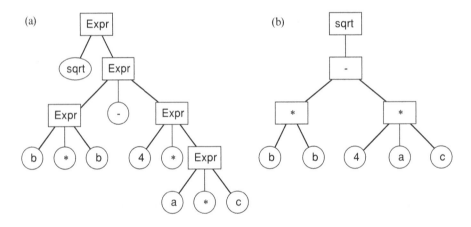

Fig. 3.19. A parse tree for the expression sqrt(b*b-4*a*c) (a), and its reduced representation (b).

The reduced version of the parse tree can be written recursively with a parenthesized expression, which is: "(sqrt(-(* b b)(* 4 a c)))". This is also the original expression translated into language LISP. This language is conceived to be able to generate directly a parse tree from a source text, after a parse process that is reduced to its simpler form. For this reason LISP became one of the favorite languages for the genetic programming.

The variation operators act on the structure of the parse trees and the contents of the internal nodes or the leaves. The set of the terminal symbols should have a small size, while being relevant for the problem under consideration. For example, it would be a priori unsuitable (though possible) to choose a set of arithmetic operators and real variables to solve a Boolean problem. Thus, all the knowledge available about the problem must be exploited to provide a set of primitives and terminals adequate for the evolutionary engine. The parse trees must moreover have a mechanism of regulation of their sizes. Otherwise, the trees of the population will tend to grow indefinitely during generations, consuming thus unnecessarily more and more memory and computing power. The regulation mechanism can be simply implemented by limiting the maximum depth of the trees, or the maximum number of nodes for any tree generated by the variation operators.

The first examples of genetic programming presented by J. Koza was restricted to the evolution of parse trees of expressions. He has empirically shown that his approach makes it possible to discover relevant programs for a great number of application examples, e.g. design of complex objects like electronic circuits, with a significantly higher efficiency than the chance could do. The approach was improved by the introduction of the "*automatically defined functions*" (ADF) which decompose a program into a set of subroutines called by a main program, all that being integrated within the same tree [Koza, 1994].

The parse trees are transformed by the action of evolutionary operators such as the crossover and the mutation. The structures of the individuals are very different from those which were described previously for other representations. It was thus necessary to propose specific variation methods.

The crossover of parse trees

Its operation is very simple. First, two trees $\mathbf{A}$ and $\mathbf{B}$ are randomly selected from a population to be mated. In the next step, a branch is selected randomly in each tree to be cut. Let $(\mathbf{A}_p, \mathbf{A}_f)$ be the cut point between the parent node $\mathbf{A}_p$ and the child node $\mathbf{A}_f$ (see figure 3.20a). Similarly, let $(\mathbf{B}_p, \mathbf{B}_f)$ be the cut point in the tree $\mathbf{B}$ (see figure 3.20b). $\mathbf{A}_f$ and $\mathbf{B}_f$ thus become the roots of two sub-trees. During the third stage of the crossover, the sub-trees are exchanged so that $\mathbf{B}_f$ is a child of $\mathbf{A}_p$ (see figure 3.20c), and $\mathbf{A}_f$ a child of $\mathbf{B}_p$ (figure 3.20d).

This general principle presented by N Cramer in 1985 [Cramer, 1985] can be refined according to various points of view. First, it is necessary to respect the limit of size assigned to the trees of the population, so that they do not become unnecessarily gigantic. If the chosen cut points do not respect it, then the crossover cannot take place. The attitude adopted in this case is a parameter of the crossover. It could be at least one of the following:

- selecting a new couple and attempting to reactivate a crossover until an offspring satisfies the size constraint;
- or attempting to choose different cut points on the two selected parents, until a satisfactory offspring is obtained.

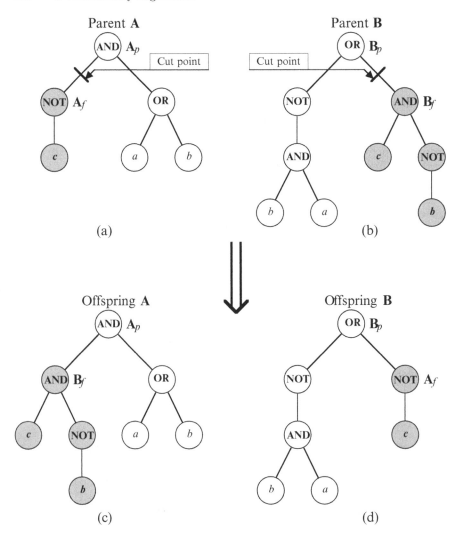

Fig. 3.20. Crossover of two parse trees.

In another point of view, the handled data could be typed. In this case, it will be necessary to take precaution in such a way that the arguments corresponding to the nodes $\mathbf{A}_f$ and $\mathbf{B}_f$ are of compatible types, which means that node $\mathbf{A}_f$ having been selected, the choice of $\mathbf{B}_f$ will be restricted among the set of the nodes of $\mathbf{B}$ such that their types are compatible with $\mathbf{A}_f$ [Montana, 1995].

In addition, J. Koza noticed that the number of terminals is of the same order of magnitude as the number of internal nodes. According to him, the crossover of leaves is similar to mutations, which does not make significant modifications to the trees, thus slowing down the evolution. He thus suggested to limit the probability to crossover a leaf by proposing a rate of about ten percent.

Mutation of parse trees

The genetic programming of J. Koza plants its roots in the domain of the genetic algorithms [Holland, 1992] which attributes only a minor role to the mutation. Within this framework, the purpose of the mutation is only to maintain diversity in a too small population, denying it of any notable role in the search for an optimal solution. This is why the mutation was not originally present in the genetic programming. However, other evolutionary models have shown the usefulness of the mutation and hence some genetic programming specialists assert today that it can have a significant role with the parse trees.

J. Koza [Koza, 1992] proposed a simple model of mutation. It consists in substituting a sub-tree whose root is chosen at random within the mutated individual, by another tree built at random. Thus, at the time of the first stage of mutation, a root M is selected (see figure 3.21a). Then, a tree of root N is generated at random (see figure 3.21b). Finally the sub-tree of root M is replaced by the tree of root N (figure 3.21c).

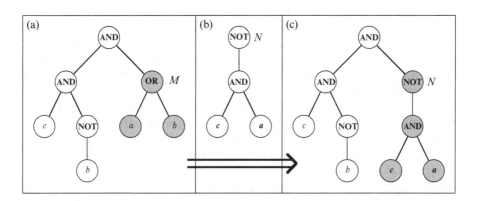

Fig. 3.21. Mutation of a parse tree.

If a tree uses typed operators and data, it is easy to adapt this mutation so that the root of the tree generated at random is of compatible type with the root of the sub-tree which will be replaced. Angeline [Angeline, 1996] proposes another type of mutation, which permutes two sub-trees chosen at random within an individual. It also suggests, as another variant, to simply substitute the symbol associated with a node by another symbol chosen randomly, the parent and offspring of this node being unchanged.

There still is, irrespective of the type of mutation, the constraint of maximum size for a mutated individual which must be satisfied. One of the two strategies enumerated above for the crossover can be directly adapted to this end. But also, as regards the mutation of J. Koza, it is possible to generate random trees with sizes controlled according to the size of the replaced sub-tree.

3.5 Particular case of the genetic algorithms

The simple genetic algorithm follows the outline of an evolutionary algorithm, such as the one presented in figure 3.1 with a notable originality: they implement a genotype – phenotype transcription that is inspired by the natural genetics. This transcription precedes the phase of fitness evaluation of the individuals. A *genotype* is often a binary symbol string. This string is decoded to build a solution of a problem represented in its natural formalism: it is viewed as the *phenotype* of an individual. This last one can then be evaluated to give a fitness value, that can be exploited by the selection operators.

The flowchart of a simple genetic algorithm is presented in figure 3.22. It is noticed that it implements a proportional selection operator (see section 3.3.3) and a generational replacement, i.e. the population of the offspring replaces that of the parents completely. Another classical version uses a steady state replacement (section 3.3.6). The variation operators work on the genotypes. As those are bit strings, the operators of crossover and mutation presented in the section 3.4.2 related to the binary representation are often used. The crossover is regarded as the essential search operator. The mutation is usually applied with a small rate, in order to maintain a minimum degree of diversity in the population. The representation being based on bit strings, the difficulty is to discover a good coding of the genotype, such as the variation operators in the space of the bit strings produce viable offspring, often satisfying the constraints of the problem. It is generally not a trivial job ...

Holland, Goldberg and many other authors worked on a mathematical formalization of the genetic algorithms based on a "Schema Theorem" [Goldberg, 1989], whose utility is controversial. A first glance enables it to justify the choice of a binary representation. However research works using this theorem did not prove finally very useful to model an evolution. Many counterexamples showed that the conclusions formulated from considerations deduced from this theorem are debatable, in particular even the choice of the binary representation.

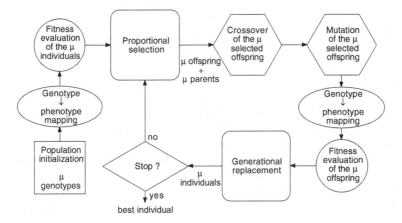

Fig. 3.22. A simple genetic algorithm.

The genetic algorithms have been subject to many modification suggestions in order to improve their performances or to extend their application
domains. Thus, the bit strings were replaced by representations closer to the
formalism of the problems dealt with, avoiding the hard question of the design of an effective coding. For example research works using the "Real Coded
Genetic Algorithms" use the real representations discussed in section 3.4.3.
In addition, the proportional selection is often replaced by other forms of
selection. These modifications are sufficiently significant so that the specific
features of the genetic algorithms disappear compared to the diversity of the
other evolutionary approaches.

3.6 Some considerations on the convergence of the evolutionary algorithms

Let $\hat{f}_n$ be the fitness of the best individual(s) obtained between generation 0
and n. It is said that an evolutionary algorithm converges towards a global
optimum if the stochastic sequence $\{\hat{f}_n\}$ converges towards f^*, the global optimal value of the fitness function. There is no theory of global convergence for
the evolutionary algorithms. It is easy to show that for operators such as the
BLX-0 crossover for the real representation, the convergence is not guaranteed
in the absence of mutation. And even in the presence of mutation, the convergence speed of an evolutionary algorithm which uses BLX-0 could be lower
by several orders of magnitude than the speed of the simple random search,
according to the landscape of the fitness function. Because the representation
is not fixed a priori, the question of the convergence of an evolutionary algorithm is reformulated in the following way: which are the properties that the

variation and selection operators must satisfy so that convergence is guaranteed [Rudolph, 1996]? And in this case, what will be the expected convergence speed? Partial answers could be put forward for particular representations and operators, or, for that matter precise answers but only for some very simple problems, the representation and the operators being fixed.

Thus, a Markovian analysis proves the convergence, as defined above, for a simple genetic algorithm for a population of bit strings, provided that the mutation rate is non-zero. But such a result is weak, because it is noticed that a simple random search in the space of the bit strings also converges in the same manner towards the optimum. Nothing is mentioned about the convergence speed and it can be noted that the crossover does not play any role in the result of convergence [Rudolph, 1994].

The convergence of the Evolution Strategies has been proven for very simple functions, like the "corridor" function and the "sphere" function, with an estimated convergence speed [Rechenberg, 1965] [Beyer, 2001]. These are the studies which made it possible to develop by extrapolation the rule of the "one fifth" enabling to adapt the value of the variance of the Gaussian mutation at any moment of the evolution.

3.7 Conclusion

This chapter has presented a set of principles and algorithmic techniques to implement the various operators that function in an evolutionary algorithm. Like building blocks, they can be chosen, configured and assembled according to the flowchart of the generic evolutionary algorithm (see figure 3.1) in order to solve a given problem as efficiently as possible. Obviously, specific choices of operators allow to reconstitute a Genetic Algorithm, an Evolution Strategy, or an Evolutionary Programming method such as designed by the pioneers of evolutionary computation in years 1960–70. However, the references to these original models, which have merged today to evolve one unifying paradigm, should not influence the engineer or the researcher in his choices. He should on the contrary concentrate on the essential and more important questions, e.g. the choice of a good representation, a fitness function corresponding well to the problem posed and finally on the efficient variation operators for the chosen representation.

The solution of industrial problems which are typically multicriteria, must satisfy constraints and which, too often, cannot be completely formalized, requires the implementation of additional mechanisms within the evolutionary algorithms. These aspects are treated in chapter 6 of this book.

3.8 Glossary

allele: within the framework of genetic algorithms: a variant of a gene, i.e. the value of a symbol in a specified position of the genotype.

chromosome: within the framework of genetic algorithms: synonymous to "genotype".

crossover: combination of two individuals to form one or two new individuals.

fitness function: function giving the value of an individual.

generation: iteration of the basic loop of an evolutionary algorithm.

gene: within the framework of the genetic algorithms: an element of a genotype, i.e. one of the symbols of a symbol string.

genotype: within the framework of the genetic algorithms: a symbol string generating a phenotype at the time of a decoding phase.

individual: an instance of solution for a problem dealt with by evolutionary algorithm.

locus: within the framework of the genetic algorithms: position of a gene in the genotype, i.e. the position of a symbol in a symbol string.

mutation: random modification of an individual.

search operator: synonymous to "variation operator".

replacement operator: determines which individuals of a population will be replaced by the offspring. It thus makes it possible to create the new population for the next generation.

selection operator: determines how much time a "parent" individual generates "offspring" individuals.

variation operator: operator modifying the structure, the parameters of an individual, such as the crossover and the mutation.

phenotype: within the framework of the genetic algorithms: set of the observable manifestations of the genotype. More specifically, it is an instance of solution for the problem dealt with, expressed in its natural representation obtained after decoding the genotype.

population: the set of the individuals who evolve simultaneously under the action of an evolutionary algorithm.

recombination: synonymous to "crossover".

3.9 Annotated bibliography

[Baeck et al., 2000a, Baeck et al., 2000b]: An "encyclopedia" of the evolutionary computation in which, as it should be, the most recognized specialists in this field have contributed. The vision offered by these two volumes is primarily algorithmic.

[Koza, 1992, Koza, 1994]: Two reference books written by the well known pioneer of the genetic programming. The first volume exposes the basic concepts of the genetic programming viewed by J. Koza. The second introduces the concept of "automatically defined functions". The largest portion of these books, which comprise of more than seven hundred pages in each volume, is devoted to the description of examples of applications resulting from a large variety of domains. They are useful to help the reader to realize the potentials of the genetic programming. There is also a third volume published in 1999 which contains a significant part dedicated to the automated synthesis of analogical electronic circuits.

[Goldberg, 1989]: The first and the most famous book in the world about the genetic algorithms. It was published in 1989, and has not been revised since then. As a result, the large part of the current knowledge on genetic algorithms, a field that evolves very quickly, is not in this book.

4

Ant Colony Algorithms

4.1 Introduction

Ant colony algorithms form a class of recently proposed metaheuristics for difficult optimization problems. These algorithms are initially inspired from the collective behaviors of trail deposit and follow-up, observed in the ant colonies. A colony of simple agents (the *ants*) communicate indirectly via dynamic modifications of their environment (*trails of pheromones*) and thus propose a solution for a problem, based on their collective experience.

The first algorithm of this type (the "Ant System") was designed for the traveling salesman problem, but the results were not very encouraging. However, it initiated the interest for the metaphor among the research community and since then several algorithms have been proposed, some of them showing very convincing results.

This chapter puts stress initially (section 4.2) on the biological aspect underlying these algorithms. In our view it is interesting to put side by side the design and the use of this metaheuristic algorithm along with the biological theories which inspired it. Section 4.3 describes in detail the first ant colony algorithm proposed and some of its principal variants. This is followed by some gradual developments which can be useful to discover the large variety of possible adaptations of these algorithms (section 4.4). Then, the operating principles of the metaheuristics are studied from section 4.5, which is immediately followed by the research perspective in this field (section 4.6). Thereafter a conclusion is presented on the whole chapter (section 4.7) and a bibliography is proposed with accompanying notes to look further and dig deeper into the subject (section 4.8).

4.2 Collective behavior of social insects

4.2.1 Self-organization and behavior

Self-organization

The *self-organization* is a phenomenon described in many disciplines, notably in the fields of physics and biology. A formal definition has been proposed [Camazine et al., 2000, p.8]:

> Self-organization is a process in which *pattern* at the global level of a system *emerges* solely from numerous interactions among *lower-level* components of the system. Moreover, the rules specifying interactions among the system's components are executed using only local information, without reference to the *global* pattern.

Two terms need clarification for a better understanding, "pattern" and "to emerge". Generally, the first one applies to an "organized arrangement of objects in space or time" (figure 4.1). Additionally, an *emerging* property of a system is a characteristic which appears unforeseen (not being *explicitly* determined), from the interactions among the components of this system.

Thus, the crucial question is to understand how the components of a system interact with each other to produce a complex pattern (in relative sense of the term, i.e. *more* complex than the components themselves). A certain number of necessary phenomena have been identified: these are the processes of *feedback* and the management of the *information flow*.

The *positive feedbacks* are processes which result in reinforcing the action, for example by amplification, facilitation, self-catalysis, etc. Positive feedbacks are able to amplify the *fluctuations* of the system, permitting the updating of even imperceptible informations. Such processes can easily lead to an explosion of the system, if they are not controlled by applying *negative feedbacks*. Hence negative feedbacks act as stabilizers for the system. When they are coupled, such feedback processes can generate powerful models.

Within the framework of biological behavior, it is easy to understand that the interactions among the components of a system will very often give rise to *communication* processes i.e. transfer of information between individuals. Generally, individuals can communicate, either by means of signals, i.e. by using a specific means to carry information, or by means of indices, where information is carried accidentally. In a similar manner, information can come directly from other individuals, or pass via the state of a work in progress. This second possibility of exchanging information, by means of modifying the environment, is called the *stigmergy*.

Generally, all these processes are more or less inter-connected, allowing a system consisting of a large number of individuals to act together to solve problems that are too complex for a single individual.

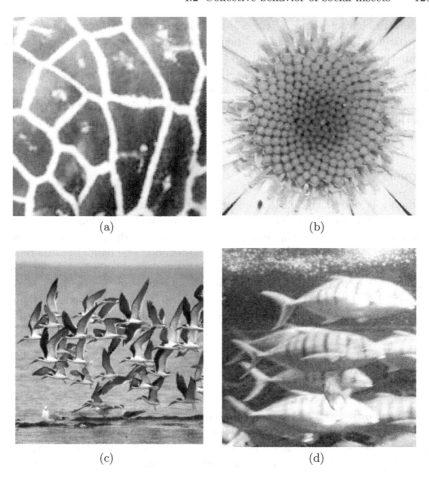

Fig. 4.1. Examples of observable *patterns* in biological systems. (a) motives for the dress of a reticulated giraffe (U.S. Fish and Wildlife Service, Gary M. Stolz), (b) double spiral of Fibonacci in the heart of a daisy, (c) birds flocking, (d) fish schooling.

Certain characteristics of the self-organized systems are very interesting, in particular their *dynamism*, or their capacity to generate *stable* patterns. Within the framework of the study of the behavior of the social insects, certain concepts related to the principle of self-organization deserve to be underlined: the intrinsic *decentralisation* of these systems, their organization in *dense heterarchy* and the recurring use of the *stigmergy*. Indeed, these concepts are sometimes used to view the same problem from different angles and partially cover the principles of self-organization.

Stigmergy

Stigmergy is one of the basic concepts for the creation of ant colony metaheuristics. It is precisely defined as a "form of communication by means of modifications of the environment", but one can utilize the term "indirect social interactions" to describe the same phenomenon. The biologists differentiate the "quantitative stigmergy" from the "qualitative" one, but the process in itself is identical. An example of the use of stigmergy is described in the section 4.2.2. The great force of stigmergy is that the individuals exchange information by means of the task in progress, to achieve the *state* of the total task in advance.

Decentralized control

In a self-organized system, there is no decision-making at a given level, in a specified order and no predetermined actions. In fact, in a decentralized system, each individual has a *local* vision of his environment, and thus does not know the problem as a whole. The literature of the multi-agent systems (see [Weiss, 1999] for an initial approach) often employs this term or that of "distributed artificial intelligence" [Jennings, 1996]. However, generally this discipline tends to study more complex behaviors patterns, founded in particular in cognitive sciences. To be precise, the advantages of decentralized control are the *robustness* and the *flexibility* [Bonabeau et al., 1999]. Robust systems are desired because of their ability to continue to function in the event of breakdown of one of their components; flexible devices are welcome, because they can be useful for dynamic problems.

Dense heterarchy

The dense heterarchy is a concept borrowed directly from biology [Wilson and Hölldobler, 1988], used to describe the organization of the social insects, and more particularly of the ant colonies. The concept of heterarchy describes a system where not only the global level properties influence the local level properties, but also the activities in the local units can influence, in return, the global levels. The heterarchy is known as *dense* in the direction in which such a system forms a highly connected network, where each individual can exchange information with any other. This concept is to some extent contrary to that of *hierarchy* where, in a popular but erroneous vision, the queen would control her subjects while passing orders in a *vertical* structure, whereas, in a *heterarchy*, the structure is rather *horizontal* (figure 4.2).

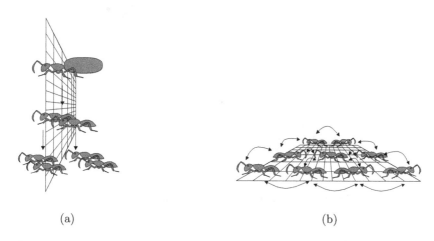

(a)   (b)

Fig. 4.2. Hierarchy (a) and dense heterarchy (b): two opposite concepts.

It should be noted that this concept not only matches with that of decentralized control, but also with that of stigmergy. This is because the concept of heterarchy describes the *manner* in which information flows through the system. However, in a dense heterarchy, any sort of communication must be taken into account, which includes the stigmergy as well as the direct exchange of information between the individuals.

4.2.2 Natural optimization: pheromonal trails

The ant colony algorithms were developed following an important observation: social insects in general, and the ant colonies in particular, can solve relatively complex problems in a natural way. The biologists studied extensively for a long time how the ants manage collectively to solve problems which are too complex for a single individual, especially the problem of choice at the time of exploitation of the sources of food.

The ants possess a typical characteristic, they employ volatile substances called *pheromones* to communicate. They perceive these substances because of the receivers located in their antennas and they are very sensitive to them. These substances are numerous and vary from species to species. The ants can deposit pheromones on the ground, utilizing a gland located in their abdomen, and thus form odorous trails, which could be followed by their fellows (figure 4.3).

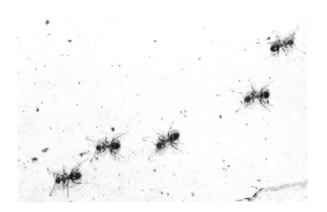

Fig. 4.3. Ants following a trail of pheromone.

The ants use the trails of pheromones to mark their way, for example between the nest and a source of food. A colony is thus able to choose (under certain conditions) the shortest path towards a source to exploit [Goss et al., 1989, Beckers et al., 1992], without the individuals having a *global* vision of the path.

Indeed, as illustrated in figure 4.4, those ants which followed the two shortest branches, arrived at the nest quickest, after having visited the source of food. Thus, the *quantity* of pheromone present on the shortest path is slightly more significant than that present on the longest path. However, a trail presenting a greater concentration of pheromones is more attractive for the ants and it has a larger *probability* to be followed. Hence the short trail will be reinforced more than the long one, and, in the long run, will be chosen by the great majority of the ants.

Here it should be noted that the choice is implemented by a mechanism of *amplification* of an initial fluctuation. However, it is possible that if, at the beginning of the exploitation, a greater quantity of pheromones is deposited on the large branches, then the colony may choose the longest route.

Other experiments [Beckers et al., 1992], with another species of ants, showed that if the ants can make half-turns on the basis of very big variation compared to the direction of the source of food, then the colony is more flexible and the risk to be trapped in the long route is weaker.

It is difficult to know precisely the physiochemical properties of the trails of pheromone, which vary from species to species and depend on a great number of parameters. However, the metaheuristics of ant colony optimization are

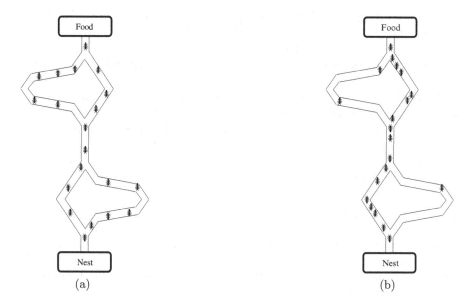

Fig. 4.4. Experiment for selection of the shortest branches by a colony of ants:
(a) at the beginning of the experiment, (b) at the end of the experiment.

mainly based on the phenomenon of *evaporation* of the trails of pheromone.
It should be noted that, in nature, the trails evaporate slower than the models
envisage it. The real ants indeed have at their disposal "heuristics" bringing a
little more information about the problem for them (for example information
on the direction). It is necessary to keep in mind that the immediate interest
of the colony (to find the shortest path towards a source of food) can be in
competition with the adaptive interest of such behaviors. If one takes into
account all the constraints which a colony of ants has to consider (predation,
competition with other colonies, etc.), a fast and stable choice can be better,
and a change of exploited site can involve too strong costs to allow the natural
selection of such an option.

4.3 Optimization by ant colonies and the traveling salesman problem

One of the earliest problems for which an ant colony algorithm was implemented was the traveling salesman problem (*TSP*): the "Ant System" (*AS*)
[Colorni et al., 1992]. The graduation of the metaphor to the algorithm is relatively easily understood and the traveling salesman problem is well known
and extensively studied.

It is interesting to dig deep into the principle of this first algorithm for better understanding the operating principle of the ant colony algorithms. There are two ways of approaching these algorithms. The first approach, most obviously in conformation with the earliest development, is that which historically led to the development of the original "Ant System"; we chose to describe it in this section. The second is a more formal description of the common mechanisms for the ant colony algorithms, it will be described in the section 4.5.

The traveling salesman problem consists in finding the shortest path connecting n cities specified, each city has to be visited only once. The problem is more generally defined like a totally connected graph (N, A), where the cities are the nodes N and the paths between these cities are the edges A.

4.3.1 Basic algorithm

In AS algorithm, in each iteration t $(1 \leq t \leq t_{max})$, each ant k $(k = 1, \ldots, m)$ traverses the graph and builds a complete path of $n = |N|$ stages (one should note that $|N|$ is the cardinality of the set N). For each ant, the path between a city i and a city j depends on:

1. the list of the already visited cities, which defines the possible movements in each step, when the ant k is on the city i: J_i^k;
2. the reciprocal of the distance between the cities: $\eta_{ij} = \frac{1}{d_{ij}}$, called *visibility*. This static information is used to direct the choice of the ants towards close cities, and to avoid the cities too remote;
3. quantity of pheromone deposited on the edge connecting the two cities, called *intensity of the trail*. This parameter defines the relative attraction of part of the total path and changes with each passage of an ant. This can be viewed as a global memory of the system, which evolves through a training process.

The rule of displacement (called "random proportional transition rule" by the authors of [Bonabeau et al., 1999]) can be stated as following:

$$p_{ij}^k(t) = \begin{cases} \frac{(\tau_{ij}(t))^\alpha \cdot (\eta_{ij})^\beta}{\sum_{l \in J_i^k} (\tau_{il}(t))^\alpha \cdot (\eta_{ij})^\beta} & \text{if } j \in J_i^k \\ 0 & \text{if } j \notin J_i^k \end{cases} \tag{4.1}$$

where α and β are two parameters controlling the relative importance of the trail *intensity*, $\tau_{ij}(t)$, and *visibility* η_{ij}. With $\alpha = 0$, only visibility of the city is taken into consideration; the city nearest is thus selected with each step. On the contrary, with $\beta = 0$, only the trails of pheromone become influential. To avoid a too fast selection of a path, a compromise between these two parameters, exploiting the behaviors of *diversification* and of *intensification* (see section 4.5.3 of this chapter), is essential. After a full run, each ant leaves a certain quantity of pheromones $\Delta\tau_{ij}^k(t)$ on its entire course, the amount of which depends on the *quality* of the solution found:

$$\Delta\tau_{ij}^k(t) = \begin{cases} \frac{Q}{L^k(t)} & \text{if } (i,j) \in T^k(t) \\ 0 & \text{if } (i,j) \notin T^k(t) \end{cases} \quad (4.2)$$

where $T^k(t)$ is the path traversed by the ant k during the iteration t, $L^k(t)$ the length of the turn and Q a fixed parameter.

However, the algorithm would not be complete without the process of *evaporation* of the trails of pheromone. In fact, it is necessary that the system should be capable of "forgetting" the bad solutions, to avoid being trapped in sub-optimal solutions. This is achieved by counterbalancing the additive reinforcement of the trails by a constant decrease of the values of the edges in each iteration. Hence, the update rule for the trails is given as:

$$\tau_{ij}(t+1) = (1-\rho) \cdot \tau_{ij}(t) + \Delta\tau_{ij}(t) \quad (4.3)$$

where $\Delta\tau_{ij}(t) = \sum_{k=1}^m \Delta\tau_{ij}^k(t)$ and m is the number of ants. The initial quantity of pheromone on the edges is a uniform distribution of a small quantity $\tau_0 \geq 0$.

The figure 4.5 presents a simplified example of the traveling salesman problem, optimized by an AS algorithm, whose pseudo code is presented in the algorithm 4.1.

For $t = 1, \ldots, t_{max}$
    For each ant $k = 1, \ldots, m$
        Choose a city randomly
        For each non visited city i
            Choose a city j, from the list J_i^k of remaining cities, according to the formula 4.1
        End For
        Deposit a trail $\Delta\tau_{ij}^k(t)$ on the path $T^k(t)$ in accordance with the equation 4.2
    End For
    Evaporate trails according to the formula 4.3
End For

Algorithm 4.1: Basic ant colony algorithm: the "Ant System".

4.3.2 Variants

Ant System & elitism

An early variation of the "Ant System" was proposed in [Dorigo et al., 1996]: the introduction of the "elitist" ants. In this version, the best ant (that which traversed the shortest path) deposits a large quantity of pheromone, with a view to increase the probability of the other ants of exploring the most promising solution.

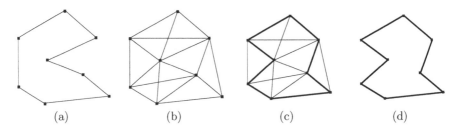

Fig. 4.5. The traveling salesman problem optimized by the AS algorithm, the points represent the cities and the thickness of the edges represents the quantity of pheromone deposited (a) example of the path built by an ant, (b) at the beginning of calculation, all the paths are explored, (c) the shortest path is reinforced more than the others, (d) the evaporation allows to eliminate the worse solutions.

Ant-Q

In this variation of AS, the rule of local update is inspired by "Q-learning[1]" [Gambardella and Dorigo, 1995]. However, no improvement compared to the AS algorithm could be demonstrated. Besides, even in the opinion of the authors, this algorithm is not more than a pre-version of the "Ant Colony System".

Ant Colony System

The "Ant Colony System" (ACS) algorithm was introduced to improve the performances of the first algorithm for problems of higher dimensions [Dorigo and Gambardella, 1997b, Dorigo and Gambardella, 1997a]. ACS is founded on the modifications proposed for the AS:

1. ACS introduces a rule of transition depending on a parameter q_0 ($0 \leq q_0 \leq 1$), which defines a balance between *diversification/intensification*. An ant k on a city i will choose a city j according to the rule:

$$j = \begin{cases} argmax_{u \in J_i^k} \left[(\tau_{iu}(t)) \cdot (\eta_{iJ})^{\beta} \right] & \text{if } q \leq q_0 \\ J & \text{if } q > q_0 \end{cases}$$

where q is a random variable uniformly distributed in $[0, 1]$ and $J \in J_i^k$ a city selected at random according to the probability:

$$p_{iJ}^k(t) = \frac{(\tau_{iJ}(t)) \cdot (\eta_{iJ})^{\beta}}{\sum_{l \in J_i^k} (\tau_{il}(t)) \cdot (\eta_{il})^{\beta}} \tag{4.4}$$

---

[1]a reinforcement based training algorithm

According to the parameter q_0, there are thus two possible behaviors: if $q > q_0$, the choice is made in the same manner as that for the AS algorithm, and the system tends to carry out a *diversification*; on the contrary, if $q \leq q_0$, then the system tilts towards an *intensification*. Indeed, for $q \leq q_0$, the algorithm exploits the information collected by the system more and it cannot choose a non explored path.

2. The management of the trails is subdivided into two levels: a local update and a global update. Each ant deposits a trail at the time of the local update according to the formula:

$$\tau_{ij}(t+1) = (1 - \rho) \cdot \tau_{ij}(t) + \rho \cdot \tau_0$$

where τ_0 is the initial value of the trail. At each passage, the visited edges see their quantity of pheromone decreasing, which supports diversification by taking into account the non explored paths. At each iteration, the total update is carried out as:

$$\tau_{ij}(t+1) = (1 - \rho) \cdot \tau_{ij}(t) + \rho \cdot \Delta\tau_{ij}(t)$$

where the edges (i, j) belong to the best turn length T^+ of length L^+ and where $\Delta\tau_{ij}(t) = \frac{1}{L^+}$. Here, only the best trail is thus updated, which takes part in an intensification by selection of the best solution.

3. The system uses a list of candidates. This list stores for each city v the closest neighbors, classified by increasing distances. An ant will consider an edge towards a city apart from the list only if this one was already explored. To be specific, if all the edges were already visited in the list of candidates, the choice will be done according to the rule 4.4, if not, then it is the closest to the not visited cities which will be selected.

ACS & 3-opt

This variant is a hybridization of the ACS and a local search algorithm of 3-opt type [Dorigo and Gambardella, 1997b]. Here, the local search is initiated to improve the solutions found by the ants thus far (and thus to bring the ants to the nearest local optimum).

Max-Min Ant System

This variant (abbreviated as $MMAS$) is founded on the basis of the AS algorithm and presents some notable differences [Stützle and Hoos, 1997, Stützle and Hoos, 2000]:

1. Only the best ant updates a trail of pheromone;
2. The values of the trails are limited by τ_{min} and τ_{max};
3. The trails are initialized with the maximum value τ_{max};

4. The updating of the trails is made in a proportional manner, the strongest trails being less reinforced than the weakest;
5. A re-initialization of the trails can be carried out.

The best results are obtained by updating the best solution with an increasingly strong *frequency*, during the execution of the algorithm.

4.3.3 Choice of the parameters

For the AS algorithm, the authors recommend that, although the value of Q has little influence on the final result, this value is of the same order of magnitude as the estimated length of the best found path. In addition, the town of departure for each ant is typically selected at random as no significant influence of specific starting point for the ants could be demonstrated.

With regard to the ACS algorithm, the authors advise to use the relation $\tau_0 = (n \cdot L_{nn})^{-1}$, where n is the number of cities and L_{nn} the length of a turn found by the nearest neighbor method. The number of ants m is a significant parameter, since it takes part in the principal positive feedback of the system. The authors suggest using as many ants as the cities (i.e. $m = n$) for obtaining good performances for the traveling salesman problem. It is possible to use only one ant, but the effect of amplifying different lengths is then lost, just as the natural parallelism of the algorithm, which can prove to be harmful for certain problems. In general, the ant colony algorithms do not seem to be very sensitive to a precise selection of the number of ants.

4.4 Other combinatorial problems

The ant colony algorithms have been extensively studied in recent past and it would take a long time to make an exhaustive list of all the applications and variations which were produced in the past few years. In the two principal fields of application (NP-difficult problems and dynamic problems), certain algorithms however gave very good results. In particular, interesting performances were noted in the case of the quadratic assignment problem [Stützle and Hoos, 2000], the planning problems [Merkle et al., 2000], sequential scheduling [Gambardella and Dorigo, 2000], the vehicle routing problem [Gambardella et al., 1999], or for the network routing problem [Di Caro and Dorigo, 1998] (see also the section 4.6.2 of this chapter for this application). A significantly large collection of literatures is available on almost all kinds of problems: traveling salesman, graph coloring, frequency assignment, generalized assignment, multidimensional knapsack, constraint satisfaction, etc.

4.5 Formalization and properties of ant colony optimization

An elegant description was proposed in [Dorigo and Stützle, 2003], which can be applied to the (combinatorial) problems where a partial construction of the solution is possible. This description, although restrictive, makes it possible to highlight the original contributions of these metaheuristics (called *ACO,* for "Ant Colony Optimization", by the authors).

> Artificial ants used in ACO are stochastic solution construction procedures that probabilistically build a solution by iteratively adding solution components to partial solutions by taking into account (i) heuristic information on the problem instance being solved, if available, and (ii) (artificial) pheromone trails which change dynamically at run-time to reflect the agents' acquired search experience.

A more precise formalization exists [Dorigo and Stützle, 2003]. It develops a *representation* of the problem on the basis of a basic *behavior* of the ants and a general *organization* of the metaheuristic under consideration. Several concepts have also been laid down to facilitate the understanding of the principles of these algorithms, in particular the definition of the trails of pheromone as an *adaptive memory*, the need for an adjustment of *intensification/diversification* and finally, the use of a *local search*. These various subjects are covered in detail hereafter.

4.5.1 Formalization

Representation of the problem

The problem is represented by a *set of solutions*, an *objective function* assigning a value for each solution and a *set of constraints*. The objective is to find the global optimum satisfying the constraints. The various states of the problem are characterized similar to a sequence of components. It should be noted that, in certain cases, a cost can be associated to the states which do not belong to the set of solutions. In this representation, the ants build solutions while moving on a graph $G = (C, L)$, where the nodes are the components of C and the set L connects the components of C. The constraints of the problem are implemented directly in the rules of displacement of the ants (either by preventing the movements which violate the constraints, or by penalizing such solutions).

Behavior of the ants

The movements of the ants can be characterized like a stochastic procedure of *building* constructive solutions on the graph $G = (C, L)$. In general, the ants try to work out feasible solutions, but if necessary, they can produce unfeasible solutions. The components and the connections can be associated with the trails of pheromone τ (establishing an adaptive memory describing the state of the system) and a heuristic value η (representing a priori information about the problem, or originating from a source other than that of the ants; it is very often the cost of the state in progress). The trails of pheromone and the value of the heuristics can be associated either with the components, or with the connections (figure 4.6).

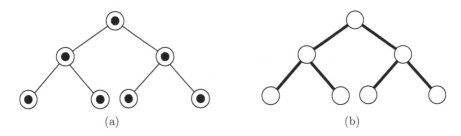

(a)   (b)

Fig. 4.6. In an ant colony algorithm, the trails of pheromone can be associated with the components (a) or connections (b) of the graph representing the problem to be solved.

Each ant has a memory to store the path traversed, an initial state and the stopping conditions. The ants move according to a probabilistic rule of decision function of the local *trails* of pheromone, *state* of the ant and *constraints* of the problem. At the time of addition of a component to the solution in progress, the ants can update the trail associated with the component or the corresponding connection. Once the solution is built, they can update the trail of pheromone components or connections used. Lastly, an ant has the capacity of at least building a solution for the problem.

Organization of the metaheuristic

In addition to the rules governing the behavior of the ants, another major process is activated: the *evaporation* of the trails of pheromone. In fact, with each iteration, the value of the trails of pheromone is *decreased*. The goal of this reduction is to avoid a too fast convergence and the trapping of the algorithm in local minima. This causes a gradual lapse in memory which helps in exploration of new areas.

According to the authors of the *ACO* formalism, it is possible to implement other processes requiring a *centralized* control (and thus not being able to be directly controlled by some ants), as additional processes. In our opinion, this is not desirable; in fact, one then loses the decentralized characteristic of the system. Moreover, the implementation of the *additional* processes with rigorous formalization becomes difficult, because one should be able to view any process there.

4.5.2 Pheromones and memory

The use of the *stigmergy* is a crucial factor for the ant colony algorithms. Hence, the choice of the method for implementation of the trails of pheromone is significant to obtain the best results. This choice is mainly related to the possibilities of *representation* of the search space, each representation being able to bring a different way to implement the trails. For example, for the traveling salesman problem, an effective implementation consists in using a trail τ_{ij} between two cities i and j like a representation of the *interest* to visit the city j after the city i. Another possible representation, less effective in practice, consists in considering τ_{ij} as a representation of the interest to visit i as the jth city. In fact, the trails of pheromone describe the state of the search for the solution by the system in each iteration and the agents modify the way in which the problem will be *represented* and perceived by the other agents. This information is shared by the ants by means of modifications of the *environment*, in form of an indirect communication: the stigmergy. Information is thus stored for a certain time duration in the system, which led certain authors to consider this process as a form of *adaptive memory* [Taillard, 1998, Taillard et al., 1998], where the dynamics of storage and of division of information will be crucial for the system.

4.5.3 Intensification/diversification

The problem of the relative use of the process of *diversification* and *intensification* is an extensively explored problem in the design and the use of a metaheuristic. By intensification, one understands the *exploitation* of the information gathered by the system at a given time. On the other hand, diversification is the *exploration* of search space areas imperfectly taken into account. Very often, it is a question of choosing where and when "to inject the random perturbation" in the system (*diversification*) and/or to improve a solution (*intensification*). In the ACO type algorithms, as in the majority of the cases, there are several ways in which these two facets of metaheuristics of optimization can be organized. The most obvious method is by adjusting the parameters α and β, which determine the relative influence of the trails of pheromone and the heuristic information. Higher the value of α, more significant will be the *intensification*, because the trails will have more influence on the choice of the ants. Conversely, lower the value of α, stronger *diversification* will take place, because the ants will avoid the trails. The parameter β acts in a similar manner. Hence both the parameters must be tuned simultaneously to have a tighter control over these aspects.

A viable alternative can also be introduced in form of modifications of the management of the trails of pheromone. For example, the use of the *elitist* strategies (the best solutions contribute more to the trails, see section 4.3.2: the AS algorithm with elitism) supports intensification, whereas a re-initialization of all the trails supports exploration (section 4.3.2, algorithm MMAS).

This choice of diversification/intensification can be undertaken in a static manner before initiating the algorithm, by using an a priori knowledge about the problem, or in a dynamic manner, by allowing the system to decide the better adjustment. There can be two possible approaches: adjustment of the parameters or introduction of new processes. These algorithms are mostly based on the concept of self-organization and these two approaches can be equivalent, a change of parameter can induce a behavior of the system that is completely different, at the global level.

4.5.4 Local search and heuristics

The ant colony metaheuristics are often more effective when they are *hybridized* with local search algorithms. These algorithms optimize those solutions found by the ants before the ants are used for updating the trails of pheromone. From the point of view of local search, the advantage of employing ant colony algorithms to generate an initial solution is undeniable. Very often hybridization with a local search algorithm becomes the important factor in differentiating an *interesting* ACO type metaheuristic from a really *effective* algorithm.

Another possibility to improve the performances is to inject more relevant heuristic information. This addition generally has a high cost in term of additional computational burden.

It should be noted that these two approaches are similar from the point of view of employing cost information to improve a solution. In fact, local search in a way is more direct than the heuristics, however the latter is perhaps more natural to use a priori information about the problem.

4.5.5 Parallelism

The structure of ant colony metaheuristics comprises of an *intrinsic parallelism*. Generally, the good quality solutions emerge as a result of the indirect *interactions* taking place inside the system, not of an explicit implementation of exchanges. Here each ant takes only the local information about its environment (the trails of pheromones) into account; it is thus very easy to parallel such an algorithm. It is interesting to note that the various processes in progress in the metaheuristic (i.e. the behavior of the ants, evaporation and the additional processes) can also be implemented independently, the user has the liberty to decide the manner in which they will interact.

4.5.6 Convergence

The metaheuristics can be viewed as modified versions of a basic algorithm: a random search. This algorithm has the interesting property to *guarantee* that the optimal solution will be found, early or late, and hence one can concentrate on the issue of convergence. However, since this basic algorithm is *skewed*, the guarantee of convergence does not exist any more.

If, in certain cases, one is sure about the convergence of an ant colony algorithm (*MMAS* for example, see section 4.3.2), the problem of convergence of an unspecified *ACO* algorithm remains unsolved. However, there is a variant of the ACO whose convergence was proven [Gutjahr, 2000, Gutjahr, 2002]: the "Graph-Based Ant System" (*GBAS*). The difference between the *GBAS* and the AS algorithm lies in the updating of the trails of pheromone, which is allowed only if a better solution is found. For certain values of parameters, and for a given small $\epsilon > 0$, the algorithm will find the optimal solution with a probability $P_t \geq 1 - \epsilon$, after a time $t \geq t_0$ (where t_0 is a function of ϵ).

4.6 Prospect

Armed with the early success of the ant colony algorithms, allied research interests started exploring many areas other than that of combinatorial optimization: for example, the use of these algorithms for *continuous* and/or *dynamic* problems, or the comparison of this type of algorithms within a framework of *swarm intelligence* and with other metaheuristics.

4.6.1 Continuous optimization

Problems of adaptation

The metaheuristics are very often employed for combinatorial problems, but there is a class of problems often encountered in engineering, where the objective function is *continuous* and for which the metaheuristics can be of great help (nonderivable function, multiple local minima, large number of variables, nonconvexity, etc.; see section 6.2). Several research efforts to adapt metaheuristic ant colonies to the continuous domain have been reported.

In addition to the traditional problems of adaptation of a metaheuristic, the ant colony algorithms pose some specific problems. Thus, the principal problem arises if one places oneself in ACO formalism with a construction of the solution composed by components. Indeed, a continuous problem can — according to the perspective chosen — have an infinite number of components and the problem of construction cannot be easily solved in this case. The majority of the algorithms are thus inspired by the characteristics of self-organization and external storage by the ant colonies, leaving aside the iterative construction of the solution.

We list here four ant colony algorithms for continuous optimization: CACO, a hybrid algorithm not baptized, CIAC and API.

The CACO algorithm

The first of these algorithms, quite naturally called *CACO* ("Continuous Ant Colony Algorithm") [Bilchev and Parmee, 1995, Wodrich and Bilchev, 1997, Mathur et al., 2000], uses two approaches: an *evolutionary* algorithm selects and crosses areas of interest, that the *ants* explore and evaluate. An ant selects an area with a probability proportional to the concentration of pheromone in that area, in an identical manner as — in the "Ant System" —, an ant would select a trail going from a city to another:

$$p_i(t) = \frac{\tau_i^\alpha(t) \cdot \eta_i^\beta(t)}{\sum_{j=1}^{N} \tau_j^\alpha(t) \cdot \eta_j^\beta(t)}$$

where N is the number of areas and $\eta_i^\beta(t)$ is used to include specific heuristics for the problem. The ants then leave the centre of the area and move in a direction chosen randomly, as long as an improvement in the objective function is observed. The displacement step used by the ant in each evaluation is given by:

$$\delta r(t, R) = R \cdot \left(1 - u^{\left(1 - \frac{t}{T}\right)^c}\right)$$

where R is the diameter of the explored area, $u \in [0, 1]$ a random number, T the total number of iterations of the algorithm and c a cooling parameter. If the ant found a better solution, the area is moved so that its centre coincides with this solution, and the ant increases the quantity of pheromone of the area proportional to the found improvement. The evaporation of the "trails" is done classically according to a coefficient ρ.

Modifications were proposed by Wodrich et al. [Wodrich and Bilchev, 1997] to improve the performances of the original algorithm. Thus, in addition to the "local" ants of CACO, the "global" ants will explore the search space (figure 4.7) so that, if required, the areas which are not very interesting will be replaced by new areas which are not previously explored. The areas are also affected by a factor called age, which increases if no improvement is discovered. Moreover, the parameter t in the search step of the ants $\delta r(t, R)$ is defined by the age of the explored area.

A remodeling of the algorithm [Mathur et al., 2000] was proposed in order to more finely associate *CACO* with the paradigm of the ant colonies and to abolish the association with the evolutionary algorithm. Thus it can be noted that, for example, the algorithm speaks about diffusion to define the creation of new areas. This algorithm was compared with some traditional algorithms and has shown average performances in its first version and better performances in its later versions.

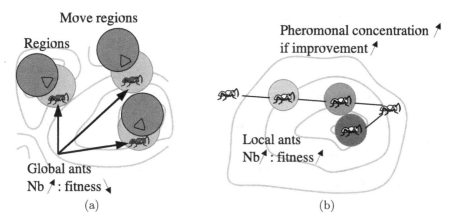

Fig. 4.7. The CACO algorithm: the global ants (a) take part in the displacement of the areas which the local ants (b) evaluate.

A hybrid method

A similar approach — with simultaneous employment of ant colonies and evolutionary algorithm — was proposed by Ling et al. [Ling et al., 2002], but few results are available at the moment when this book is written. The principal idea of this method is to consider the differences between two individuals in each dimension as many parts of a path where the pheromones are deposited. The evolution of the individuals is dealt with by employing the mutation and the crossover operators. From a certain point of view, this method thus tries to reproduce the construction mechanism of the solution, using components.

The method proceeds precisely as described in the algorithm 4.2. Each ant x_i of the population containing m individuals is considered as a vector with n dimensions. Each element $x_{i,e}$ of this vector can thus be regarded as a candidate with the element $x^*_{i,e}$ providing the optimal solution. The idea is to use the path between the elements $x_{i,e}$ and $x_{j,e}$ — given (i,j) — to deposit a trail of pheromone whose concentration is given as $\tau_{ij}(t)$ at the time step t.

The authors proposed an "adaptive" version where the probabilities of mutation and crossover are variable quantities. Unfortunately this algorithm is not yet completely tested, its performances are thus doubtful and need validation.

The CIAC algorithm

Another algorithm was developed by two of the co-authors of this book, which focused on the principles of *communication* of the ant colonies. It proposes to

1. At each iteration, each ant selects an initial value in the group of candidate values with the probability:

$$p_{ij}^k(t) = \frac{\tau_{ij}(t)}{\sum \tau_{ir}(t)}$$

2. Use the mutation and the crossover operators on those m values in order to obtain m new values;
3. Add these new values to the group of candidate values for the component $x_{i,e}$;
4. Form m solutions of the new generation;
5. Calculate the "fitness" of these solutions;
6. When m ants traversed all the edges, update the trails of pheromone of candidate values of each component by:

$$\tau_{ij}(t+1) = (1 - \rho)\tau_{ir}(t) + \sum \tau_{ij}^k$$

7. If the kth ant chooses the jth candidate value of the group of components, then $\delta\tau_{ij}^k(t+1) = Wf_k$, if not $\delta\tau_{ij}^k = 0$. With W a constant and f_k the "fitness" of the solution found by the kth ant;
8. Erase the m values having the lowest intensities of pheromone in each group of candidates.

Algorithm 4.2: A hybrid ant colony algorithm for the continuous case.

add the *direct* exchanges of information [Dréo and Siarry, 2002] to the stigmergic processes, being inspired by a similar action adopted in "heterarchic approach" described previously in the 4.2.1. Thus, a formalization of the exchange of information is proposed, based on the concept of communication channels. Indeed, there are several possible ways to pass information between two groups of individuals, for example either by deposits of trails of pheromone or by direct exchanges. One can define various types of *channels of communication* representing the set of the characteristics of the transmission of information. From the point of view of metaheuristics, there are three principal characteristics (see figure 4.8):

Range: the number of individuals involved in the exchange of information. For example, information can be emitted by an individual and received by several others, and vice-versa.

Memory: the persistence of information in the system. Information can remain within the system for a specific time duration or can be only transitory.

Integrity: the modifications generated by the use of the channel of communication. Information can vary in time or be skewed during its transmission.

Moreover, information passing through a communication channel can be of varied interest, such as for example the value and/or the position of a point on the search space.

The *CIAC* algorithm (acronym for "Continuous Interacting Ant Colony") uses two communication channels:

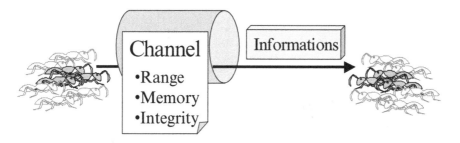

Fig. 4.8. Structural characteristics of a channel of communication for transmission of information : range, memory and integrity.

1. The stigmergic channel is formed by the spots of pheromone, deposited on the search space, which will be more or less attractive for the artificial ants, according to their concentrations and their distances. The characteristics of the stigmergic channel are thus the following: the range is at its maximum, all the ants can potentially take information into account, there is use of memory since the spots persist on the search space, finally, information evolves with time as the spots evaporate. The information carried by a spot implicitly contains the position of a point and explicitly the value of the improvement found by the ant, having deposited the spot.
2. The direct channel is implemented in the form of message exchange between two individuals. An artificial ant has a stack of received messages and can send some to another ant. The range of this channel is unity since only one ant receives the messages, the memory is implemented in form of the stack of messages which the ant memorizes and finally, information (here the position/value of a point) does not fade with passage of time.

The algorithm showed some interesting characteristics, it utilizes the self-organization properties of the ant colony algorithms, in particular a certain capacity to be oscillated between a process of intensification and a process of diversification when the two communication channels (stigmergic and direct) are used in synergy. The figure 4.9 illustrates this behavior of oscillations: the ordinate shows the standard deviation of the distribution of the objective function values, a high standard deviation corresponds to a high dispersion of the ants on the axis of the values (*diversification*) whereas a low value corresponds to a gathering of the ants (*intensification*). It should be noted that this behavior is not observed when only one channel is in use; hence there is synergy between the two channels.

However, the results are comparable only with those produced by the other ant colony algorithms implemented for the continuous domain, therefore better results should be obtained by employing other metaheuristics adapted for the continuous case.

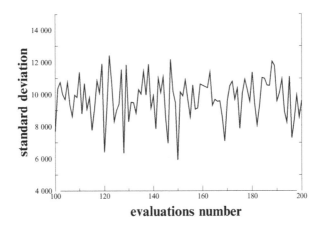

Fig. 4.9. Oscillations observed during the simultaneous use of the two channels of communication in CIAC algorithm.

This approach gave rise to a hybridization with the Nelder-Mead algorithm for local search [Dréo and Siarry, 2003]. This modification of the original $CIAC$ algorithm, called $HCIAC$, thus uses two channels of communication, adds a local search methodology and stochastic decision-making processes. The last feature is implemented by using the stimulus/response type functions, which facilitates us to define a threshold of choice for an action. To be precise, one can use a sigmoid function $p(x) = \frac{1}{1+e^{\delta\omega-\omega x}}$ to test the function for choice of a state x of an ant where a threshold δ determines the position of the point of inflection and the power ω characterizes the inflection of the sigmoid function. If we draw a random number r from an uniform distribution, one can have two possible choices: $r < p(x)$ or $r > p(x)$. Considering $\delta = 0.5$ and $\omega = +\infty$, one can obtain a simple binary choice. Using this type of function one can dispense with a delicate parameter setting procedure, for example by distributing the thresholds according to a normal law on the entire population. In a similar manner, one can initiate by this way a simple training procedure, while varying the thresholds.

HCIAC algorithm is described in the figure 4.10. Hybridization has — as often with the ant colony algorithms — facilitated to reach comparable results with those obtained from other metaheuristic competitors for the continuous problems.

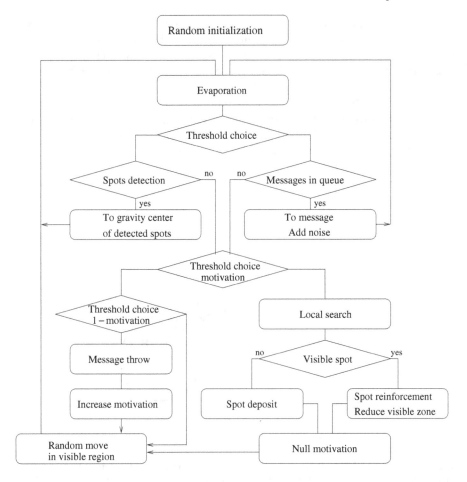

Fig. 4.10. The HCIAC algorithm.

The API algorithm

In all these algorithms adapted for continuous problems, the term "ant colonies" could be utilized as all of them use processes very similar to stigmergy for information exchange.

However, there is one algorithm which can be adapted to the continuous case [Monmarché et al., 2000] that utilizes the behavior of primitive ants (which does not mean *not-adapted*) of the *Pachycondyla apicalis* species as a starting point, and that does *not* utilize the indirect communication by trails of pheromone: the *API* algorithm.

In this method, one can start by positioning a nest randomly on the search space, and then ants are sent at random in a given perimeter. These ants then locally explore the "hunting site" by evaluating several points in a given perimeter (see figure 4.11). Each ant memorizes the best-found point. If during the exploration of its hunting site it finds a better point, then it will reconsider this site, if not after a certain number of explorations, it will choose another site. Once explorations of the hunting sites are completed, randomly peeked ants compare, on two by two basis (as can be the case for the real ants when they exhibit the behavior of "tandem-running"), their best results and then they memorize the best two hunting sites. After a specified time period, the nest is re-initialized at the best point found, the memory of the sites of the ants is reset and the algorithm executes a new iteration.

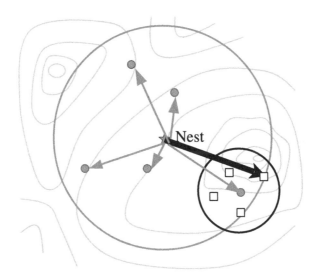

Fig. 4.11. The API algorithm: a method with multiple starting inspired by a species of primitive ant. The ants (full circles) explore hunting sites (small squares) within a perimeter (large circle) around the nest. The nest is moved to the best point when the system is re-initialized (arrow in thick feature).

Conclusion in the continuous domain

It should be noted that out of these four algorithms, two were in fact more or less hybridized with an evolutionary algorithm, and a third one did not utilize the "classic" metaphor for ant colonies. Generally, it can be opined that research in this domain is still at its primitive stage and the proposed algorithms are not fully matured, and are thus not yet really competitive compared to the other established metaheuristic classes for the continuous problems.

4.6.2 Dynamic problems

A problem is known as a dynamic one if it *varies* with time, i.e. the optimal solution does not have the same characteristics during the time of optimization. These problems give rise to specific difficulties, owing to the fact that it is necessary as well as possible to approach the best solution at *each* instant of time.

The first application of the ant colony algorithms for dynamic problems was proposed for optimization of the routing of the telephone networks [Schoonderwoerd et al., 1996]. However the proposed algorithm was not intensively studied in the literature and hence it is difficult to learn some lesson from it. Another application on similar problems was proposed by White et al. [White et al., 1998, Bieszczad and White, 1999]. An application for problems of routing of Internet networks (see figure 4.12) has also been presented: the AntNet algorithm [Di Caro and Dorigo, 1997]. This metaheuristic was the subject of several studies (see in particular [Di Caro and Dorigo, 1998]) and seems to have proven its effectiveness for several test problems.

Fig. 4.12. The network example used to test the AntNet algorithm: *NFSNET* (each edge represents an oriented connection).

To update probabilistic tables of routing, each of these algorithms uses ants to explore the network so that the relevant information is the frequency of passage of the ants over each node. Generally, the distributed and the flexible aspects of the ant colony algorithms seem to adapt well for the dynamic problems.

4.6.3 Metaheuristics and ethology

Very often the metaheuristics originate from *metaphors* drawn from nature, and in particular from biology. The ant colony algorithms are inspired by the behavior of social insects, but they are not the only algorithms which evolved from the study of the animal behavior (*ethology*). For example, optimization by particle swarms ("Particle Swarm Optimization" [Eberhart et al., 2001], see 5.6) originated from an analogy with the collective behaviors of animals displacements, as observed in fish schooling or bird flocking; there are other algorithms also which are inspired by the behaviors of the bees [Choo, 2000, Panta, 2002]. Moreover some algorithms can be found in literatures which consider some aspects of the behavior of the social insects as the starting point, although they do not make use of the *classic characteristics* of the ant colony algorithms (see for example [De Wolf et al., 2002, Nouyan, 2002] as well as the section 5.12 of this book).

Hence, all doors remain open to believe that ethology can be a source of interesting inspiration for the design of new metaheuristic algorithms.

4.6.4 Links with other metaheuristics

The metaheuristics form a wide class of algorithms, where many concepts are found across several categories. Moreover, many variations of a specific category of algorithms make the borders between different metaheuristics fuzzy.

An example of overlapping between two metaheuristics can be cited by the term "swarm intelligence", which is used not only to describe the operating mode of the ant colony algorithms [Bonabeau et al., 1999], but also of other algorithms like the "particle swarm" [Eberhart et al., 2001] (see section 5.6 for a detailed description). Generally, this term refers to any system (normally artificial) having self-organization properties — similar to those described in the section 4.2.1 — that is able to solve a problem by utilizing only the forces of interactions at the individual level.

A broader attempt for unified presentation has also been made: the framework of the "adaptive memory programming" [Taillard et al., 1998] (see section 7.5), in particular including the ant colonies, the tabu search and the evolutionary algorithms. This framework insists on the use of a form of *memory* in these algorithms, and on the use of the *intensification* and the *diversification* phases (see section 4.5.3 for this aspect of the artificial ant colonies).

Thus several metaheuristic algorithms can be brought closer to the ant colony algorithms and vice-versa. One feature that strongly supports this overlapping is the fact that the ant colony algorithms are very often effective only with a local search (see section 4.5.4). Hence, from a certain point of view, an ant colony algorithm strongly resembles the *GRASP* [Feo and Resende, 1995, Resende, 2000] ("Greedy Randomized Adaptive Search Procedures", see section 5.8) algorithm with a specific *construction* phase.

Similarly, the "Cross-Entropy" [Rubinstein, 1997, Rubinstein, 2001] (see section 5.9) method has two phases: initially generate a *random* data file, then change the parameters which *generate* this data file to obtain a better performance for the next iteration. Still, this method can be considered to be close to the ant colony algorithm [Rubinstein, 2001]. Some works have even aimed at using these two methods jointly [Wittner and Helvik, 2002].

One can also point out the similarities of these algorithms with particle swarm optimization [Kennedy and Eberhart, 1995, Eberhart et al., 2001] (described in section 5.6), which also strongly utilizes the attributes of distributed systems. Here, large groups of particles are traversing the search space with a displacement dynamic that make them gathering each other.

Another very interesting overlapping of ant colony algorithms can be observed with the estimation of distribution algorithms (*EDA*, [Larranaga and Lozano, 2002], described section 5.7). Indeed, these algorithms — derived from the evolutionary algorithms in the beginning — are based on the fact that in each iteration, the individuals in the search space are chosen at random according to a *distribution*, built from the states of the preceding individuals. Schematically, for a better individual, the probability of creation of other individuals in the neighborhood is higher. One can observe that the similarity of these EDA algorithms to the *ACO* algorithms is remarkable [Monmarché et al., 1999].

One can thus draw a parallel between evolutionary algorithms (see chapter 3) and ant colonies, that both use a population of "agents" selected on the basis of memory-driven or probabilistic procedures. One can also harp on the idea, supported by some biologists, that the phenomenon of self-organization has an important role to play in the evolutionary processes [Camazine et al., 2000]... which the evolutionary algorithms consider as a starting point.

A new approach — less related to the metaheuristics — consists in considering a particular class of ant colony algorithms (the class called "Ant Programming") and can be placed in between the optimal control theories and the reinforcement learning [Birattari et al., 2002].

It is well observed that many interactions and overlapping do exist and the relations between evolutionary algorithms, evolution of distribution algorithms and ant colonies do iterate the fact that each one can finally reveal the characteristics of the others. It is thus difficult to study ant colony metaheuristic as a homogeneous, stand-alone algorithm which in itself is a separate class from the others. However, the power of the metaphor utilized and the combination of a whole group of relatively well-known characteristics (see section 4.5) make it possible to clarify its definition.

4.7 Conclusion

The metaheuristic which is inspired by the ant colonies is initiated to be well described and formalized. The entire set of properties required for its description is known: probabilistic construction of a solution (by addition of components in the *ACO* formalism), heuristics on the specific problem, use of indirect memory form and a structure comparable with that of a self-organized system. The ideas underlying the ant colony algorithms are powerful; one can describe this metaheuristic like a *distributed* system where the *interactions* between basic components, by means of *stigmergic* process, facilitate the emergence of a coherent global behavior so that the system is able to solve difficult optimization problems.

The ant colonies have been successfully applied to many combinatorial problems and research initiations have been undertaken to adapt them for continuous problems. The importance of the choice of a local search has been emphasized to produce competitive algorithms against other older and often more specialized metaheuristics. It seems that these algorithms can become natural choices for dynamic problems as they are based on a self-organized structure, especially when only local information is available.

4.8 Annotated bibliography

[Hölldobler and Wilson, 1990]: This book presents an impressive collection of knowledge on the biology of the ants. A bible on the subject, which received the Pullitzer price.

[Camazine et al., 2000]: One can find here a complete description of the self-organization principles in the biological systems, accompanied by many examples. Descriptions of patterns make it possible to understand the theoretical bases of the self-organization.

[Bonabeau et al., 1999]: This work treats ant colony algorithms as systems showing swarm intelligence. The book is articulated around biological and algorithmic concepts, in particular around metaheuristics of ant colonies. A reference on the ACO algorithms.

[Dorigo and Stützle, 2003]: A chapter specifically dedicated to the ant colony algorithms in a book which provides general descriptions on several metaheuristics. Less rich than the preceding one, but more recent.

[Dorigo et al., 2002]: Proceedings of the last *ANTS* congress on the "ant algorithms", a fast view on the most recent research in this field. The congress is held every two years since 1998.

Part II

Variants, Extensions and Methodological Advices

5

Some Other Metaheuristics

5.1 Introduction

The metaheuristics form an extremely promising field of research, where several ideas are coined, almost everyday. The well-known techniques, which were presented in the first part of this book, should not shadow the existence of many other methods. The development of such algorithms often follows a regular, predictable pattern. In particular, often the initial versions of these algorithms are innovative but not very effective and there remains enough scope for further improvements, followed by proposals of several improved variants, and finally they are hybridized with other approaches. The origin of inspiration for each such algorithm may be useful, however the induced classification of metaheuristics is often arbitrary.

We propose in this chapter a sample collection — inevitably incomplete and partial — of the metaheuristics, from the "simple" variant of the simulated annealing to the more innovative concept of the algorithms based on estimation of distribution. It will be interesting to draw analogy between these methods and those described in the first four chapters of the book, or the approaches of unification presented in the chapter 7.

Hereafter we present, in a largely arbitrary order:

- some variants of simulated annealing;
- the noising method;
- the method of distributed search;
- the Alienor method;
- particle swarm optimization;
- algorithms with estimation of distribution;
- the GRASP method;
- the "Cross-Entropy" method;
- artificial immune systems;
- the differential evolution;
- algorithms inspired by the social insects.

Some other metaheuristics have also been separately discussed within the scope of the chapters 1 to 4. The annotated bibliography placed at the end of this chapter 5 finally aims at pointing out recently published books, which give a detailed presentation of certain techniques described in this book, or can even describe other methods which are left beyond the scope of this book.

5.2 Some variants of simulated annealing

5.2.1 Simulated diffusion

The idea in the method of the simulated diffusion [Aluffi-Pentini et al., 1985, Geman and Hwang, 1986], is to introduce random fluctuations authorizing the degradations of the objective function, while preserving the descent along the directions of the gradients. The minimum of the function f is localized starting from the asymptotic behavior of the solutions of the ordinary differential equation of the gradient:

$$\dot{x} = -\nabla f(x) \tag{5.1}$$

for which the minimum is a stable state. However, there remains a major risk: the algorithm may get trapped in a local minimum of f, rather than converging towards the global minimum x^*. In order to overcome this difficulty, a stochastic disturbance is added to the relevant equation (eq.5.1) which is then written as:

$$dx_t = -\nabla f(x_t)\, dt + \sigma(t) \cdot dW_t \tag{5.2}$$

where $\{W_t, t \geq 0\}$ is a standard Brownian motion (i.e. a process of "random walk"), for a suitable choice of the scalar coefficient of diffusion $\sigma(t)$.

It can be assumed that ∇f allows a "Lipschitz constant" (when a function g allows a Lipschitz constant K, it checks the following inequality: $\|g(X) - g(Y)\| \leq K \|X - Y\|$, in all the points X and Y in the domain where g is defined, where $|\ldots\|$ denotes the Euclidian norm). It can also be assumed that ∇f satisfies the limit of growth:

$$|\nabla f(x)|^2 \leq K\left(1 + \|x\|^2\right) \tag{5.3}$$

for a given constant K, and all $x \in \mathbb{R}^n$.

Then, for $\sigma(t) = \frac{c}{\sqrt{\log(t+2)}}$, with $c > 0$, it can be shown that the probability distribution $p(x_t)$ converges towards a limit of Gibbs density proportional to $e^{\frac{-f(x)}{T}}$ when the " absolute temperature" $T = \sigma^2(t) \to 0$ when $t \to \infty$. This limit density is "concentrated" around the global minimum x^* of f. Other choices of $\sigma(t)$ can lead to convergence towards a local minimum of f with a probability larger than that for convergence towards the global minimum.

In fact, the following significant result:

$$E\left(\|x_t - x^*\|^2 \log t\right) \geq \gamma \tag{5.4}$$

can be established for a certain value of $\gamma > 0$ and for t sufficiently large. There is a lower limit for this relation to converge, within the framework of least squares, from x_t towards x^*.

The procedure using the solutions x_t of (eq. 5.2) to locate x^* is also called "stochastic annealing" or " stochastic gradient". In practice, it requires the employment of a numerical method to solve the stochastic differential equation (eq. 5.2).

The simulated diffusion method was very successful in the treatment of certain applications in the area of electronics. However, it could enjoy limited applications, because it requires the gradient information of f for its successful operation (see also the section 6.2).

5.2.2 Microcanonic annealing

Let us consider an isolated system, i.e. a system that does not have any heat transfer with its environment. A "microcanonic" analysis of this system will reveal that the principal property of this physical system, in this case, is that its total energy is constant, irrespective of its dynamic evolution. In accordance with this analysis, Creutz proposed a variant of simulated annealing, the "microcanonic annealing" [Creutz, 1983]. The total energy of the system, which is the sum of the potential energy and the kinetic energy, remains preserved during the process.

$$E_{total} = E_p + E_c \qquad (5.5)$$

For the optimization problem, the potential energy E_p can be considered as the objective function, to be minimized. The kinetic energy E_c plays a role similar to that of the temperature in simulated annealing; it is forced to remain positive. E_c allows to cut off or add energy to the system, according to the disturbance imposed. The algorithm accepts all those disturbances which cause moves towards the lower energy states, by adding $-\delta E$ (lost potential energy) to the kinetic energy E_c. The moves towards higher energy states are only accepted when $\delta E < E_c$, and the energy acquired in the form of potential energy is cut off from the kinetic energy. Thus, the total energy remains constant.

The algorithm is described in detail in algorithm 5.1.

At each energy stage, the "thermodynamic equilibrium" is reached as soon as the ratio $r_{eq} = \frac{\langle E_c \rangle}{\sigma(E_c)}$ of the average kinetic energy observed to the standard deviation of the distribution of E_c is in the "neighborhood" of 1.

The equation (eq. 5.6) involving the kinetic energy and the temperature establishes a bond between simulated annealing and microcanonic annealing.

$$k_B T = \langle E_c \rangle \qquad (5.6)$$

(k_B = Boltzmann constant)

1. Choose, at random, an initial solution x for the system to be optimized and evaluate the value of the objective function $f = f(x)$;
2. Perturb this solution to obtain a new solution $x' = x + \Delta x$;
3. Calculate $\Delta f = f(x') - f(x)$;
4. *If* $\Delta f < E_c$
   Then accept the new solution x'; make $x \leftarrow x'$ and $E_c \leftarrow E_c - \Delta f$;
   Else refuse the solution x';
5. Save the best point met;
6. *If* the thermodynamic "equilibrium" of the system is reached,
   Then decrease the kinetic energy E_c;
   Else go to step 2;
7. *If* the kinetic energy E_c is close to 0,
   Then go to step 8;
   Else go to step 2;
8. Solution=best point found; terminate the program.

Algorithm 5.1: Algorithm for microcanonic annealing.

This algorithm has several advantages compared to simulated annealing. It neither requires the evaluation of the transcendent functions like e^x, nor any random number is required to be drawn for the acceptance or the refusal of a configuration. From execution point of view, this can achieve greater speed. However, Creutz noted a disadvantage, in the case of the problems of "small dimension". This method exhibited higher probability of getting trapped in metastable states than simulated annealing [Hérault, 1989].

1. Choose, at random, an initial solution x for the system to be optimized and evaluate the value of the objective function $f = f(x)$;
2. Choose an initial threshold T;
3. Perturb this solution to obtain a new solution $x' = x + \Delta x$;
4. Calculate $\Delta f = f(x') - f(x)$;
5. *If* $\Delta f < T$,
   Then accept the new solution x'; *make* $x \leftarrow x'$;
6. Save the best point met;
7. *If* the quality of the optimum does not improve for a "certain duration", or if a given number of iterations were reached,
   Then lower the threshold T;
8. *If* the threshold is close to 0,
   Then go to step 10;
9. Go to step 3;
10. Solution = best point found; terminate the program.

Algorithm 5.2: Algorithm for the threshold method.

5.2.3 The threshold method

The principal difference between the threshold method [Dueck and Scheuer, 1989] [Bertocchi and Odoardo, 1991] and the simulated annealing technique lies in the criterion for acceptance of the solutions attempted: simulated annealing accepts those configurations which cause deterioration of the objective function f only with a certain probability; on the other hand, the threshold method accepts a new configuration, if the (possible) degradation of f does not exceed a certain threshold T, which is a function of the iteration k.

The algorithm is presented in detail in the algorithm 5.2.

The method compares favorably with simulated annealing for the combinatorial optimization problems like the traveling salesman problem. An adaptation of this method to solve problems involving continuous variables can be carried out, similarly to continuous simulated annealing.

5.2.4 "Great deluge" method

The method of the "great deluge" [Dueck, 1993], and the method of the "record to record travel", presented in the paragraph 5.2.5, are methods for maximization of the objective function (an adaptation of the initial objective function is thus necessary). These methods are variants of the simulated annealing technique and the threshold method. The differences lie in the laws of acceptance of the solutions, which degrade the objective function. Moreover, compared to the simulated annealing method (which requires a complex, delicate choice of several parameters), these two methods are simpler to use, as they comprise of less parameters (only two in each method).

The principle of the algorithm of the "method of the great deluge" is presented in the algorithm 5.3.

The metaphor of the great deluge facilitates to understand the intuitive mechanism of this method: to keep the feet dry, the hiker will visit the peaks of the explored area. If the water level always goes up, an immediate disadvantage due to the separation of the "continents" will appear, which should trap the algorithm in local maxima. However, for combinatorial problems, the author presented results which were completely at par with those obtained with other methods of global optimization [Dueck, 1993].

5.2.5 Method of the "record to record travel"

The other variant, entitled "record to record travel" [Dueck, 1993], is presented in the algorithm 5.4. In this method, any solution can be accepted, if it is not "much worse" than the best *record* value obtained previously. A certain similarity with the preceding method can be found, the difference being between the *record* and the variation (*deviation*) corresponding to the water level WATER-LEVEL.

1. Choose, at random, an initial solution x for the system to be optimized and evaluate the value of the objective function $f = f(x)$;
2. Initialize the "quantity of rain" UP> 0;
3. Initialize the "level of water" WATER-LEVEL> 0;
4. Perturb this solution to obtain a new solution $x' = x + \Delta x$;
5. Evaluate the new value of f;
6. *If* $f >$WATER-LEVEL,
   Then accept the new solution x'; make $x \leftarrow x'$;
   increase the level WATER-LEVEL of the quantity UP;
7. Save the best point met;
8. *If* the function did not improve for a long time, or if there were too many function evaluations,
   Then go to step 9;
   Else go to step 4;
9. Solution = best point found; terminate the program.

Algorithm 5.3: Algorithm for the method of the great deluge.

In this method, as in the preceding one, there are only two parameters to be adjusted (quantity of water UP for the preceding method, or the variation *deviation* for this method, and the termination criterion in both methods). The choice of the first parameter is significant, because it realizes a compromise between the convergence speed and the quality of the maximum obtained.

The author specifies that the results of these two methods for the traveling salesman problem of dimension higher than 400 cities are better than those obtained using simulated annealing.

1. Choose, at random, an initial solution x for the system to be optimized and evaluate the value of the objective function $f = f(x)$;
2. Initialize the authorized "deviation" DEVIATION> 0;
3. Evaluate the initial RECORD : RECORD$= f(x)$;
4. Perturb this solution to obtain a new solution $x' = x + \Delta x$;
5. Evaluate the new value of f;
6. *If* $f >$RECORD$-$DEVIATION,
   Then accept the new solution; make $x \leftarrow x'$;
7. Save the best point met;
8. *If* $f >$RECORD,
   Then RECORD$= f(x)$;
9. *If* the function did not improve for a long time, *or* if there were too many function evaluations,
   Then go to step 10;
   Else go to step 4;
10. Solution= best point found; terminate the program.

Algorithm 5.4: Algorithm for the method of the record to record travel.

5.3 Noising method

The noising method [Charon and Hudry, 2002] uses a descent-based algorithm, i.e. an algorithm which, starting from an initial solution, carries out iterative improvements until obtaining a local minimum. On the basis of an unspecified point x in the domain S, the data are "disturbed", i.e. the values taken by the function f are modified in a certain manner; then the descent-based algorithm is applied to the disturbed function. At each iteration, the amplitude of the noising of f decreases until it is zero. The best solution met is regarded as the global minimum.

The algorithm for this method is proposed in the algorithm 5.5. The authors proposed and analyzed various possible functions of noising. They showed that, according to the noising carried out, the method can be made exactly identical with the threshold method described above, or with the simulated annealing: in this direction, the noising methods represent a generalization of simulated annealing and acceptance with threshold.

1. Choose, at random, an initial solution x for the system to be optimized;
2. Initialize the "amplitude of noise";
3. Update the objective function, using noising of a given amplitude;
4. Apply, for the "disturbed" objective function, a descent-based method, starting from the current solution;
5. Decrease the amplitude of noise;
6. *If* the amplitude of noise is zero,
   Then go to step 7;
   Else go to step 3;
7. Solution=best point found; terminate the program.

Algorithm 5.5: Algorithm for the noising method.

The noising method compares favorably with the simulated annealing on certain combinatorial optimization problems, like the problem of the partitioning of graphs with a non-fixed number of cliques. On the other hand, its adaptation for the problems with continuous variables remains a subject of further study.

5.4 Method of distributed search

The algorithm of "distributed search" [Courrieu, 1991] evolves a probability distribution of visit in the search domain. This distribution converges towards a stable state, where the probability density is maximum in the neighborhood of the extrema to be searched. Put in another way, the probability of visit is "concentrated" on those zones where the objective function takes the required extreme values.

The basic idea is very simple: let x_1 and x_2 be two points in the search domain S such that $f(x_1) < f(x_2)$. Then, there is a collection of points x, in the neighborhood of x_1 (whose size varies during the process), such that $f(x) < f(x_2)$. The algorithm samples the domain S according to a probability distribution, controlled by an estimate of the adequate size of the neighborhood, which varies in various stages of the process. The generation of the points to be visited is done independently in each dimension, by means of a generation function, known as "law of filiation". This law is given as:

$$x_i = s_i \tan(\pi u_i) + m_i \tag{5.7}$$

where:

- u_i is a uniform random number in $\left] \frac{-1}{2}, \frac{+1}{2} \right[$
- m_i are the components of the distribution center
- s_i are the scale parameters, i.e. the parameters which define the size of the neighborhood of the minimum, or the quartiles (i.e. the values of x for which the value of the distribution function is $\frac{1}{4}$ or $\frac{3}{4}$).

The random variable thus defined obeys an n-dimensional Cauchy law of density:

$$g(x; m, S) = \prod_{i=1}^{n} \left[\frac{1}{\pi s_i} \frac{1}{1 + \left(\frac{x_i - m_i}{s_i}\right)^2} \right] \tag{5.8}$$

The Cauchy law does not have moment and its variance is infinite (the density decreases slowly and is never negligible). In the opinion of the author, the Cauchy variables have properties which are particularly suitable for this method.

Practically, the method can be clearly understood while following the algorithm described in the algorithm 5.6.

The application of this algorithm does not require any condition to be fulfilled, it is capable of evaluating the objective function in any point of the domain. The author showed the convergence of the method, and compared its performances with those of three other methods: a uniform random search, a simulated annealing algorithm and a genetic algorithm.

5.5 "Alienor" method

The   *Alienor*   method   [Cherruault, 1986b,   Cherruault, 1986a] [Cherruault, 1989], proposed by Cherruault *et al.*, is based on a succession of reducing transformations, which can transform any function of several variables to a function of a single variable: the polar angle of an Archimedes spiral. Then one can choose from a pool of powerful methods, usually implemented

1. Generate, at random, a population of N points of the domain S;
   Calculate, for each one of the points, the value of the function f;
   Calculate the total of these values in the population;
2. Make N times:
   - Choose, at random, two points from the current population;
   - Take the better of the two points as "father";
   - Replace the other point by a "son" of the first, generated according to the current "law of filiation" (equation 5.7);
   - Calculate the value of the function f associated with the new point, and update the total of f in the population;
3. *If* the total of f in the population improved,
   Then go to step 2;
   Else go to 4;
4. *If* the termination criterion is reached,
   Then go to step 6;
   Else go to 5;
5. Adjust the law of filiation, by reducing the scale parameters; Go to 2;
6. Solution=best point found; terminate the program.

Algorithm 5.6: Algorithm for the distributed search method.

for solving monovariable problems, to solve a multivariable optimization problem.

For example, in the case of a problem with 4 variables, let us replace the vector components $x = (x_1, x_2, \ldots, x_n)^T$ by:

$$x_1 = r_1 \cos \theta_1 \quad x_2 = r_1 \sin \theta_1 \quad r_1 = a_1 \theta_1$$
$$x_3 = r_2 \cos \theta_2 \quad x_4 = r_2 \sin \theta_2 \quad r_2 = a_2 \theta_2$$

This first transformation of variables gives:

$$f(x_1, x_2, x_3, x_4) = f(a_1\theta_1 \cos \theta_1, a_1\theta_1 \sin \theta_1,$$
$$a_2\theta_2 \cos \theta_2, a_2\theta_2 \sin \theta_2)$$

$$f(x_1, x_2, x_3, x_4) = g(\theta_1, \theta_2)$$

The function f of 4 variables is now transformed to a function g of 2 variables. A second transformation of variables:

$$\theta_1 = r \cdot \cos \theta, \theta_2 = r \cdot \sin \theta \text{ with } r = a\theta$$

leads to a function of a single variable θ which implies:

$$f(x_1, x_2, x_3, x_4) = f(a_1 a\theta \cdot \cos \theta \cdot \cos (a\theta \cdot \cos \theta),$$
$$a_1 a\theta \cdot \cos \theta \cdot \sin (a\theta \cdot \cos \theta),$$
$$a_2 a\theta \cdot \sin \theta \cdot \cos (a\theta \cdot \sin \theta),$$
$$a_2 a\theta \cdot \sin \theta \cdot \sin (a\theta \cdot \sin \theta))$$

$$f(x_1, x_2, x_3, x_4) = G(\theta)$$

It is equivalent to seek for the global optimum of f, or that of G. However, the advantage of using G is that it does not comprise of any variable other than θ. Indeed, it can be shown that min G tends towards min f, when a, a_1, and a_2 tend towards 0 [Cherruault, 1989].

Once the absolute minimum of G is determined, the absolute minimum of f can be easily obtained by performing the following transformations, for the example with 4 variables:

$$x_1 = a_1 a\theta \cdot \cos\theta \cdot \cos(a\theta \cdot \cos\theta)$$
$$x_2 = a_1 a\theta \cdot \cos\theta \cdot \sin(a\theta \cdot \cos\theta)$$
$$x_3 = a_2 a\theta \cdot \sin\theta \cdot \cos(a\theta \cdot \sin\theta)$$
$$x_4 = a_2 a\theta \cdot \sin\theta \cdot \sin(a\theta \cdot \sin\theta)$$

The minimization of $G(\theta)$ can be carried out employing a traditional function minimization algorithm for a single variable.

There are two disadvantages associated with the use of this method:

- on one hand, the mandatory computational burden associated with the trigonometric functions becomes huge, as soon as the number of variables increases;
- in addition, it is very difficult to ensure the satisfaction of the constraints of the type:

$$a_j \leq x_j \leq b_j, 1 \leq j \leq n$$

It is necessary to evaluate many expressions of the type:

$$x_j = a_j a\theta \cdot \cos\theta \cdot \cos(a\theta \cdot \cos\theta)$$

to check the preceding inequalities.

The figure 5.1 represents the global exploration of the plane (objective function with 2 variables x_1 and x_2) parameterized by a single parameter θ.

The *Alienor* method in particular made it possible to numerically solve the partial differential equations used to model biological systems. However, its effectiveness is not yet established for the problems involving a large number of variables.

5.6 Particle swarm optimization method

The particle swarm optimization ("Particle Swarm Optimization", *PSO*) [Kennedy and Eberhart, 1995, Eberhart et al., 2001] evolved from an analogy drawn with the collective behavior of the animal displacements (in fact, the metaphor was largely derived from socio-psychology). Indeed, for certain groups of animals, e.g. the fish schools, the dynamic behavior in relatively complex displacements can be observed, where the individuals themselves

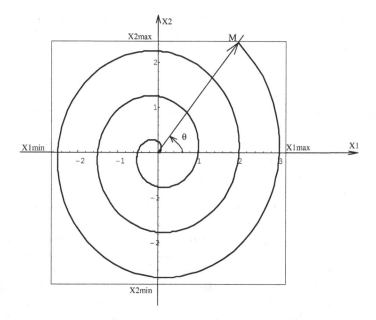

Fig. 5.1. Global exploration of the plane parameterized by θ (the Archimedes spiral).

have access only to limited information, like the position and the speed of their closer neighbors. For example, it can be observed that a fish school is able to avoid a predator in the following manner: initially it gets divided into two groups, then the original school is reformed (see figure 5.2), while maintaining the cohesion among the school.

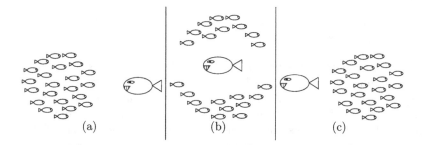

Fig. 5.2. A schematic of a fish school avoiding a predator. (a) the school forms only one group, (b) the individuals avoid the predator by forming a "fountain" like structure, (c) the school is reformed.

These collective behaviors completely conform to the theory of the self-organization described in the beginning of the chapter 4. To summarize, each individual uses the local information regarding the displacement of his closer neighbors, which are reachable by him, to decide on his own displacement. Very simple rules like "remain relatively close to the other individuals", "go in the same direction", "at same speed" etc. are good enough to maintain the cohesion among the entire group, and to allow complex and adaptive collective behaviors.

The authors, who proposed the method of particle swarm optimization, drew their original inspiration by first comparing the behaviors in accordance with the theory of socio-psychology for data processing and the decision-making in social groups, side by side. It is an exceptional and remarkable achievement that this metaheuristic was originally conceived for the continuous domain, and, till date, majority of its applications are in this domain. The method conceives a large group of *particles*, in the form of vectors, moving in the search space. Each particle i is characterized by its *position* $\mathbf{x}_i$ and a vector of change in position (called *velocity*) $\mathbf{v}_i$. In each iteration, the movement of the particle can be characterized as: $\mathbf{x}_i(t) = \mathbf{x}_i(t-1) + \mathbf{v}_i(t-1)$. The core of the method consists in the manner in which $\mathbf{v}_i$ is chosen, after each iteration. Socio-psychology suggests that the movements of the individuals (in a socio-cognitive chart) are influenced by their last behavior and that of their neighbors (closely placed in the social network and not necessarily in space). Hence, the updating of the position of the particles is dependent on the direction of their movement, their speed, the best preceding position $\mathbf{p_i}$ and the best position $\mathbf{p_g}$ among the neighbors:

$$\mathbf{x}_i(t) = f\left(\mathbf{x}_i(t-1), \mathbf{v}_i(t-1), \mathbf{p}_i, \mathbf{p}_g\right)$$

The change in position, in each iteration, is thus implemented according to the following relation:

$$\begin{cases} \mathbf{v}_i(t) = \mathbf{v}_i(t-1) + \varphi_1\left(\mathbf{p}_i - \mathbf{x}_i(t-1)\right) + \varphi_2\left(\mathbf{p}_g - \mathbf{x}_i(t-1)\right) \\ \mathbf{x}_i(t) = \mathbf{x}_i(t-1) + \mathbf{v}_i(t-1) \end{cases}$$

where the φ_n parameters are drawn randomly from the discourse $U_{[0,\varphi_{max}]}$ and are influential in striking a balance between the relative roles of the *individual experience* (governed by φ_1) and of the *social communication* (governed by φ_2). Uniform random selection of these two parameters is justified from the fact that it does not give any a priori importance to any of the two sources of information. The algorithm also employs another parameter, V_{max}, to limit the rapidity of movement in each dimension, so that it can prevent any "explosion" of the system, in case there are too large amplifications of the oscillations. The pseudo code for the generalized version of the algorithm — in the continuous version — is presented in the algorithm 5.7.

This basic algorithm can be further improved, if the problem is addressed from the point of view of controlling the divergence. In the initial version,

n =number of individuals
D =dimensions of the problem
Until **reaching** the termination criterion:
    For $i = 1$ to n :
      If $F(\mathbf{x}_i) > F(\mathbf{p}_i)$ then :
        For $d = 1, \ldots, D$:
          $p_{id} = k_{id}$ $//$ p_{id} is thus the best found individual
        end d
      end if
      $g = i$
      For j =index of the neighbors:
        If $F(\mathbf{p}_j) > F(\mathbf{p}_g)$ then :
          $g = j$ $//$ g is the best individual in the neighborhood
        end if
      end j
      For $d = 1, \ldots, D$:
        $v_{id}(t) = v_{id}(t-1) + \varphi_1 \left(p_{id} - x_{id}(t-1)\right) + \varphi_2 \left(p_{gd} - x_{id}(t-1)\right)$
        $v_{id} \in (-V_{max'} + V_{max})$
        $v_{id}(t) = x_{id}(t-1) + v_{id}(t)$
      end d
    end i
end

Algorithm 5.7: Particle swarm optimization (in continuous domain).

the parameter V_{max} is employed to prevent any "explosion" of the system, in the process of amplification of the positive feedback. One such improvement proposed to use constriction factors to exercise better control over this behavior [Clerc and Kennedy, 2002]. The algorithm proposes to limit rapidity by constricting the velocity update relation such that the user can control intensification/diversification dynamics of the system. In the "simple constriction" version, only one factor is multiplied with two component members of the velocity update relation which are influential in calculating rapidity or swiftness. This method of constriction enables the algorithm to converge (the amplitudes of the movements of the individuals are gradually decreased until they get cancelled). The algorithm could implement an effective compromise between intensification and diversification. The only problem arises when the points p_i and p_g move apart, in that case the particles will continue to oscillate between these two points without converging. An interesting characteristic of this algorithm is that, if a new optimum is discovered after the algorithm converged (i.e., after a phase of intensification), the particles will explore the search space around the new point (i.e. a phase of diversification).

In the same idea, a version sets up an *inertia weight* by multiplying each member of the equation of swiftness by a different coefficient [Shi and Eberhart, 1998]. To summarize, the inertia weight decreases according to time, which causes a controllable convergence by this parameter. General dynamics remains the same as in the version with coefficient of constriction, except for the impossibility of setting out again in a dynamics of diversification if a new better point is found.

Another significant parameter can be the introduction of the concept of neighborhood for the particles. It seems to be a well established fact that a *social* neighborhood (for example, an individual x_2 is the neighbor of the individuals x_1 and x_3, irrespective of the locations of x_1, x_2, x_3 in space) gives better results than a *spatial* neighborhood (a function of the proximity of the individuals in the search space, for example).

Other variations were also proposed, e.g. modifying the concept of the *better preceding position* of a particle by that of the better position of the centre of gravity of the *"clusters"* selected in the population [Kennedy, 2000]. The influence of the initial distribution of the particles was also studied [Shi and Eberhart, 1999].

The readers are redirected to read [Eberhart et al., 2001] to obtain a detailed, state of the art, understanding of the particle swarm optimization and the concepts associated with it and [Clerc, 2005] to read a synthesis.

5.7 The estimation of distribution algorithm

The algorithms with estimation of distribution ("Estimation of Distribution Algorithms", *EDA*) were originally designed as an alternative for the evolutionary algorithms [Mühlenbein and Paaß, 1996]. However, in the *EDA* type methods, there are no crossover or mutation operators. In fact, the population of the new individuals is drawn randomly according to a distribution estimated with the help of the information gathered from the preceding population. In the evolutionary algorithms, variables do maintain an implicit relationship, whereas in the *EDA* algorithms, the basic essence of the method precisely consists in estimating these relations, through the estimation of the probability distribution associated with each selected individual.

The best way of understanding the principle of *EDA* is to study the simplest possible example. Let us consider a simple problem of a function seeking to maximize the number of 1s in the three dimensions: the objective is to maximize $h(x) = \sum_{i=1}^{3} x_i$ with $x_i = \{0, 1\}$ (the "OneMax" problem).

The first stage consists in generating the initial population, hence M individuals are randomly drawn according to the probability distribution: $p_0(x) = \prod_{i=1}^{3} p_0(x_i)$ where the probability that each element x_i is equal to 1 is $p_0(x_i)$. In other words, the probability distribution, according to which the individuals are randomly drawn, is factorized like a product of three univariant marginal probability distributions (here, since the variables are binary, they are the Bernouilli distributions of parameter 0.5). The population thus sampled is named D_0. Let us consider, for example, a population of six individuals (illustrated in figure 5.3).

i	x_1	x_2	x_3	$h(x)$
1	1	1	0	2
2	0	1	1	2
3	1	0	0	1
4	1	0	1	2
5	0	0	1	1
6	0	1	0	1

Fig. 5.3. EDA algorithm employed for optimization of the OneMax problem: the initial population D_0.

The second stage consists in selecting individuals among this population; hence the second population $D_0^{S_e}$ is built in a probabilistic manner, drawn from the best individuals in D_0. The method of selection is unrestricted, here one can, for example, select the three best individuals (see figure 5.4).

i	x_1	x_2	x_3	$h(x)$
1	1	1	0	2
2	0	1	1	2
4	1	0	1	2
$p(x)$	0.3	0.3	0.3	

Fig. 5.4. EDA algorithm employed for optimization of the OneMax problem: selected individuals $D_0^{S_e}$.

The third stage consists in estimating the parameters of the probability distribution represented by these selected individuals. In this example, the variables are assumed to be independent. Hence three parameters will be necessary to characterize the distribution. Thus each parameter is estimated as $p(x_i \mid D_0^{S_e})$ on the basis of its relative frequency in $D_0^{S_e}$. Hence, this results in: $p_1(x) = p_1(x_1, x_2, x_3) = \prod_{i=1}^{3} p(x_i \mid D_0^{S_e})$. By sampling this probability distribution $p_1(x)$, one can obtain a new population D_1 (figure 5.5).

i	x_1	x_2	x_3	$h(x)$
1	1	1	1	3
2	0	1	1	2
3	1	0	0	1
4	1	0	1	2
5	0	1	1	2
6	1	1	0	2
$p(x)$	0.3	0.3	0.3	

Fig. 5.5. EDA Algorithm for optimization of the OneMax problem: the new population D_1.

This process has to be continued until a predefined termination criterion is reached ... A generalized version of the estimation of distribution algorithm is presented in the algorithm 5.8.

$D_0 \leftarrow$ Generate M individuals at random
$i = 0$
Until reaching the termination criterion:
$\quad i = i + 1$
$\quad D_{i-1}^{Se} \leftarrow$ Select $N \leq M$ individuals in D_{i-1} following the method of selection
$\quad p_i(x) = p\left(x \mid D_{i-1}^{Se}\right) \leftarrow$ Estimate the probability distribution of an individual
$\quad\quad$ belonging to the selected individuals
$\quad\quad D_i \leftarrow$ Sample M individuals since $p_i(x)$
end loop

Algorithm 5.8: Algorithm for estimation of distribution.

The main difficulty associated with the estimation of distribution algorithm is the estimation of the probability distribution. In practice, it is necessary to determine the parameters of the distribution in accordance with a selected model. Hence, many approximations for this estimation have been proposed for problems of both continuous and combinatorial optimization. Various algorithms proposed can be classified according to the complexity of the model utilized to evaluate the dependency among the variables. Hence three such categories can be determined:

1. Models *without* dependence: the probability distribution is factorized starting from univariant independent distributions for each dimension. The main drawback of such a choice is that it is not a practicable assumption for difficult optimization problems, where a strong inter-dependence of the variables is a common occurrence;
2. Models with *bivariant* dependences: the probability distribution is factorized starting from bivariant distributions. In this case, the training of the distribution can be wide until the concept of *structure*;
3. Models with *multiple* dependences: the factorization of the probability distribution is carried out starting from the statistics of the order *superior* than two.

I should be noted that in the continuous domain, the model of distribution is generally a normal distribution. Some important alternatives have also been proposed. These include the use of "data clustering" for multimode optimization and similar alternatives for combinatorial problems. Convergence theorems have also been formulated, in particular with the help of modeling by Markov chains or dynamic systems.

Unfortunately it is beyond the scope of this book to detail all such algorithms proposed, however an interested reader is redirected to the exhaustive book of reference [Larranaga and Lozano, 2002] in this field.

5.8 GRASP method

The *GRASP* ("Greedy Randomized Adaptive Search Procedure" [Feo and Resende, 1995, Resende, 2000]) algorithm is a metaheuristic which is characterized by multiple initializations. Basically, it operates in the following manner: first a feasible solution is obtained, which is then further improved by a local search technique . The main objective is to repeat these two phases in an iterative manner and to preserve the best found solution (see the algorithm 5.9).

cut = dimension of the problem
For $k = 1,\ldots$, **iterations_max make:**
    Construction:
        solution $= 0$
        Evaluate the incremental cost of each individual candidate
        If **find** Dimension(solution) $\neq$ **cut:**
            Build **LLC:** the limited list of the candidates
            Select randomly an element e from **LCR**
            solution $=$ Add(solution, e)
        end
    *Search **local**:*
        If **find** the **solution** locally nonoptimal:
            Find $s' \in the$Neighborhood of (solution) such that $f(s') < f(solution)$
            solution $= s'$
        end
    better_solution $=$ Min(solution, **better_solution**)
end
return **better_solution**

Algorithm 5.9: The *GRASP* algorithm (the minimization problem).

In the construction phase, a limited list of the candidates (LLC) is drawn as follows:

1. Choose all the feasible elements.
2. Choose among those the best elements according to a given performance index.
3. Randomly draw, among these best elements, those which will belong to the limited list of candidates.

Generally, the performance index is chosen as the degradation in the value of the objective function due to the incorporation of the element in the solution, in the construction phase. This process gives the *greedy* dimension to the algorithm, whereas the *probabilistic* dimension arises due to the phenomenon of random selection. The important point to be noted is that the list of candidates is updated after *each* incorporation of a new element, which provides the *adaptive* flavor to the algorithm.

As long as the construction phase is continued, the solution found is gradually improved while a local search is implemented to transform this solution into the local optimum.

The most interesting aspect of *GRASP* is its simplicity in implementation as well as the limited number of parameters that are required to be tuned. It can also be quickly realized that the local search phase can utilize any relevant algorithm that is found effective for a given problem; same logic can also be applied for the construction phase.

The *LLC* can be limited according to the number of elements, or according to the quality of these similar elements. In the later case, the use of a parameter to regulate the threshold of choice, $\alpha \in [0, 1]$, facilitates easy adaptation of the algorithm between a phase of greedy construction ($\alpha = 0$, only the elements equal to the minimal value are accepted) and a phase of random construction ($\alpha = 1$, all the elements are accepted). A similar mechanism can be designed to directly regulate the size of the *LLC* with the help of a similar coefficient.

In a more formal manner, *GRASP* can be viewed as a technique of sampling an unknown distribution, which makes it similar to other algorithms that use similar techniques, e.g. the estimation of distribution algorithms (see section 5.7 of this chapter) or the ant colony based metaheuristics (see chapter 4), to name a few.

In the "reactive" version of *GRASP*, the parameter α is dynamically chosen during the execution of the algorithm, according to various strategies. This version produced better results that the first. Other versions use different procedures for the construction phase, implementing a specific form of memory by modifying the probabilities of choice for each element of the *LLC*. It was also proposed to use the "Path-Relinking" method in combination with *GRASP*, in particular to implement them in parallel.

All these alternatives are elaborately described in [Resende, 2000], accompanied by several results and comparisons.

5.9 "Cross-Entropy" method

The "Cross-Entropy[1]" (*CE*) method was initially conceived to estimate the probability of rare events in complex stochastic networks [Rubinstein, 1997],

---

[1] originally synonymous with the *Kullback-Leibler* distance, a well-known measurement of information in neuro-informatics

which was later adapted for difficult combinatorial optimization problems [Rubinstein, 1999, Rubinstein, 2001].

The algorithm comprises of two phases: firstly, a solution is produced at random according to a specified mechanism, then the parameters of the mechanism are modified, on the basis of the solution obtained, in order to obtain a better solution in the next iteration.

The *CE* method uses a graph-based formulation of the problems and views the *deterministic* optimization problems like the problems of *stochastic* optimization. In this direction, the random component lies either on the edges, or on the nodes of the graph.

The simplest manner in which the operation of the basic algorithm can be understood will be the description of its operation to solve problems of searching for the shortest path in a network. Schematically, a probability matrix represents the "memory" system, and the new solutions are produced by a process of selection of Markov chains. The algorithm deteriorates the transition matrix and amplifies the probabilities in the Markov chains by sampling during iterations. Thus the objective is to ensure an effective deterioration of the matrix, which is the essence of the CE method.

The algorithm 5.10 gives a step-by-step description of the method. It utilizes a performance function, which calculates the quality of a solution equivalent to the objective function of the problem. An amplification (controlled by ρ) of the best solutions is carried out in step 2. Step 3 will attempt to maximize the proximity with the optimum. Indeed, the solution for this problem minimizes the entropy crossed between $f\left(\cdot; v_{t-1}\right)$ and $f\left(\cdot; v_t\right)$.

1. Initialize the iteration $t = 0$ with a first uniform solution $\hat{v}_0$.
2. Generate a sample of solutions with the probability density $f\left(\cdot; v_{t-1}\right)$ (by using a strategy of selection) and calculate $(1 - \rho)^{\text{th}}$ quantile $\hat{\gamma}_t$ of the performance function.
3. By using the same sample of solutions, solve the "stochastic program", memorize the solution $\hat{v}_t$.
4. If for a fixed number of d iterations:

$$\hat{\gamma}_t = \hat{\gamma}_{t-1} = \ldots = \hat{\gamma}_{t-d}$$

Then terminate
Else $t = t + 1$ and return to step 2

Algorithm 5.10: The "Cross-Entropy" method.

There does not exist any synthesis book yet on this method at the time when this book is written. However a Web page is maintained on the subject [De Boer, 2002].

5.10 Artificial immune systems

The term "artificial immune systems" (*AIS*) is applicable for a vast range of different systems, in particular for metaheuristic optimization, inspired by the operation of the immune system of the vertebrates. A great number of systems have been conceived in several varied fields e.g. robotics, the detection of anomalies or optimization (see [De Castro and Von Zuben, 2000] for a detailed exploration of various applications).

The immune system is responsible for the protection of the organism against the "aggressions" of external organisms. The metaphor from which the *AIS* algorithms originate harps on the aspects of *training* and *memory* of the immune system known as *adaptive* (in opposition to the system known as *innate*), in particular by discriminating between *self* and *non-self*. Indeed, the alive cells have on their membranes some specific molecules, called "antigens". Each organism thus has a single identity, determined by the whole of the antigens present on its cells. The *lymphocytes* (a type of white globule) are those cells of the immune system which have *receivers* that are able to bind specifically to a single antigen, thus enabling to recognize a foreign cell at the organism. Once a lymphocyte recognizes a cell of non-self, it will be stimulated to proliferate (by producing clones of itself) and to be different in cell enabling to keep the antigen in memory or cell enabling to fight the aggressions. In the first case, it will be able to react more quickly, when exposed to a new antigen: in fact, that is also the principle behind the effectiveness of the vaccines. In the second case, the fight against the aggressions is possible with the help of the production of the antibody. The figure 5.6 summarizes these principal stages. The diversity of the receivers in the entire population of the lymphocytes should also be noted as it is produced by a mechanism of *hyper-mutation* of the cloned cells.

The principal ideas used for the design of this metaheuristic are the selections operated on the lymphocytes accompanied by the positive feedback, allowing the multiplication and the implementation of memory by the system. Indeed, these are the chief characteristics to maintain the self-organized characteristics of the system (see the beginning of the chapter 4 for a more precise definition of the self-organization).

The approach used in the *AIS* algorithms is very similar to that of the evolutionary algorithms but was also compared with that of the neural networks. Within the framework of difficult optimization, the *AIS* can be regarded to take the shape of evolutionary algorithm, introducing particular operators. To operate the selection, it has to be based, for example, on a measurement of affinity (i.e. between the receiver of a lymphocyte and an antigen). The process of mutation takes place through an operator of hyper-mutation, resulting directly from the metaphor. In the final analysis, the algorithm developed is very close to a genetic algorithm (see algorithm 5.11).

For example, algorithms which are interesting for dynamic optimization can be elaborated [Gaspar and Collard, 1999].

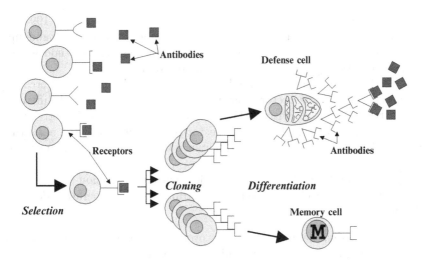

Fig. 5.6. Selection by cloning: lymphocytes, presenting specific receivers of an antigen, are different in cell memory or cell taking part in active defence against the organism with the help of the antibody.

1. Generate a collection of solutions P composed of an entire collection of cell memories P_M added to the present population P_r: $P = P_M + P_r$;
2. Determine the n best cells P_n from the population P, which is based on the measure of affinity;
3. Clone n individuals to form a population C. The number of clones produced for each cell is a function of affinity;
4. Implement a hyper-mutation process for the clones, which thus generates a population C^*. The mutation is proportional to affinity;
5. Select the individuals C^* to form the memory population P_M;
6. Replace the worst individuals in P to form P_r;
7. If a termination criterion is not reached, return to 1.

Algorithm 5.11: A simple example of the algorithm of *artificial immune system*.

A description of the basic theory and many applications of the artificial immune systems can be found in [De Castro and Von Zuben, 1999], [De Castro and Von Zuben, 2000] and in [Dasgupta and Attoh-Okine, 1997], and also in a book of reference [Dasgupta, 1999].

5.11 Method of differential evolution

The differential evolution technique belongs to the class of the evolutionary algorithms. In fact, it is founded on the principles of mutation, crossover and selection. However, it was originally conceived [Storn and Price, 1995] for continuous problems and it uses a weighted difference between two randomly selected individuals as the source of random variations. The method is very effective and recently has become increasingly popular.

Thus, the core of the method is based on a particular manner of creating new individuals. A new vector is created by adding the weighted difference between two individuals with a third; if the resulting vector is better than a predetermined individual, the new vector replaces it. Thus, the algorithm extracts information of direction and distance to produce its random component.

Let us consider a D dimensional problem with a population of N individuals, evolving with each generation t, according to three operators designed as follows:

Mutation: an individual mutant $v_{i,t+1}$ is produced starting from an individual $x_{i,t}$ relating to three other individuals conforming to:

$$v_{i,t+1} = x_{r_1,t} + F \cdot (x_{r_2,t} - x_{r_3,t})$$

where the three other individuals are indicated by r_1, r_2, r_3, different indices are chosen at random and by employing $F \in [0, 2]$, a real number called "the amplification factor". The disturbance will become smaller and smaller as the difference $(x_{r_2,t} - x_{r_3,t})$ diminishes.

Crossover: the individual $x_{i,t}$ is "mixed" with the mutant, thus creating the "test vector" $u_{i,t+1}$:

$$u_{i,t+1} = (u_{1i,t+1}, u_{2i,t+1}, \ldots, u_{Di,t+1})$$

where

$$u_{ji,t+1} = \begin{cases} v_{ij,t+1} & \text{if } (r(j) \leq C_R \text{ or } j = rn(i)) \\ x_{ji,t} & \text{if } (r(j) > C_R \text{ and } j \neq rn(i)) \end{cases}$$

with:

$j = 1, 2, \ldots, D$;

$r(j) \in [0, 1]$: the jth evaluation of a uniform random distribution;

$C_R \in [0, 1]$: the crossover constant;

$rn(i) \in (1, 2, \ldots, D)$: a random index ensuring that $u_{i,t+1}$ is at least an element obtained from $v_{i,t+1}$.

Selection: an individual is selected if the test individual $u_{i,t+1}$ allows an improvement in the objective function compared to $x_{i,t}$. A selected individual takes part in the next generation; in the opposite case, it is only retained to be used as parent during the next generation.

Generate an initial population in accordance with an uniform distribution.
For $t = 0, \ldots, T$ generations
>   For $i = 0, \ldots, N$ individuals
>   Choose three individuals at random $x_{r_1,t}, x_{r_2,t}, x_{r_3,t}$
>   Mutation:
>>   Create a mutant $v_{i,t}$ starting from the individual $x_{i,t}$ and from
>> $x_{r_1,t}, x_{r_2,t}, x_{r_3,t}$
>
>   Crossover:
>>   Create the test individual $u_{i,t}$ starting from the mutant $v_{i,t}$ and from
>> $x_{i,t}$
>
>   Selection:
>>   If $f(u_{i,t}) < f(x_{i,t})$
>> $x_{i,t+1} \leftarrow u_{i,t}$
>>   Else
>> $x_{i,t+1} \leftarrow x_{i,t}$
>
>   End individuals

End generations

Algorithm 5.12: The differential evaluation method for the minimization problem.

The method (summarized by the algorithm 5.12) thus has only three parameters: N, F and C_R, which facilitates a relatively simple adjustment, studied in detail in [Gämperle et al., 2002]. Many improvements were proposed, for example by changing the choices of the vectors to be mutated, by hybridizing a *DE* method with a local search algorithm [Rogalsky and Derksen, 2000] or combining with particle swarm optimization [Tsui and Liu, 2003]. . . . A further study in this field should be undertaken with the help of detailed descriptions presented in [Storn and Price, 1995, Storn, 1997], and the supplementary bibliography presented in [Lampinen, 2001], which describes many applications in particular.

5.12 Algorithms inspired by the social insects

These are those metaheuristics which are inspired by the behavior of the social insects which are not explicitly related to the — more known — ant colony algorithms. Indeed, the behaviors of these species are complex and rich; and the self-organized characteristics possessed by these systems present an interesting source of inspiration.

A good example can be the creation of an algorithm, inspired by the models of the organization of work in the ants [Campos et al., 2000] [Cicirello and Smith, 2001, Nouyan, 2002]. The task sharing among certain species reveals that specialized individuals are employed to achieve specific tasks, which facilitates to avoid the costs (in time and energy for example) related to the reallocation of the tasks. However, such specializations of the individuals are not rigid, which could be prejudicial with the colony, but adapts according to many internal and external stimuli perceived by the individuals. Thus, the colony presents a specialization which is an adapted and flexible function of the parameters perceived by the individuals [Wilson, 1984].

The behavioral models proposed an explanation of the mechanisms involved in this phenomenon [Bonabeau et al., 1996, Bonabeau et al., 1998, Theraulaz et al., 1998]. These models are responsible for creating thresholds of response for each type of tasks. These thresholds represent the levels of specializations of the individuals and they are either fixed [Bonabeau et al., 1998], or updated according to the achievement of the tasks by the individuals [Theraulaz et al., 1998].

These models are inspired by the algorithms employed for the dynamic allocation of tasks, where each machine can be viewed as associated with an individual having a set of thresholds of answer θ_a, where $\theta_{a,j}$ represents the threshold of the agent a for the task j. The tasks send a stimulus S_j to the agents, representing the latency of the task. The agent a will have a probability of carrying out the task j of:

$$P\left(\Theta_{a,j}, S_j\right) = \frac{S_j^2}{S_j^2 + \Theta_{a,j}^2}$$

The algorithm then employs update rules for the thresholds and the decision rules whether two agents would try to carry out the same task (see [Cicirello and Smith, 2001] for more details). Later, improvements for this basic algorithm have also been proposed [Nouyan, 2002] which addressed the issues of increasing the speed and the effectiveness of the algorithm for complex problems.

5.13 Annotated bibliography

[Saït and Youssef, 1999]: Several metaheuristics are described in detail in this book. In particular, two metaheuristics, which are not discussed here, are explained in detail: "Simulated Evolution" and "Stochastic Evolution".

[Pham and Karaboga, 2000]: This book concentrates on elaborating several metaheuristics. Special attention could be drawn towards a chapter devoted to the neural networks, utilized for a problem of placement of electronic components.

[Teghem and Pirlot, 2002]: This recent book is a collection of the contributions of a dozen authors. This collection can be very useful in supplementing our very brief presentation of the noising methods and the *GRASP* method. This book also includes a chapter devoted to the utility of the neural networks in combinatorial optimization, and another chapter exploring the possible integration of the techniques of operations research in constraint programming.

[Resende, 2000]: A synthesis of the *GRASP* algorithms, accompanied by some results and comparisons.

[Eberhart et al., 2001]: A very complete book on particle swarm optimization, the most obvious reference in this domain.

[Larranaga and Lozano, 2002]: A reference book on the algorithms of estimation of distribution, comprising of a collection of articles, detailed and rather clearly explained.

[Dasgupta, 1999]: A book on the artificial immune systems with a detailed view of their various aspects.

[Reeves, 1995]: Main metaheuristics are presented in this book.

6

Extensions

6.1 Introduction

Our interest in this chapter will be mainly focused to four types of extensions of the metaheuristics, which have been proposed to encounter optimization problems of following particular nature:

- adaptation for the problems with continuous variables;
- multimodal optimization;
- multiobjective optimization;
- evolutionary optimization with constraints.

6.2 Adaptation for the continuous variable problems

In this section, we initially present the general framework of "difficult" continuous optimization. Then some typical applications are described in detail, followed by discussions on specific difficulties that arise in continuous problems. A few pitfalls encountered with the adaptation of metaheuristics for the problems of continuous variables are underlined in the following section.

The second part of the chapter describes as an illustration the methods, which we proposed to adapt some metaheuristics: simulated annealing, tabu search and genetic algorithms. The adaptation of the ant colony algorithms for the continuous case has been treated in detail in the 4.6.1 section.

6.2.1 General framework of "difficult" continuous optimization

Some typical problems

The "difficult" problems usually encountered in the presence of continuous variables can be classified into three families:

- problems of *optimization* strictly speaking,
- problems of *identification* or *characterization*,
- problems of *learning*.

We illustrate each of these families by an application example.

Example of a strictly speaking optimization problem: the optimization of the performances of an electronic circuit.

Let us assume that the electrical diagram of the circuit is known, which fulfils a given function: for example, a constant current generator, or a wide frequency band amplifier. This circuit comprises of n "parameters": for example, resistance values, geometries of the transistors, etc. For each of them, generally the discourse of the physically acceptable values is known. The problem of the "optimization of the performances of the circuit" consists in seeking the numerical value that is necessary to be assigned to each of the n parameters available so that the circuit can realize a given objective, as far as practicable.

Example: optimization of a constant current generator. The circuit under consideration is represented in the figure 6.1. It shows a constant current generator, comprising of 4 *MOS* transistors, denoted as M_1 to M_4. The 8 unknown factors of the problem are the dimensions (width W and length L) of the channels of each transistor. The specifications imposed for the circuit are the following: the delivered current i must be constant and equal to $50\mu A$, irrespective of the continuous output voltage V_1, which is between $-3V$ and $-2V$.

The objective function f is formed by means of a circuit simulator, e.g. *SPICE*. In "continuous simulation" mode, the simulator evaluates the current i delivered by the generator, when the voltage V_1 takes a specified value. In this application, carried out in the *CEA*, Center of Bruyères-le-Châtel, f is formed from the results of 6 continuous simulations, corresponding to the 6 specified values of V_1 : -3 ; -2.8 ; -2.6 ; -2.4 ; -2.2 and $-2V$. The corresponding currents obtained in the simulation are denoted as: $i_1, \ldots, i_6$ (in μA).

Then f can be formed as:

$$f = \sum_{k=1,6} (i_k - 50)^2$$

For a fixed range of the channel dimensions of the transistors, 6 continuous simulations are thus necessary to lead to an evaluation of the objective function. Let us note that the choice of the number of simulations results from a compromise between the computing time (proportional to this number of simulations) and a better adequacy of the electronic circuit sought with the specified objective.

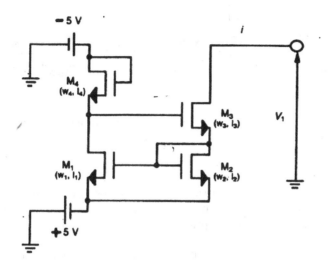

Fig. 6.1. Current constant generator: optimization problem with 8 variables.

Example of identification or characterization problem: the "tuning" of a model of a synchronous electrical motor.

In an application implemented in the Centre de Génie Electrique de Lyon (*CEGELY*) [Clerc et al., 2002], the objective was to bring the calculated curves (obtained from the numerical model) and the experimental curves (based on the measures taken from the motor) as close as possible. The experimental test bench employed to take measurements from the synchronous motor (*MS*) is shown in figure 6.2. There is 19 unknown factors for this optimization problem and all are parameters of the numerical model. The objective function to be minimized is based on the difference (absolute value or least squares) between the experimental data and the theoretical model based calculations.

Example of training problem. This type of problem is usually encountered in neural networks, or fuzzy rule bases. We illustrate it with the "analog neural network" of the 6.3, used by the *CEA* to reproduce the sine function in wired form, as accurately as possible [Berthiau et al., 1994]. This multi-layer neural network, of "feedforward" type, is used to calculate:

$$Y = 0.8 \sin X, \text{for } -\pi \leq X \leq \pi$$

The network takes two inputs: the argument X of the *sine* function and the initial constant X_0 (which allows to adjust the activation function of each neuron in the network). The network comprises of two layers: an input layer of 5 neurons, each one receiving two inputs; an output layer of only one neuron,

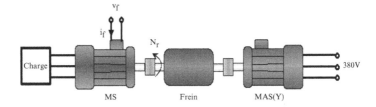

Fig. 6.2. Experimental test bench for the extraction of measurements on a synchronous motor.

which receives 6 inputs. Each " synaptic coefficient" of the neural network is implemented by a resistance in the circuit, as shown in the figure 6.3a. The training of the neural network can thus be expressed as an optimization problem of 16 variables: the 16 synaptic resistances, whose discourse of variation is fixed in $[1k\omega, 1M\Omega]$.

Going through these few preceding examples, one can guess the abundance of the continuous applications in most of the engineering fields: in particular, in the domains of electronics and mechanics. Moreover, there are many "mixed" problems, which are simultaneously discrete and continuous, such as — that we described in the chapter 1 (paragraph 1.5.3) — search for an equivalent diagram in electronics.

Specific difficulties in the continuous problems

To ascertain these difficulties, let us consider the following case:

- mono-objective problem;
- objective function f, to be minimized;
- decision variables accumulated in a vector x ;
- only constraints: "box constraints": $x_i^{MIN} \le x_i \le x_i^{MAX}$.

The continuous optimization problems, like those referred above, often present specific difficulties. The name "difficult problem" is again used, even if this term does not refer here to the theory of complexity, relevant for the discrete problems.

The principal sources of these difficulties are the following:

1. There does not exist any analytical expression of f.
2. f is "noised". This noise can be of experimental nature, if the calculation of f is carried out with the help of measurements. It can also be a "numerical calculation noise", for example when f is evaluated with the help of an electronic circuit simulator (which, in particular, makes use of methods for numerical integration).
3. f comprises of non-linearities.

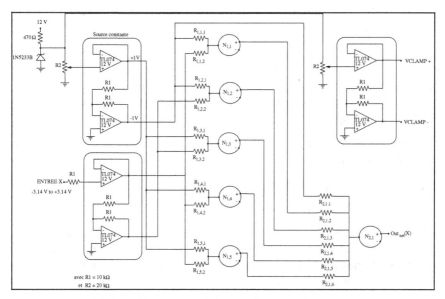

(a) Overall diagram of the "analog neural network"

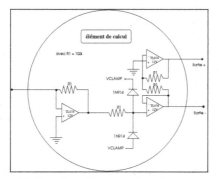

(b) Detailed diagram of each neuron

Fig. 6.3. Electric diagram of the "analog neural network" used for the approximation of the *sine* function.

4. There exists correlations — non precisely localized — between certain variables of the problem.

The difficulties (1) and (2) prohibit to access the gradients of f. As for the difficulties (3) and (4) are concerned, they involve the existence of a tormented "energy landscape", comprising of many local minima. Consequently, those methods which attempt to solve such difficult continuous problems effectively must have two properties:

- They must be *"direct"*, i.e. do not require calculation of the gradients: this condition prohibits the use of powerful traditional methods, of "Newton type".
- They must be *"global"*, in the following sense: methods that do not get trapped, in theory, in a local minimum.

Remark. It is to be noted that the terms "global" and "local" above do not characterize the type of "movement" utilized . It follows that, for example, simulated annealing can be described as local, by its mechanism, but also as global, by its finality...

This twofold requirement justifies the utility of employing the metaheuristics, as all metaheuristics are simultaneously "direct" and "global".

One can put stress on the point that the "direct " aspect of metaheuristics — that is related to their combinatorial origin — is not specially attractive for discrete problems. On the contrary, it is a determining advantage in the difficult continuous domain.

The preceding considerations explain the significant interest raised in the metaheuristics in the context of continuous optimization. On the other hand, the majority of metaheuristics were conceived within a discrete framework (a notable exception being the "particle swarm optimization" method, described in the chapter 5), from where the need arose for an "adaptation" for the continuous problems.

Some pitfalls of the adaptation of metaheuristics for the problems with continuous variables

These pitfalls are initially of "cultural" order: the continuous applications are generally the domain of the engineers, specially the electronics and the mechanical engineers. On the other hand, the know-how concerning the metaheuristics lies with the computer scientists, who are less interested in the continuous problems, that are very little standardized. This gap probably explains the lack of theoretical results developed in this field.

There also exist the pitfalls of "technical" order, like:

- The heterogeneity of the definition domains of different variables;
- The presence of variables with very wide definition ranges (sometimes more than 10 decades).

In addition, confrontation between the competing empirical studies is perilous; indeed:

- The competitive codes are seldom available in public.
- It is delicate to program all the competitive methods.
- The results published are not easily comparable, due to the following reasons:
  - Different sets of test functions, and of evolution domains of the variables are utilized;
  - Different methods for the selection of the initial point, and for the number of averaged executions per result are implemented;
  - Different definitions of the "success" (approach to the global optimum x^*), of the "final error" (in f? in x?), of the computing time.

One of the authors of this book recommends an objective comparison between the competing algorithms, via the use of statistical tests [Taillard, 2003b]: this procedure could solve the problem of the comparison, if the authors of the competing algorithms agreed to conform.

Many methods were proposed in the literature to adapt the metaheuristics to the continuous case. It is not possible to deduce general considerations from these methods, except for the following ones:

- The majority of the techniques developed to adapt a metaheuristic are not applicable for the other metaheuristics...
- As in the discrete domain, no metaheuristic seems to surpass its competitors...

To clarify the domain, and to present the reader some examples of operational methods, we now describe the techniques which we proposed to adapt some metaheuristics for the continuous case.

6.2.2 Some continuous metaheuristics

Simulated annealing

The main problem relates to the management of the "discretization" of the search space. It is necessary " to maintain almost the same effectiveness" for the process of optimization, throughout the descent in temperature. To this effect, it is necessary to use the length of the discretization step as an additional controllable parameter, which adapts automatically, for example according to the rate of acceptance of the modifications attempted, or average length of displacements already carried out (this length constituting, to some extent, a measurement of the local topography of the configuration space).

In practice, for the development of continuous simulated annealing, one can proceed as follows:

1. The discretization step, a priori different for each variable x_i of the problem, is called $STEP_i$. It represents the maximum variation that x_i can undergo in a single "movement".
2. For a fixed step $STEP_i$, we must specify the design rule F for calculating the movement of a variable, passing from the value x_i to the value $x_i' = F(x_i, STEP_i)$.
3. As mentioned before, $STEP_i$ must be adjusted periodically, to maintain the efficiency of optimization almost constant, when T is decreased; it is thus necessary to choose:
   a) the mode of evaluation of this effectiveness:
      i. using the rate of acceptance of the movements attempted?
      ii. using the average amplitude of the movements already carried out?
   b) the frequency of the adjustment of the step;
   c) the law of the adjustment of the step.
4. We must decide which variables x_i are concerned with a given movement:
   a) all variables of the problem?
   b) only one at a time?
   c) a subset of variables?

Unfortunately, only a few theoretical results are available to validate this intuitive procedure, and to specify its various options. Consequently, many authors justify their choices by following a systematic *empirical* approach: the program is evaluated and tuned by using, like objective functions, various published analytical test functions, comprising up to 100 variables.

The a priori knowledge of the global and local minima of the problem facilitates the study of the influence of the main parameters of the method on its convergence speed. For the purpose of illustration, we define and present three classical test functions in the figure 6.4.

The tuning that we utilized [Siarry and Berthiau, 1997] is specified in the algorithm 6.1.

Remarks.

For strongly correlated variables, simultaneous displacement of several of them enables us to approach the optimum more than that can be achieved by employing individual displacements. An estimation of the gradients and the "hessian" of the objective function f would make it possible to determine the more "sensible" variables (i.e. those variables whose variations produce the most significant effect on f), and to analyze the correlations between the variables. However, as indicated before, the evaluation, by finite differences, of the gradients and the hessian of f frequently leads to severe numerical instabilities. It is thus necessary to give up a priori partitioning of the problem based on the regrouping of the most correlated variables.

Michalewicz (MZ) $(n$ variables):

$$MZ(x) = -\sum_{i=1}^{n} \sin(x_i) \cdot \left(\sin\left(\frac{ix_i^2}{\pi}\right)\right)^{20} ;$$

search space: $0 \leq x_i \leq \pi$, $i = 1, \ldots, n$;
$n = 2$, 1 global minimum: $MZ(x^*) = -1.80$;
$n = 5$, 1 global minimum: $MZ(x^*) = -4.687$;
$n = 10$, 1 global minimum: $MZ(x^*) = -9.68$.

Goldstein-Price (GP) (2 variables):

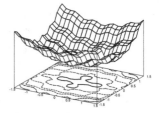

$$GP(x) = \left(1 + (x_1 + x_2 + 1)^2 \cdot \left(19 - 14x_1 + 3x_1^2 - 14x_2 + 6x_1x_2 + 3x_2^2\right)\right) \cdot$$
$$\left(30 + (2x_1 - 3x_2)^2 \cdot \left(18 - 32x_1 + 12x_1^2 + 48x_2 - 36x_1x_2 + 27x_2^2\right)\right) ;$$
search space: $-2 \leq x_i \leq 2$, $i = 1, 2$;
4 local minima;
1 global minimum: $x^* = (-1, 0)$; $GP(x^*) = 3$.

B2 $(B2)$ (2 variables):

$$B2(x) = x_1^2 + 2x_2^2 - 0.3 \cdot \cos(3\pi x_1) - 0.4 \cdot \cos(4\pi x_2) + 0.7$$
search space: $-1.5 \leq x_i \leq 1.5$, $i = 1, 2$
25 local minima;
global minimum: $x^* = (0, 0)$; $B2(x^*) = 0$

Fig. 6.4. Examples of classical test functions.

Tabu search

We describe, as an illustration, the algorithm which we proposed: $ECTS$ ("Enhanced Continuous Tabu Search") [Chelouah and Siarry, 2000a], which is desired to be a faithful transposition of the combinatorial method.

The various elements of ECTS are the following:

- The tabu list comprises of "balls" centered on the last adopted solutions.
- The concept of neighbourhood is specified in figure 6.5: the space around a current solution is deconstructed into concentric hyper-rectangular "crowns". The list of the neighbours of this current solution is formed by randomly selecting a solution inside each "crown".

- Initial $STEP_i : \frac{1}{4}$ of the domain of variation of x_i ;
- $STEP_i$ modified at the end of each temperature step;
- The rate of acceptance A_i of the movements attempted, during the stage, for x_i is analyzed:
  - if $A_i > 20\%$, $STEP_i$ is doubled,
  - if $A_i < 5\%$, $STEP_i$ is divided by 2,
- *If* the domain of variation of x_i is "not very wide":

$$x_{i'} = x_i \pm y \cdot STEP_i$$

where y indicates a real random number, drawn in $[0, 1]$.
Else (in case of several decades), the perturbation is operated according to a logarithmic law;
- The n variables of the problem are modified in groups of p, formed at random: typically, $p \cong \frac{n}{3}$.
The frequencies of movement of the various variables are equalized.
- Several complementary criteria are used for the automatic termination of the program.

Algorithm 6.1: Simulated annealing in continuous variables.

- The general structure of the algorithm is presented in the figure 6.6. The core of the tabu method, not described in this figure, is classical:
  - N neighbours, not tabu, of the current solution are generated;
  - the best of these neighbours becomes the new current solution;
  - the old current solution occupies position in the tabu list.
- ECTS comprises of two successive phases (figure 6.6):
  - a *diversification* phase, intended to locate the "promising zones" in the solution space, where the global minimum of f may be located. The "promising" status is initially attributed to some solutions. The solution x is called promising, if it "clearly" surpasses (within the context of a threshold fixed dynamically by ECTS) its N neighbours, evaluated in turn in the core of tabu search. Then the promising list is made of "balls" centered on the promising solutions.
  - an *intensification* phase: a new tabu search is initiated inside the best promising zone. It operates in an even finer manner (reducing the dimension of the search space and the size of the tabu list).

To illustrate this mechanism, we present, in figure 6.7, the path followed by ECTS during the optimization of a test function. ECTS located 6 promising zones, including the one where is situated the global optimum $(0, 0)$.

The principal drawback of this approach resides in the large number of parameters required to be tuned:

- size of the tabu list and promising list;
- radius of the tabu and promising balls;

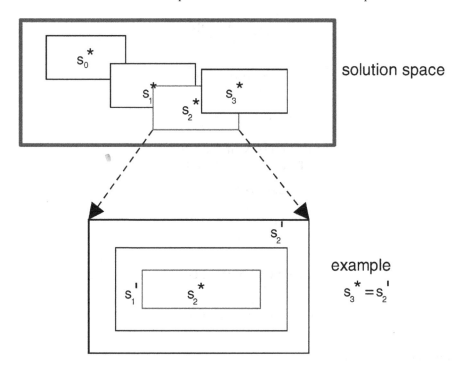

Fig. 6.5. The concept of neighborhood of a current solution in continuous tabu search.

- dimensions of the neighbourhood of a current solution;
- number of neighbours of a current solution;
- ...

The majority of the parameters in ECTS are calculated automatically, or are "at best" fixed empirically. The detailed experimental validation of ECTS, elaborated in [Chelouah and Siarry, 2000a], shows that the computing time grows slowly (quasi linearly) with the number of variables in f. Convergence can still be accelerated, by hybridizing ECTS with the polytope algorithm of Nelder & Mead. This last one is a "local descent" algorithm, which has the advantage — like the metaheuristics — of not requiring to calculate the gradients of the objective function. The descent is implemented by successive deformations of a "polytope", via geometrical transformations described in figure 6.8. In the hybrid variant of ECTS, the diversification and intensification phases are alternate, as shown in figure 6.9.

Fig. 6.6. General structure of the continuous tabu algorithm ECTS.

Genetic algorithms

Here we describe, as an illustration, the algorithm which we have proposed: *CGA* (Continuous Genetic Algorihm) [Chelouah and Siarry, 2000b].

The characteristics of CGA are the following:

- alternating phases of diversification and intensification, specified later;
- dynamic reductions of the population size and the search space;
- real coding of the individuals; the name given to an AG exploiting this type of coding is twofold: RCGA (Real Coded Genetic Algorithm) or EA (Evolutionary Algorithm);
- specific crossover and mutation operators, described later;
- dynamic reduction of the probability of mutation.

The alternation between diversification and intensification in CGA is illustrated in the figure 6.10:

- The diversification phase is carried out using classical GA, employing "opulent" individuals. It leads to the identification of a "promising solution" x^*.
- The intensification phase is carried out in a new restricted search space, around x^*. It is carried out using a new GA, comprising a reduced population, formed by thinner individuals.

The crossover operator, often called "recombination" in real coding, is inspired, in CGA, by the mono-point crossover in binary coding:

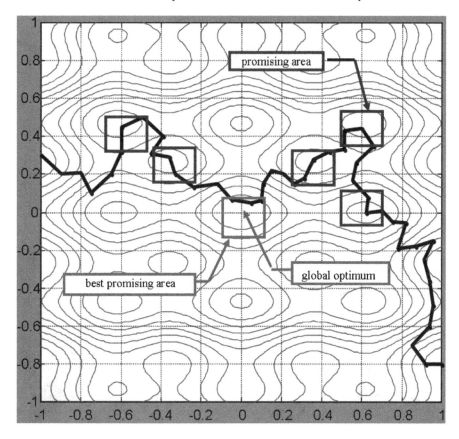

$$B_2(x) = x_1^2 + 2x_2^2 - 0.3 \cdot \cos\left(3\pi x_1\right) - 0.4 \cdot \cos\left(4\pi x_2\right) + 0.7$$

search space: $-1 \leq x_i \leq 1$, $i \in \{1, 2\}$
global minimum: $(x_1, x_2)^* = (0, 0)$ et $B2(0, 0) = 0$

Fig. 6.7. The path followed by the ECTS algorithm (diversification) during the optimization of the B2 test function.

- Each individual is a vector x of dimension n.
- The ith component of x is the ith variable of optimization.
- The position i of the crossover is selected at random.
- The components of index higher than i are exchanged between the two parents.
- The components of index i undergo opposite variations, as indicated in an example with $(n = 2, i = 2)$, in the figure 6.11.

The mutation operator affects the value of only one variable. The maximum amplitude of the perturbation and the probability of mutation are

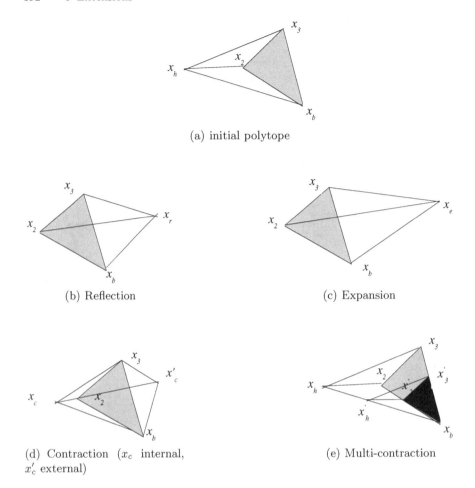

(a) initial polytope

(b) Reflection

(c) Expansion

(d) Contraction (x_c internal, x'_c external)

(e) Multi-contraction

Fig. 6.8. Operating principle of the geometrical transformations applied by the polytope algorithm of Nelder & Mead. x_b represents the vertex where the value of the objective function is lowest; x_h represents the vertex where the value of the objective function is highest.

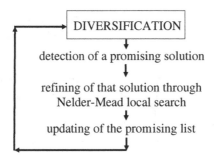

Fig. 6.9. Operating principle of the hybrid variant of ECTS algorithm.

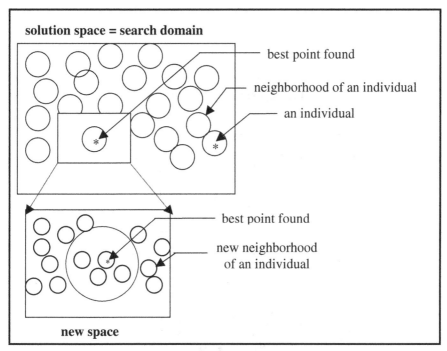

effect of successive "zooming"

Fig. 6.10. Alternation between diversification and intensification in CGA.

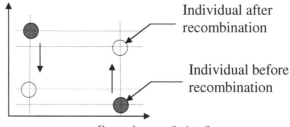

Example: $n = 2,\ i = 2$

Fig. 6.11. Opposite variations of the components of the two parents located at the crossover point.

Typical example of result

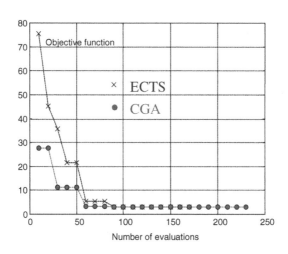

Fig. 6.12. Typical example of performances obtained with CGA and ECTS algorithms.

gradually diminished: the later factor is more delicate and is discussed in detail in [Chelouah and Siarry, 2000b].

A typical example of performance is presented in figure 6.12. In the case of the optimization of the Goldstein-Price function, we compare the evolutions obtained towards the optimum with two algorithms: the CGA algorithm and the continuous tabu ECTS algorithm described before. It can be observed that CGA performs better at the beginning, but then it "runs out of steam", which suggests that CGA should be hybridized with a method of local descent. We proposed to carry out the intensification with the polytope algorithm of Nelder & Mead [Chelouah and Siarry, 2003]. The effect of this transformation is illustrated in the figure 6.13. From the example presented, it should be noted that:

- the CGA algorithm converges slowly after a few generations.
- the polytope algorithm only (*SS*) led to a local optimum.
- the hybrid method (*CHA*), obtained by hybridization of CGA and SS, is very effective, provided the transition from one approach to the other is not premature.

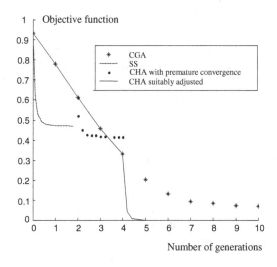

Fig. 6.13. Effect of the hybridization of CGA with the polytope algorithm of Nelder & Mead: optimization of the B_2 function.

The automated transition between these algorithms is discussed in [Chelouah and Siarry, 2003].

The hybrid variant was exploited successfully, in the CEA, for optimal design of a sensor with eddy currents intended for non destructive testing of metal tubes [Chelouah et al., 2000]. The algorithm in particular enables us to obtain a sensor, in plate form, very wide spread: the "pancake".

6.3 Multimodal optimization

6.3.1 The problem

The industrial problems can be seldom completely formalized. Many decisions depend on the image that a company desires to project, the policy it wants to adopt vis-a-vis its customers and its competitors, its economic or legal environment, etc. Its decisions regarding the design of a new product, its manufacturing, its launching, depend on dialogues, negotiations, with several players. All these factors make it difficult to formalize such problems with the aim of solving them by means of a computer. In the context of optimization, a problem often presents several optimal solutions of equivalent value. Theoretically only one solution is enough. However, this does not offer any degree of freedom to the decision makers. On the other hand, if a set of optimal solutions is taken into account, this will enable the decision makers to freely choose one of these solutions according to factors which can not be formalized.

Multimodal optimization consists in locating multiple global optima, and possibly the best local optima, of an objective function. The evolutionary algorithms are good candidates to achieve this task because they handle a population of solutions that can be distributed among various optima. Let us note that there are methods to search for several optima, like the sequential niching [Beasley et al., 1993], which do not require a population based algorithm to succeed. But they present poor performances. This is why this section is entirely devoted to the evolutionary methods. However, if a multimodal objective function is subjected to a standard evolutionary algorithm, the experimentations and the theory show that the population is attracted by only one of the maxima of the fitness function, and it is not necessarily a global maximum. For example, let us consider a function comprising of two peaks of equal height. An initial population is built where the individuals are already evenly located, on the two optima. After a few generations, the balance will be broken because of the genetic drift. From this point, the crossover amplifies imbalance until the majority of the population is localized only on one peak. The problem of multimodal optimization would be correctly solved if a mechanism could stabilize subpopulations located on the highest peaks of the fitness function. It is about *speciation*, which makes it possible to classify the individuals of a population into different *subpopulations*, and *niching* that stabilizes subpopulations within *ecological niches* containing the optima of the objective function. There are several methods of speciation and niching. The most common or the most effective ones are described below.

6.3.2 Niching with the sharing method

The concept of "sharing of limited resources within an *ecological niche*", suggested by J.H. Holland [Holland, 1992], constitutes one of the most effective approaches to create and maintain stable subpopulations around the peaks of the objective function with an evolutionary algorithm. The concept of ecological niche originates from the study of the population dynamics. It was formalized by Hutchinson in 1957 [Hutchinson, 1957], who defined it as a hyper-volume in an n dimensional space, each dimension representing a living condition (e.g. quantity of food, temperature, size of the vital domain, etc). An ecological niche cannot be occupied by several species simultaneously. It is the empirical principle of *competitive exclusion*. The resources within a niche being limited, the size of the population that occupies it stabilizes.

In 1987, Goldberg and Richardson [Goldberg and Richardson, 1987] proposed an adaptation of this concept for the genetic algorithms, which can be generalized for any evolutionary algorithm. The technique is known under the name of *sharing method*. It is essential that a concept of dissimilarity among the individuals be introduced. For example, if the individuals are bit strings, the Hamming distance can be appropriate. If they are vectors of $\mathbb{R}^n$, the Euclidean distance is a priori a good choice. The value of dissimilarity is the criterion to decide whether two individuals belong to the same niche or not. The method consists in attributing a *shared fitness* to each individual, which is equal to its raw fitness divided by a quantity that increases with the number of individuals resembling it. The shared fitness can be viewed as representing a quantity of resource available for each individual in a niche. The selection is ideally proportional, so that the number of offspring of an individual is proportional to its shared fitness, although the method has also been employed with other selection models. Thus, with the same raw fitness, an isolated individual will definitely have more offspring than an individual having many neighbors in the same niche. At equilibrium, the number of individuals located on each peak becomes proportional, at first approximation, to the fitness associated with this peak. This appears to give rise to a stable *subpopulation* in each niche. The shared fitness of an individual i can be expressed as:

$$\tilde{f}(i) = \frac{f(i)}{\sum_{j=1}^{\mu} \mathrm{sh}(d(i,j))}$$

where sh is of the form:

$$\mathrm{sh}(d) = \begin{cases} 1 - \left(\frac{d}{\sigma_s}\right)^{\alpha} & \text{if } d < \sigma_s \\ 0 & \text{otherwise} \end{cases}$$

with:

sh :   the sharing function;
$d(i,j)$:   the distance between the individuals i and j, that depends on the chosen representation;
σ_s :   the *niche radius*, or dissimilarity threshold;
α :   the "sharpness" parameter;
μ :   the population size.

Let us assume that α is chosen very large, tending towards infinity, then $(d/\sigma_s)^\alpha$ tends towards 0 and $\text{sh}(d)$ is 1 if $d < \sigma_s$, or otherwise equal to 0. Then $\sum_{j=1}^\mu \text{sh}\,(d(i,j))$ is the number of individuals located within a ball of radius σ_s centered on the individual i. The shared fitness is thus, in this case, the raw fitness of the individual i, divided by the number of its neighbors. This type of niching performs well, at the condition that the distances between the peaks are less than the niche radius σ_s. However, for a given optimization problem, barring a few rare cases, the distances between the peaks are not known a priori. Then, if the radius is selected too large, all optima cannot be discovered by the individuals of the population. An imperfect solution for this problem consists in defining niches as balls with a fuzzy boundary. Thus, the individuals j from which the distances to the individual i are close to σ_s have a weaker contribution to the value of $\text{sh}(d(i,j))$ than the others. In this way, if unfortunately the niche already presumably centered on a peak contains another peak close to its boundary, it will be less probable that the later one will perturb the persistence of the presence of individuals on the central peak. The "sharpness" of the niche boundaries is controlled by the parameter α, which is assigned a default value of 1.

Now, let us consider the case where the radius σ_s is selected too small compared to the distances between the peaks. Then there will be several niches for each peak. In theory, this is no trouble, but in practice it implies to put much more individuals among the niches than necessary and thus it will require a population size larger than it would be necessary. This will cause wastage of computation resources. If the population is not of sufficient size, it is very much possible that all the global optima of the population will not be discovered. Hence an accurate estimation of σ_s is of prime importance. Hence suggestions will be subsequently made allowing to come close to this objective.

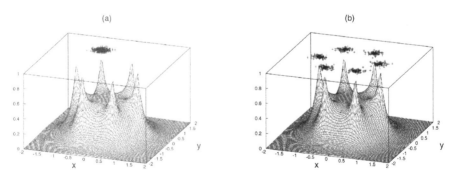

Fig. 6.14. (a): selection without sharing: the individuals converge towards only one of the optima (b): selection with sharing: the individuals converge towards several optima.

The figures 6.14a and 6.14b show the distribution of the individuals on the peaks of a multimodal function defined in $\mathbb{R}^2$, after convergence of the evolutionary algorithm with and without sharing of the fitness function. The individuals are projected on the plane located at the height of the optima, parallel with the axes x and y, so that they are more visible.

Genetic drift and the sharing method

Let us assume that the individuals are distributed on all the global peaks of the fitness function after a sufficient number of generations. Let N be the population size and p be the number of peaks, each one of them will be occupied by a subpopulation of approximately N/p individuals. Also let us assume that the fitnesses of all the individuals are close to the fitness of the global optima. As an equilibrium situation is reached, the subpopulations for the next generation will have approximately the same size. Consequently, each individual is expected to have a number of offspring close to unity. In this case, the effective number of offspring of an individual obtained by employing a stochastic selection technique can be zero with a not negligible probability. Even with a sampling of minimal variance like the SUS selection, an individual will be able to have zero or one effective offspring if the expected number of offspring is slightly lower than unity. Hence, there is a possibility, which becomes more significant for a small population, that a subpopulation covering a peak may disappear because of stochastic fluctuations. To reduce this possibility to an acceptable level, it is necessary to allot to each peak a high number of individuals, so that the sharing method requires a priori big population sizes.

Advantages and difficulties for the application of the method

The basic sharing method enjoys an excellent stability if the population size is large enough to counter the genetic drift. With the help of variation operators capable to ensure a good diversity, the distribution of the population after some generations does not depend on the initial population. The main difficulty of this method lies in the appropriate choice of the niche radius σ_s. Another drawback relates to the algorithmic complexity which is given as $\mathcal{O}(\mu^2)$, where μ is the population size. As the method requires big population sizes, this can be seriously disadvantageous except when the calculation of the fitness function is very long. The basic sharing method is not compatible with the elitism. Lastly, it is well suited to be used with a proportional selection technique. Various authors proposed solutions to overcome these disadvantages. The seniority of the sharing method and its effectiveness in the maintenance of diversity make it, still today, the most known and the most used niching technique.

6.3.3 Niching with the deterministic crowding method

The first niching method by *crowding* was presented by De Jong in 1975 [De Jong, 1975]. It utilizes a value of distance, or at least of dissimilarity, between individuals, like the sharing method, but it operates at the level of the replacement operator. De Jong suggests that for each generation the number of offspring be of the order of ten times less that the number of parents. A higher value decreases the effectiveness of the method. A lower value would favor too much the genetic drift. All the offspring find themselves in the population of the parents for the next generation, and hence the parents that they replace have to be chosen. The replacement operator selects a parent who must "die" for the offspring that resembles it closest. Nevertheless, the similarity comparisons are not systematic, and an offspring will be compared only with one small sample C_F of parents randomly drawn from the population. C_F is the crowding factor. De Jong showed, for some test functions, that a value of C_F fixed at two or three gives interesting results. Hence the individuals tend to be distributed among the various peaks of the fitness function, thus preserving preexistent diversity in the population.

However the method makes frequent replacement errors due to the low value of C_F, which is prejudicial to the niche effect. But a high value of C_F produces a too strong reduction of the selection pressure. Indeed, the parents which are replaced, being similar to the offspring, have almost the same fitnesses if the function is continuous. Their replacement thus improves the fitnesses within the population very little. On the contrary, the selection pressure is stronger if efficient offspring replace less efficient parents, i.e. if errors in replacement are made, which implies that C_F must be weak. In 1992, S.W. Mahfoud [Mahfoud, 1992] *proposed the deterministic crowding method* as a major improvement over the method of De Jong. The main idea is that a pair of offspring e_1 and e_2 obtained after crossover and mutation enters in competition only with its two parents p_1 and p_2. There are two possibilities of replacement:

(a): e_1 replaces p_1 and e_2 replaces p_2
(b): e_1 replaces p_2 and e_2 replaces p_1

The choice (a) is selected if the sum of dissimilarities $d(p_1, e_1) + d(p_2, e_2)$ is weaker than $d(p_1, e_2) + d(p_2, e_1)$; otherwise the choice (b) is carried out. Lastly, the replacement of a parent by an offspring is effective only if the parent is less efficient than the offspring. This can be described as a deterministic tournament. This implies that the method is elitist, because if the best individual is in the population of the parents and not in that of the offspring, it will not be able to disappear from the population in the next generation.

Advantages and difficulties for the application of the method

Deterministic crowding does not require determination of appropriate parameter values depending on the problem such as a radius of niche. In fact, only the population size is significant and is chosen according to a very simple criterion: the larger the number of optima to find, the larger the population. The number of calculations of distances to be carried out is of the order of the population size, which is lower by an order of magnitude compared to the sharing method. Deterministic crowding is a replacement operator that favours the best individuals. Thus, the selection for the reproduction may be absent, i.e. reduced to its simplest expression: a parent always produces only one offspring, irrespective of its fitness. In this case, the selection operators involve only computational dependencies between couples of offspring and their parents. Thus the parallelization of the method is both simple and efficient. All these qualities are interesting, but deterministic crowding does not reduce the genetic drift significantly compared to an algorithm without niching. From this point of view, this method is less powerful than the sharing method. This implies that, if the peaks are occupied during a certain number of generations by individuals, the population will finally converge towards only one optimum. This disadvantage often leads us to the conclusion that the methods with low genetic drift are preferred, even if their use is less simple.

6.3.4 The clearing procedure

The *clearing procedure* was proposed in 1996 by A. Pétrowski [Petrowski, 1996]. It is based on limited resource sharing within ecological niches, according to the principle suggested by J.H. Holland, with the particularity that the distribution of the resources is not equitable among the individuals. Thus the clearing procedure will assign all the resources of a niche, typically to the best individual, designated as the dominant. The other individuals of the same niche will not have anything. This means that only the dominant individual will be able to reproduce to generate a subpopulation for the next generation. The algorithm thus determines the subpopulations in which the dominant individuals are identified. The simplest method consists in choosing a distance d significant for the problem and to assimilate the niches with balls of radius σ_c centered on the dominants. The value of σ_c must be lower than the distance between two optima of the fitness function so that they can be distinguished to maintain individuals on all of them. Thus the problem now consists in discovering all the dominant individuals of a population. The population is initially sorted according to the decreasing fitnesses. A step of the algorithm is implemented in three phases to produce a niche:

1. The first individual of the population is the best individual. This individual is obviously a dominant.
2. The distances of all the individuals from the dominant are computed. The individuals located at a distance closer than σ_c belong to the niche centered on the dominant individual. Hence, they are dominated and thus their fitnesses are assigned to zero.
3. The dominant and the dominated individuals are withdrawn virtually from the population. The procedure is then reapplied from step 1 to the new population thus reduced.

The operator comprises of as many steps as the algorithm finds dominants. These preserve the fitness which they obtained before the application of the niching. The operator is applied just after the fitness evaluation and before application of the selection.

Elitism and genetic drift

The clearing procedure lends itself easily to implement an elitist strategy: it suffices to preserve the dominant individuals from the better subpopulations to inject them in the population for the next generation. If the number of optima to be discovered is known a priori, the same number of dominants is preserved. In the opposite case, a simple strategy, among others, consists in preserving in the population the dominant individuals whose fitness is better than the average fitness of the individuals in the population before clearing. Nevertheless it will be necessary to take precaution so that the number of preserved individuals is not too large compared to the population size.

If the dominant individuals located the optima of the function in a given generation, the elitism will indefinitely maintain them on the peaks. This algorithm is perfectly stable, contrary to the methods discussed before. The genetic drift does not have a detrimental effect in this context! This enables us to reduce the required population sizes compared to other methods.

Niche radius

Initially, the estimation of the niche radius σ_c follows the same rules as for the sharing method. Theoretically it should be lower than the minimum distance between all the global optima considered two by two, so that all of them will be discovered. However, the choice of a too large niche radius does not have the same effects as with the sharing method, where this situation leads to instabilities with an increased genetic drift. If this occurs with the clearing procedure, certain optima will be ignored by the algorithm, without disturbing its convergence towards those which are located. Hence, the criterion of determination of the radius can be different. Indeed, the user of a multimodal optimization algorithm does not require to know all the global optima, which is impossible when those are in infinite number in a continuous domain, but rather a representative sample of the diversity of these optima. Locating the global optima corresponding to the instances of almost identical solutions will not be very useful. On the other hand, it will be more interesting to determine the instances of optimal solutions distant from each other in the search space. Thus, the determination of σ_c depends more on the minimum distance between the desired optimal solutions, an information independent of the fitness function, than the minimum distance between the optima, which strongly depends on fitness, and which is generally unknown. If however the knowledge of all the global optima is required, there are techniques which enable estimation of the niche radius by estimating the width of the peaks. It is also possible to build niches which are not balls, with the help of an explicit speciation (see section 6.3.5).

Advantages and difficulties for application of the method

The principal quality of the method lies in its great resistance to the loss of diversity by genetic drift, especially in its elitist version [Sareni and Krahenbbuhl, 1998]. Therefore it can work with relatively modest population sizes, which results in reduced computing time. The niche radius is a parameter which can be defined independent of the landscape of the fitness function, unlike the sharing method, and rather according to the desired diversity of the multiple solutions.

The clearing procedure requires about $\mathcal{O}(c\mu)$ distance calculations where c indicates the number of niches and μ the population size. It is weaker than the sharing method, but higher than the deterministic crowding method.

If it was found during the process of evolution that the number of dominant individuals is of the same order of magnitude as the population size, this would indicate:

- either that the population size is insufficient to discover the optima with the sampling step fixed by the niche radius;
- or that this step is too small, compared to the computation resources assigned for the resolution of the problem. It is then preferable to increase the niche radius, so that the optima found are distributed as widely as possible in the entire search space.

The method performs unsatisfactorily with the condition of a restricted mating, using a restriction radius lower or equal to the niche radius (see chapter 3, section 3.4.1). In that case, the crossover will be useless, because it will be applied only to similar individuals: the selected individuals, which are the clones of the same dominant individual. To overcome this problem, there are at least two solutions. One solution can be to carry out a mutation at a high rate before the crossover, in order to restore diversity within each niche. The other can be to increase the restriction radius. In the later case, the effect of exploration of the crossover becomes more significant. Indeed, it may be that between two dominant individuals around two peaks are located some areas of interest that the crossover is likely to explore. But this can also generate a high rate of lethal crossovers, reducing the convergence speed of the algorithm.

6.3.5 Speciation

The main task of the speciation is to identify the existing niches in a search space. So far in our discussions only one species can occupy a niche, it is then assumed that the individuals of a population who occupy it belong to a species or a *subpopulation*. Once determined by speciation, a subpopulation can be used in several ways. For example, it can be stabilized around a peak by employing a niching technique. The restricted mating can also be practiced inside subpopulations, which, in addition to the improvement due to the reduction of the number of lethal crossovers, thus conforms to the biological metaphor, which requires that two individuals of different species cannot mate and procreate.

The balls used in the techniques of niching described above can be viewed as niches created by an implicit speciation. The sharing method, and the clearing procedure, also perform satisfactorily if the niches are provided to them by the explicit and prior application of a speciation method. For that, such a method must provide a partition of the population $\mathbf{S} = \{\mathbf{S}_1, \mathbf{S}_2, \ldots, \mathbf{S}_c,\}$ in c subpopulations $\mathbf{S}_i$. Hereafter, it is then easy to apply, for example:

- a niching by the sharing method, by defining the shared fitness as:

$$\tilde{f}(i) = \frac{f(i)}{\mathrm{card}(\mathbf{S}_j)}, \forall i \in \mathbf{S}_j$$

  for all subpopulations $\mathbf{S}_j$;
- a niching by the clearing procedure, by preserving the fitness of the best individual of any subpopulation $\mathbf{S}_j$ and by forcing fitnesses of other individuals to zero;
- a restricted mating, which only operates between the individuals of any subpopulation $\mathbf{S}_j$.

Moreover, an explicit speciation technique is compatible with the elitism: the individuals of a subpopulation being clearly identified, it is possible to preserve the best one from each subpopulation in a generation for the next one.

Label based speciation

In 1994, W.M. Spears proposed [Spears, 1994] a simple speciation technique using *tag-bits*, where an integer number belonging to a set $\mathbf{T} = \{T_1, T_2, \ldots, T_c\}$ is associated with each individual in a population. The value of the label T_i signifies the subpopulation $\mathbf{S}_i$ to which all the individuals labeled by T_i belong. c is the maximum number of subpopulations which can exist in the population. The method was so named because originally Spears had proposed his method within the framework of the genetic algorithms, and the labels were represented by bit strings. During the construction of the initial population, the labels attached to each individual are drawn randomly in the set $\mathbf{T}$. During the evolution, the labels can mutate, by selecting randomly a new value in $\mathbf{T}$. The mutation corresponds in this case to a *migration* from a subpopulation towards another. After some generations, the subpopulations are placed on the peaks of the fitness function because of the selective pressure. However, there is no guarantee that each peak containing a global optimum is maintained by one and only one subpopulation. Some of them can be forgotten, while others can be occupied by several subpopulations. Hence the method is not a reliable one. It is quoted here because it is well-known in the world of evolutionary computation.

Island model

The *island model* is also a classical concept in the field of evolutionary computation. This model evolves several subpopulations S_i through a series of epochs. During each epoch, the subpopulations evolve independent of each other, over a given number of generations G_i. At the end of each epoch, the individuals move between the subpopulations during a phase of *migration*, followed by a possible phase of assimilation. The later phase is employed to carry out integration operations of the migrants in their host subpopulations, for example by stabilizing their sizes. This iterative procedure continues until the user-defined termination criterion for the algorithm is satisfied. The migration does not take place arbitrarily, but according to a relation of neighbourhood defined between the subpopulations. The proportion of the migrating individuals is determined by migration rates chosen by the user.

Originally, the model was developed as a parallelization model of a genetic algorithm. That enables them to be efficiently implemented in a distributed memory multiprocessor computer, where each processing unit deals with a subpopulation [Cohoon et al., 1987]. It can be noticed that, logically, this process is similar to a label based speciation, with a mutation of the labels constrained by the neighborhood relations. The label mutation takes place only at the end of each epoch. As the label based speciation, this method lacks in reliability in the distribution of the subpopulations on the peaks of the fitness function. However, the fact that the subpopulations evolve independently during each epoch, offers the advantage of a more accentuated local search for optima.

Speciation by *clustering*

During an evolution, the individuals of a population tend to gather in the areas of the search space showing high fitness under the action of the selection pressure. These areas are good candidates to contain global optima. The application of a classical clustering method (e.g. K-means algorithm, LBG algorithm, etc.) partitions the search space in as many areas as accumulations of individuals are detected. Each detected area is assimilated with a niche, and the individuals located there constitute a subpopulation [Yin and Germay, 1993b]. The method is reliable with large population sizes, because a niche can be identified only if it contains a large enough cluster. This number can be significantly reduced if the speciation algorithm exploits the fitness values of the individuals in each area, in order to recognize better the existence of possible peaks in those regions [Petrowski and Girod Genet, 1999]. It is interesting to combine a clustering based speciation with an island model, in order to profit from the advantages of both methods: a reliable global search of the highest peaks, which occurs during the migration phases (exploration), and an improved local search for the optima (exploitation), during the epochs [Bessaou et al., 2000].

6.4 Multiobjective optimization

The multiobjective, or multicriterion, optimization treats the case of the simultaneous presence of several objectives, often contradicting with each other. In a recent work devoted to this subject at Springer [Collette and Siarry, 2003], the authors underline the significant role of the metaheuristics, which largely contributed to the renewal of the multiobjective optimization. A whole chapter of this book is devoted to the multiobjective optimization methods exploiting a metaheuristic.

6.4.1 Formalization of the problem

Let $\mathbf{f}(\mathbf{x})$ be a vector of c objectives associated with an instance of solution $\mathbf{x}$ of a multiobjective optimization problem. Each of its components $f_i(\mathbf{x})$ is equal to the value of the ith objective for the solution associated with $\mathbf{f}$. A simple solution consists in aggregating all the objectives to a single one by the method of weighted summation. Thus, the problem is transformed by calculating a new objective function:

$$G(\mathbf{x}) = \sum_{i=1}^{c} w_i f_i(\mathbf{x})$$

where w_i are the weights that must be chosen according to the importance attached to each objective. But this solution is not very satisfactory for several reasons:

- On one hand, the choice of the weights is a difficult exercise, because objectives are often of different, incommensurable nature.
- On the other hand, rather than obtaining a single solution, one would rather like to know some representative samples of solutions that can achieve at best the optimization of the possibly conflicting objectives. The argument that can be put forward for the interest in having several solutions is similar to that which was developed before in connection with multimodal optimization.

To solve this problem in a more adequate manner, a technique different from the simple combination of the objective functions must be employed.

Pareto optimum

It is always possible to configure a multiobjective optimization problem as a minimization problem for all its objectives. Indeed, when a problem does not conform to this condition, it is enough to change the signs of those objectives which must be maximized. In the context of minimization, let us consider two vectors of objectives $\mathbf{v}$ and $\mathbf{u}$. If all the components of $\mathbf{v}$ are lower or equal to the components of **u**, with at least one strictly lower component, then the vector **v** corresponds to a better solution than **u**. In this case, it is said that **v** *dominates* **u** in the Pareto sense. In a more formal way, it can be written as: $\mathbf{v} \overset{P}{<} \mathbf{u}$.

$$\mathbf{v} \overset{P}{<} \mathbf{u} \iff \forall i \in [1;c], v_i \leq u_i \text{ and } (\exists j \in [1;c] : v_j < u_j)$$

The collection of the objective vectors which cannot be dominated constitutes the optimal values of the problem in the Pareto sense. These vectors belong to the *Pareto front*, or *trade-off surface*. The *Pareto-optimal set* is defined as the collection of the solutions in the search space Ω whose objective vectors belong to the Pareto front. Multiobjective optimization consists of finding one or more solutions of this set, or at least close to this set.

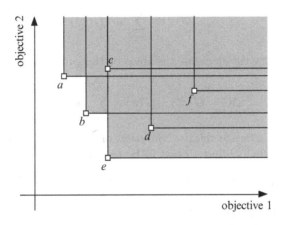

Fig. 6.15. Dominations in the Pareto sense in an objective space of dimension 2.

The figure 6.15 represents the relations of domination between 6 objective vectors in a two dimensional space. *c* is dominated by *a*, *b* and *e*. *d* is dominated by *e*. *f* is dominated by *b*, *d* and *e*.

Unquestionably the most employed metaheuristic class in the multiobjective optimization is that of the evolutionary algorithms. Indeed, they are well suited for simultaneous searching of a collection of optimal solutions, because they deal with populations of solution instances. The second part of this section is devoted to them. In the first part, we briefly present the P.A.S.A. and M.O.S.A. methods, employing simulated annealing.

6.4.2 Simulated annealing based methods

P.A.S.A. ("**Pareto Archived Simulated Annealing**") method

This method, developed at *Electricite de France* [Engrand and Mouney, 1998], uses a function of "aggregation" of objective functions, coupled with a system of archiving non dominated solutions.

Presentation of the method.

Let us assume that the objective functions to be minimized, f_i, $i = 1, 2, \ldots, n$, of the multiobjective optimization problem, are positive. This assumption makes it possible to define the problem as that of a mono-objective minimization, with the following aggregation function:

$$G\left(x\right) = \sum_{i=1}^{n} \ln\left(f_i\left(x\right)\right)$$

where x gathers the decision variables of the problem. Thus, the following expression:

$$\Delta G = G\left(x'\right) - G\left(x\right) = \sum_{i=1}^{n} \ln\left(\frac{f_i\left(x'\right)}{f_i\left(x\right)}\right)$$

represents the average relative variation of the objective functions between the current point (x) and a point to be tested (x'):

- If $\delta G > 0$, the new solution x' deteriorates, on a relative average, the set of objective functions.
- If $\delta G < 0$, the new solution x' improves, on a relative average, the set of objective functions.

In the first case, the solution x' is accepted with the usual probability of simulated annealing:

$$p = e^{\frac{-\Delta G}{T}}$$

where T denotes the temperature.

The method comprises an archiving of the "non-dominated" solutions. Before stressing on this aspect, two traditional definitions in multiobjective optimization are recalled:

relation of dominance: the vector x_1 is known to dominate the vector x_2 if:
- x_1 is at least as good as x_2 for all the objectives;
- x_1 is strictly better than x_2 for at least one objective.

non dominated solutions: the non dominated solutions, or optimal solutions in the Pareto sense, are those which dominate the other solutions, but do not dominate amongst themselves.

The P.A.S.A. method uses an "archive" of non-dominated solutions, whose size can vary between 0 and N_{max}, managed in the following manner:

- If the solution x' is dominated by at least one solution of the archive, it is not placed in the archive.
- If the solution x' dominates a solution y in the archive, it replaces y in the archive.
- If the solution x' is not dominated by the members of the archive, it is placed in the archive, and the solutions dominated by x' are withdrawn.

For the optimization to achieve an approximation of the whole compromise surface, it is necessary to start the search again regularly starting from a point chosen at random, within the archive. This stage is called the "return to base". The algorithm of P.A.S.A. method is described in the algorithm 6.2.

Discussion.

The P.A.S.A. method enables us to take into account a preexistent expertise on the problem under consideration: the "good" solutions known a priori can be placed in the initial archive population. This flexibility is also apparent in the level of the management of the archive, which may integrate heuristic rules suitable for the problem at hand. A restriction of the method appears in the definition of the aggregation function, which imposes that the objective functions must be positive in the entire search space.

1. Choose an initial archive population, possibly reduced to only one element x_0, and choose the initial temperature T_q
2. To indicate that x_n' a neighbor of x_n
3. Calculate $\delta G = G\left(x_n'\right) - G\left(x_n\right)$
4. If $\delta G \leq 0$, then $x_{n+1} = x_n'$
5. If $\delta G > 0$, calculate $p = e^{\frac{-\delta G}{T_q}}$:
   - $x_{n+1} = x_n'$ with the probability p,
   - $x_{n+1} = x_n$ with the probability $(1-p)$
6. If the "thermodynamic equilibrium" at the temperature T_q is reached, then:

$$T_{q+1} = \alpha \cdot T_q$$

   with $\alpha < 1$
7. Principle of non-dominance for the archiving of x_{n+1} :
   - if x_{n+1} is dominated by at least one solution of the archive, it is not archived;
   - if x_{n+1} dominates a solution y_p of the archive, it replaces y_p in the archive;
   - if x_{n+1} is not dominated by the members of the archive, it is placed in the archive, and the solutions dominated by x_{n+1} are withdrawn.
8. "Return to base": periodically make $x_{n+1} = y_i$, where y_i is a member of the archive chosen at random.
9. If the termination criterion is not reached, return to 2.

Algorithm 6.2: Algorithm of the P.A.S.A. method.

M.O.S.A. method ("Multiple Objective Simulated Annealing")

This method was proposed in [Ulungu et al., 1999].

Presentation of the method.

Let us consider that x_n be the current solution in the iteration n, and y be a solution under consideration at the time of the iteration n. We can start by defining a series π_k of functions expressing, for each objective function f_k, the probability of acceptance of a solution y degrading the objective:

$$\pi_k = \begin{cases} e^{\frac{-\Delta f_k}{T_n}} & \text{if } \Delta f_k > 0 \\ 1 & \text{if } \Delta f_k \leq 0 \end{cases}$$

where: T_n indicates the temperature in the iteration n,

$$\Delta f_k = f_k\left(y\right) - f_k\left(x_n\right)$$

Then the aggregation of the functions π_k is carried out, by applying one of the two following formulae:

$$t\left(\Pi, \lambda\right) = \prod_{k=1}^{N} \left(\pi_k^{\lambda_k}\right)$$

or:

$$t\left(\Pi, \lambda\right) = \min_{k=1}^{N}\left(\pi_k^{\lambda_k}\right)$$

with:

π	collection of π_k, $k = 1, \ldots, N$;
λ	collection of λ_k, $k = 1, \ldots, N$;
λ_k	a weighting coefficient related to the objective function f_k.

In addition, it can be formulated as a mono-objective minimization problem by building the equivalent objective function f_{eq} :

$$f_{eq}\left(x\right) = \sum_{i=1}^{N} w_i \cdot f_i\left(x\right)$$

The function $t\left(\Pi, \lambda\right)$ then defines the probability of acceptance of a perturbation degrading the function f_{eq}.

Remarks.

The M.O.S.A. method should not be confused with the direct application of simulated annealing for the function f_{eq} above. Indeed, in the later case, a perturbation degrading the objective will be accepted with the probability $e^{\frac{-\Delta f_{eq}}{T_n}}$.

An example studied in [Collette and Siarry, 2003] shows that M.O.S.A. method is more effective, in this direction in particular when it can achieve solutions which would not be accessible by the direct application of simulated annealing for the mono-objective problem f_{eq}.

Y. Collette in his thesis [Collette, 2002] discusses the employment of several simulated annealing based metaheuristics, for the treatment of a multiobjective problem of particularly difficult nature: that of optimization at ElectricitEde France of the nuclear fuel loading schedules.

6.4.3 Multiobjective evolutionary algorithms

This part is devoted to the methods which use the concept of Pareto dominance because of their good reliability. The purpose of these is to provide a "uniform" sampling of the Pareto front or the Pareto optimal set. The other methods available now also find solutions that are indeed Pareto-optimal, but they may "ignore" parts of the Pareto front, like its concave parts, or they can give too much advantage to some parts like those that correspond to solutions that minimize one of the objectives at the cost of the others.

In the context of solving such problems by evolutionary algorithms, the individuals correspond to instances of solutions in the search space. They are affected by a scalar fitness value, calculated from the objective vectors associated with the individuals, such that the non-dominated individuals will be more often selected than the others.

That being stated, there exists a difficulty in applying the techniques based on the Pareto dominance, relative to the dimensionality of the objective space. More the number of objectives to optimize, and thus larger the dimension of the objective space, more the Pareto front is vast, and less are the chances that individuals are dominated by others. If in this case, a maximum fitness is assigned to the individuals not dominated in a population, in order to favour their reproduction. Then many individuals will have this fitness, thus generating a low selection pressure, and consequently a slow convergence of the algorithm. The strategies using the Pareto dominance will have, consequently, to take this problem into account. The most popular or the most effective methods are described below.

Essential components

An evolutionary algorithm, which tends to sample a Pareto-optimal set uniformly, needs two essential ingredients to perform satisfactorily, in addition to the standard operators (selection, crossover, mutation, replacement):

- a method which assigns fitness values to the individuals, according to the relations of dominance which exist within a population;
- a speciation/niching method, which maintains a high level of diversity within the population, so that it covers the Pareto-optimal unit uniformly, or at least, its nearest possible neighborhood.

A third essential ingredient, but which does not belong to the evolutionary algorithm, is an archive of the non-dominated solutions discovered during a complete evolution. Indeed, there is no guarantee that at the end of the evolution, the solutions which approached the Pareto optimal set at best, were preserved in the population. Thus, at the end of each generation of the evolutionary algorithm, the population is copied in the archive, then the dominated individuals are eliminated from it. This archive is not exploited a priori by the multiobjective optimization algorithm, except notably for the most recent ones.

The Goldberg's "Pareto ranking"

An algorithm of this class of methods was described for the first time by D.E. Goldberg in his famous book [Goldberg, 1989]. However, the author did not describe any concrete implementation of this algorithm, and obviously any performance result. The idea however inspired many researchers in the years that followed.

Calculation of the individual fitnesses.

In the original proposal of Goldberg, the calculation is based on the ranking of the individuals according to the domination relation existing between the solutions which they represent. First of all, rank 1 is assigned to the non-dominated individuals of the complete population: they belong to the non-dominated front. These individuals then are fictitiously withdrawn from the population, and the new non-dominated individuals are determined, who are assigned rank 2. It can be said that they belong to the rank 2 dominated front. One can proceed in this manner until all the individuals are ranked. The fitness value of each individual is then calculated like a function of the rank of each individual in a way similar to the technique described in the paragraph 3.3.3 of chapter 3, by keeping in mind to assign to each equally placed individual the same fitness.

Niching.

It is the sharing method (section 6.3.2), possibly reinforced by a restricted mating (section 3.4.1). Goldberg does not specify whether the niching is implemented in the search space, or the objective space.

Several alternatives of this approach then appeared, differing mainly in the mode of calculation of the individual fitnesses. Thus, Fonseca and Fleming proposed the MOGA algorithm in 1993 [Fonseca and Fleming, 1993]. In this contribution, each individual is assigned a rank equal to the number of individuals who dominate it. Then a selection according to the rank is applied, in accordance with the idea of Goldberg. The niching is carried out in the objective space, which allows a uniform distribution of the individuals in the neighborhood of the Pareto front, but not in the Pareto-optimal set. This does not permit to perform multimodal and multiobjective optimization at the same time. The niche radius σ_s should be calculated so that the distribution of μ individuals of the population is uniform on the whole Pareto front. Fonseca and Fleming proposed a method to estimate its value [Fonseca and Fleming, 1993].

The "Non Dominated Sorting Genetic Algorithm" method

The "Non Dominated Sorting Genetic Algorithm" method was presented in 1994 by Srinivas and Deb [Srinivas and Deb, 1994] and is inspired directly by the idea of Goldberg. It uses the same Pareto ranking. On the other hand, it carries out a niching different from the one used by MOGA. Indeed the sharing method is applied, front by front, in the search space with a sharpness parameter α equal to two. The method is a classical one. It is recognized as an effective one in the approximation of the Pareto front.

The "Niched Pareto Genetic Algorithm" method

For the "Pareto ranking" methods, the selection according to the rank can be replaced by a tournament selection between the ranked individuals. Horn *et al.* (1994) [Horn et al., 1994] proposed the "Niched Pareto Genetic Algorithm" (NPGA) method which performs the tournaments directly according to the relations of dominance, thus avoiding a computation expensive preranking of the entire population. Applying a simple binary tournament (section 3.3.4) is not satisfactory because of the low selective pressure in this context. To increase it, the authors conceived an unusual binary tournament: the *domination tournament*.

Let two individuals x and y be drawn randomly from the population to take part in a tournament. Those are compared with a comparison sample γ, that is drawn at random and comprises of t_{dom} individuals. The winner of the tournament is x if it is not dominated by at least one of the individuals of γ and if y is dominated. In the opposite case, the winner is y. If now x and y are in the same situation: either dominated, or non-dominated, the winner of the tournament is that which has less neighbors within a ball of radius σ_s in the objective space. This last operation gives rise to a form of niching, with an aim of reducing the genetic drift which would be induced by the choice of a winner at random. Indeed, a significant genetic drift will be harmful with a regular distribution of the non-dominated individuals, which are a priori close to the Pareto-optimal set.

The parameters t_{dom} and σ_s must be fixed by the user. t_{dom} is an adjustment parameter of the selection pressure. Horn *et al.* noticed in some case studies that if t_{dom} is too weak, smaller than one percent of the population, there are too many dominated solutions and the solutions close to the Pareto optimal set have less chances to be discovered. If it is larger than twenty percent, premature convergences are frequent, due to a too high selection pressure. A value of about ten percent would be adequate to place the individuals near the Pareto front at the best. The parameter σ_s proves to be a relatively robust one. It can be smaller and smaller as the population becomes larger and larger, and vice versa, as far as the objective is only to cover the areas close to the Pareto front as regularly as possible. An estimate of its value is given in [Fonseca and Fleming, 1993] or [Horn et al., 1994]. The NPGA method is one of the most widely used one.

The "Strength Pareto Evolutionary Algorithm " (SPEA) method

This method was presented in 1999 by E Zitzler and L Thiele [Zitzler and Thiele, 1999]. Its originality lies in the utilization of the archive of the non-dominated solutions during the evolution of a population. The purpose of it is to intensify the search for new non-dominated solutions, and thus to approach more from the Pareto front by implementing a form of elitism (section 6.3.4). Moreover, the authors proposed a new niching technique dedicated to multiobjective optimization.

Calculation of the fitnesses of the individuals.

In each generation, the fitness of the individuals in the population **P** and the archive **P'** is determined in the following way:

Stage 1: The fitness f_i of any individual i in **P'** is equal to the negative of its
strength s_i :

$$f_i = -s_i \quad \text{and} \quad s_i = \frac{n}{\mu + 1}$$

where n is the number of solutions dominated by i in the population **P**, and μ is the size of **P**, s_i necessarily lies between 0 and 1.

Stage 2: The fitness f_j of any individual j in **P** is equal to the negative of the sum of the strengths of the individuals in **P'** who dominate it, added to unity:

$$f_j = -\left(1 + \sum_{\substack{P \\ i,i<j}} s_i\right)$$

f_j is thus smaller than or equal to -1, and consequently smaller than the fitnesses of the solutions of **P'**.

Thus, the fitness of an individual is all the more weak as there are more individuals of **P'** who dominate it. The figure 6.16 illustrates the calculation of the fitness with the help of an example.

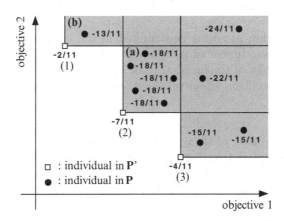

Fig. 6.16. Example of calculation of the fitness of the solutions in **P** and **P'** with SPEA method.

Niching.

The niching is carried out implicitly by the fitness calculations described above. Each rectangle in the figure 6.16 is regarded as a niche defined not in terms of distance, but in terms of Pareto dominance. Let us consider the marked niche (a). It contains a relatively high number of individuals compared to the others. Thus its contribution to the strength of the individual (2) is high. This high strength implies in return a relatively weak fitness of the individuals in the niche (a). The niche (b) contributes little to the strength of the individual (1). The fitness of the individual that it contains is found higher than those of the individuals in the niche (a). At the time of the selection stage, the overcrowded niches will thus have less number of occasions to produce offspring than the sparsely crowded niches. This produces the required stabilization effect for the subpopulation size in the niches.

The algorithm also employs a selection by binary tournaments operated on the concatenation of the populations $\mathbf{P}$ and $\mathbf{P'}$. Moreover, the authors proposed that the size of the archive $\mathbf{P'}$ can be reduced by clustering, if it exceeds a fixed threshold. The clustering procedure allows to replace a group of neighboring individuals by a single one, which is the centroid of the group. In this way, the computing power is not wasted in redundant comparisons.

The SPEA method proved to be by far the most effective one in the approximation of the Pareto front, compared to the other techniques which were presented here for a set of test functions. The enhancement in performance of this method is primarily explained by the elitism supported by the utilization of the archive $\mathbf{P'}$.

6.5 Constrained evolutionary optimization

The optimization problems encountered in the industrial world must often satisfy a certain number of constraints. These give rise to a set of relations that the variables of the function to be optimized must satisfy. While solving a numerical problem, these relations are generally expressed as a set of q inequalities:

$$g_i(x) \leq 0, \text{ for } i = 1, \ldots, q$$

where x is a solution of the optimization problem. The possible equality constraints are replaced by two inequality constraints of the type mentioned above. In case of the evolutionary algorithms, the vector x is an individual. When the constraints are satisfied by an individual, it is known as a *feasible*. The feasible region $\mathcal{F}$ is the set of the feasible solutions. The complement of $\mathcal{F}, \mathcal{U}$ in the search space Ω is the infeasible region.

However, the standard variation operators described in chapter 3, for the real or binary representation, generate individuals in a blind manner, not taking the constraints into account, and they may correspond to infeasible solutions. To impose the necessity of satisfaction of the constraints on a problem, several approaches, acting on the various operators of the evolutionary algorithm, can be employed.

A first rudimentary approach consists in calculating the fitness function only in the feasible region $\mathcal{F}$. The individuals in the infeasible region are affected with a null fitness which prevent their reproduction. This is called the *death penalty method*. Although this method is a simple one, it has proved far from being powerful, because the topology of $\mathcal{F}$ seldom satisfies good properties, like the convexity or the connectivity. Even if these properties are satisfied, $\mathcal{F}$ can have zero measure. In this case, random drawings employed by the variation operators, which are independent of the constraints, are not likely to produce individuals corresponding to feasible solutions.

In addition, people involved in the field of constrained optimization problems noted that many feasible global optima are on the boundary of $\mathcal{F}$. Having solution instances on both sides of this boundary can help much to discover these optima. Thus, when the objective function is defined also in the infeasible region, [1] then it is useful to exploit infeasible instances of solutions to help to find a feasible global optimum. In case of objective functions defined on zero measure feasible regions, it will not be possible, in the general case, to obtain feasible solutions. Within this framework, $\mathcal{F}$ should be approached as closely as possible.

Many researchers over the years have attempted to address the difficulties mentioned above. Here we present some of these approaches which have become classical.

6.5.1 Penalization methods

The principle is simple: the fitness of an infeasible individual is reduced by subtraction of a penalty.

$$f_p(x) = f(x) - p(x)$$

where $p(x)$ is positive, increasing with the measurements of the constraint violations $\eta_i(x)$, such that:

$$\begin{cases} \eta_i(x) > 0 \text{ if the } i^{\text{th}} \text{ constraint is violated,} \\ \eta_i(x) = 0 \text{ otherwise} \end{cases}$$

Typically,

$$p(x) = P\left(\sum_{i=1}^{q} \alpha_i \eta_i^{\beta}(x)\right)$$

where:

---

[1]This is not always possible because a region can be infeasible precisely because the objective function is not defined inside it.

P is an increasing function;

α_i is a positive coefficient whose value becomes larger as more im-
         portance is given to the satisfaction of the ith constraint;

β is fixed typically at 1 or 2.

Static penalties

The simplest approach is described below:

$$p(x) = \sum_{i=1}^{q} \alpha_i \eta_i(x) \quad \text{with} \quad \begin{cases} \eta_i(x) = 1 \text{ if the } i^{\text{th}} \text{ constraint is violated,} \\ \eta_i(x) = 0 \text{ otherwise} \end{cases}$$

The constants α_i must be determined in such a way that the global optima
of the restriction of f defined in $\mathcal{U}$ correspond to local optima of f_p. Their
determination is difficult when the value of the optimum is not known a priori,
which is the general case. Indeed, the values of these constants must be large
enough, but if they are too large, the penalized individuals will have only a
few chances to reproduce, and the method closely resembles that of the "death
penalty". These constants thus depend on the problem under consideration.
Moreover, this method is brutal because it does not offer a means of more
penalizing the infeasible individuals which are away from the boundary of $\mathcal{F}$
compared to the others. This property complicates the search for the feasible
optima which are located on the boundary. Also, the approach is improved
by taking into account the values of g_i in the expression of f_p, when the
constraints are violated:

$$p(x) = \sum_{i=1}^{q} \alpha_i \eta_i^{\beta}(x) \text{ with} \begin{cases} \eta_i(x) = g_i(x) \text{ if the } i^{\text{th}} \text{ constraints is violated,} \\ \eta_i(x) = 0 \quad \text{ otherwise} \end{cases}$$

There too, the determination of the constants α_i is difficult, more than in the
preceding case. The methods of static penalties can be sophisticated as that
presented by Homaifar *et al.* in 1994 [Homaifar et al., 1994].

Dynamic penalties

It can be advantageous to vary the coefficients α_i according to the number of t
generations accomplished. Indeed, in the beginning, the penalties can be weak,
thus supporting diversity in the population, both in the feasible and in the
infeasible region. Then, when the evolution is quite advanced, the individuals
must concentrate on the most promising feasible peaks, in order to locate the
global optima precisely. Then the penalty coefficients must take higher values,
to increase the selection pressure in favour of the feasible individuals.

To sight an example, the following penalty function was proposed by Joines
and Houck in 1994 [Joines and Houck, 1994]:

$$p(x,t) = t^k \sum_{i=1}^{q} \alpha_i \eta_i^{\beta}(x)$$

where:
  t represents the number of generations accomplished;
  k takes a constant value, typically 1 or 2.

Adaptive penalties

With this method, the idea is to modify the weights α_i according to the state of the population during the last generations. The method of Hadj-Alouane and Bean proposed in 1992 and published in 1997 [Hadj-Alouane and Bean, 1997] is one of the earliest methods which proposed adaptation of penalties and is known for its striking simplicity. Its modification criterion of α_i depends on the feasible or infeasible feature of the best individual x^* in the population, during G preceding generations:

$$p(x) = \sum_{i=1}^{q} \alpha_i(t)\eta_i^{\beta}(x)$$

where

$$\alpha_i(t+1) = \begin{cases} \alpha_i(t)\gamma_1 & \text{if } x^* \in \mathcal{U} \text{ on } G \text{ generations} \\ \alpha_i(t)/\gamma_2 & \text{if } x^* \in \mathcal{F} \text{ on } G \text{ generations} \\ \alpha_i(t) & \text{otherwise} \end{cases}$$

γ_1 and γ_2 are larger than 1. The update of the α_i is carried out after each period of G generations.

The "Segregated Genetic Algorithms" method

This method suggested by Leriche *et al.* in 1995 [Leriche et al., 1995] is remarkable, in the sense that it uses two populations: one made up of individuals whose fitness is strongly penalized if they are infeasible, the other containing individuals that are slightly penalized. The selections in each population are carried out independently, but the variation operators work on the reunion of the two populations. In this way, the efficiency of the method is less sensitive to the determination of the weights, compared to an evolutionary algorithm employing a simple penalty function.

6.5.2 Superiority of the feasible individuals

With the type of methods discussed now, a feasible individual has always more chances to be selected than an infeasible individual. This property was not guaranteed in case of the methods of "penalties" examined before. The simplest method in this class is that of Deb published in 2000 [Deb, 2000]. It proposes a tournament selection. If, for two individuals who are selected to take part in a tournament,

- both are feasible, the individual which has the better fitness wins;
- one is feasible and the other infeasible, the feasible individual wins;
- both are infeasible, the individual which violates the constraints less wins. The measurement of the constraint violation can simply be the sum of the values of the functions $\max(0, g_i(x))$.

This method has the advantage that it does not require the computation of the objective function in the infeasible region $\mathcal{U}$. In this way, computation resources can be saved and, moreover, the method can be applied when the objective function is not defined in $\mathcal{U}$.

To maintain diversity in the feasible region, Deb uses a niching compatible with the tournament selection: a feasible individual i being selected, a feasible individual j takes part in the tournament with i if the distance between i and j is smaller than a given threshold σ. If no individual j is found after n_f drawing, then i wins the tournament. σ and n_f are fixed by the user. They are the only parameters of the method.

The method gives interesting results for the test problems where it was applied. To produce satisfactory result, it requires that the measure (the "volume") of the feasible region is not small compared to that of the infeasible region.

There are other methods where a feasible individual is always better than an infeasible individual. The method of the *Stochastic ranking* by Runarsson and Yao [Runarsson and Yao, 2000] can be mentioned in this context.

6.5.3 Repair methods

A repair method transforms an infeasible individual into a feasible one. A "repaired" individual can be used only for the evaluation of its fitness, or it can be reintroduced in the population. The difficulty of this class of methods is that they are strongly dependent on the representation and the problem. The example of GENOCOP III makes it possible to have an idea of such an approach.

Repair algorithm of GENOCOP III

GENOCOP III [Michalewicz and Nazhiyath, 1995] is employed to solve numerical problems whose solutions must satisfy nonlinear constraints. GENOCOP III is an extension of the GENOCOP (*GEnetic algorithm for Numerical Optimization of COnstrained Problems*) algorithm which can solve problems with linear constraints.

With GENOCOP III, two populations $\mathbf{P}_s$ and $\mathbf{P}_r$ coexist and coevolve. $\mathbf{P}_s$ contains infeasible individuals and possibly feasible individuals, they are the *search individuals*. $\mathbf{P}_r$ contains individuals who satisfy all the constraints of the problem: they are the *reference individuals*. If an individual x of $\mathbf{P}_s$ selected by the selection operator does not satisfy all the constraints of the problem, individuals are generated by BLX-0 crossovers of x with selected reference individuals, until a feasible individual z is obtained. Then the fitness of z is evaluated independent of the constraints. The fitness of x is chosen equal to that of z. The algorithm 6.3 formalizes this description.

Action Repair (Individual x)
    Individual z
    Individual r
    Real a
    Repeat
        $r \leftarrow$ select$(\mathbf{P}_r)$
        $a \leftarrow$ random$(0, 1)$
        $z \leftarrow ax + (1 - a)r$
    While Non feasible(z)
    $x.f \leftarrow$ evaluate(z)
End Action

Algorithm 6.3: Description of the repair algorithm of GENOCOP III.

Instead of being stochastic, as specified for the BLX-0 crossover operator, the sequence of the coefficients a can be deterministic, for example $a = (1/2, 1/4, 1/8 \ldots)$. The selection of a reference individual r can be uniform, or according to a probability distribution which depends on the fitnesses. Once z is obtained, it can be introduced into the population $\mathbf{P}_s$ with a probability p_r to replace x. It is also introduced into $\mathbf{P}_r$ to replace the reference individual which was used for generating it, provided that its fitness is higher. The algorithm of co-evolution of $\mathbf{P}_s$ and $\mathbf{P}_r$ is not discussed here, as it is beyond the framework of this section. The interested reader can refer to [Michalewicz and Nazhiyath, 1995] for a more detailed discussion.

For proper functioning of the method, it is necessary to find feasible solutions (at least one which is replicated) to build the initial population $\mathbf{P}_r$, which can be difficult, especially when the measure of the feasible region is very small compared to the infeasible region. An advantage of GENOCOP III, compared to the penalty methods, is that it does not require to evaluate the individuals in the infeasible region.

6.5.4 Variation operators satisfying the constraint structures

When a variation operator satisfies the constraints of a problem, it generates certainly feasible offspring if the parents on which it works are feasible. As in the preceding section, where the repair algorithms are described, these operators necessarily depend on the problem dealt with and the chosen representation. In general, it is difficult to design such operators, but they improve a lot the efficiency of the optimization algorithm.

It was already shown variation operators satisfying the constraints of a problem in the section dedicated to the representations of the permutation problems in chapter 3 (section 3.4.4). Indeed, these operators transform individuals representing some valid sequences into other valid sequences, thus satisfying the unicity constraint of each element.

GENOCOP

The GENOCOP algorithm of Michalewicz and Janikow [Michalewicz and Janikow, 1994] deals with the nonlinear optimization problems with linear constraints. The possible equality constraints are eliminated by equation solving. The remaining variables define a feasible region which is a convex polytope.

It is simple to design a crossover in real representation which satisfies these constraints. The BLX-0 crossover offers this guarantee. It is used in GENOCOP under the designation of "arithmetic crossover". Michalewicz and Janikow also proposed the use of two other types of crossovers: the uniform crossover and the heuristic crossover (see [Michalewicz and Janikow, 1994]). These last two crossover techniques do not guarantee the satisfaction of the constraints in only one application of the operator. Several crossover attempts may be necessary with possibly different parameters to obtain feasible offspring. The three crossovers are applied with rates fixed by the user within the population.

GENOCOP also combines the action of three types of mutations, applied with fixed rates, that modify only one component x_k, chosen at random, of an individual x:

- the *uniform mutation*, where x_k is modified according to an uniform random distribution in the interval delimited by the boundaries of the feasible polytope;
- the *boundary mutation*, which replaces x_k either by its minimum value in the feasible polytope, or by its maximum value, with a probability $1/2$;
- the *non uniform mutation* which generates smaller variations in x_k as the number of generations carried out becomes large, thus allowing to approach an optimum with precision.

Search on the boundary of the feasible region

This search is based on the observation that the global optima are often on the boundary of the feasible region. The idea is then to design specialized variation operators that, for parents on the boundary, generate offspring who are also on the boundary. Schoenauer and Michalewicz [Schoenauer and Michalewicz, 1996] presented a development of this approach.

6.5.5 Other methods dealing with constraints

Behavioral memory method

The *behavioral memory method*, proposed by Schoenauer and Xanthakis [Schoenauer and Xanthakis, 1993], gradually deals with the constraints of the problem, during several stages and by modifying the fitness function at each stage. There are $q + 1$ stages for q constraints to be satisfied. An order of taking the constraints into account must be defined. For the first q stages, the fitness at the stage i is a function $M(g_i(x))$ which is maximum when the constraint $g_i(x) \leq 0$ is not violated. The individuals which do not satisfy the constraints g_1 to g_{i-1} disappear from the population as they are assigned zero fitness ("death penalty method"). The algorithm passes from one stage to the next one when a sufficiently high rate of the population is in the feasible region. In the last stage, the fitness depends only on the objective function in the feasible region. The infeasible individuals disappear from the population. A niching is used to maintain diversity in the population at each stage.

Converting the constraints into additional objectives

A mono-objective optimization problem which must satisfy q constraints $g_i(x) \leq 0$ can be transformed into an unconstrained problem comprising $q + 1$ objectives $(f, f_1, f_2, \ldots f_q)$, such that:

$$f_i(x) = \max(0, g_i(x)), \forall i \in [1; q]$$

In the implementation of Surry *et al.* [Surry et al., 1995], a Pareto ranking $r(x)$ is carried out on the objectives related to the constraints. The fitness f is evaluated for all the individuals. A binary tournament selection is implemented such that the tournament criterion is either f with a probability p, or the Pareto rank. The variation operators are then applied. The value of p is adapted according to the rate of the feasible individuals in the last generations. If the rate is lower than a desired value, then p is decreased, otherwise it is increased.

6.6 Conclusion

In this chapter we presented probable answers to some highly important questions raised by the modern optimization problems: how to adapt the metaheuristics, often of discrete origin, for the problems with continuous variables, many of which arising in the practical field of engineering? How to obtain several diverse solutions, but of equivalent values, to further facilitate in taking finer decisions, according to one or more possible criteria which cannot be formalized? How to discover the best trade-off solutions when several criteria must be optimized? How to take account of the constraints in the search for the optima? These have proved to be enormously interesting research topics which culminated in several fruitful studies over the years. Indeed this chapter showed that new methods of continuous, multimodal, multiobjective or constraint handling optimization are emerging at a brisk pace. The purpose of these works is to widen their applicability, to improve their effectiveness, to facilitate their implementation by increasing the robustness of their parameters, or by reducing their number.

6.7 Annotated bibliography

[Collette and Siarry, 2003]: An ideal book to look further into multiobjective optimization in multiple contexts.

[Deb, 2001]: A valuable reference book in the domain of multiobjective optimization by evolutionary algorithms.

[Michalewicz, 1996]: After an introduction to the genetic algorithms, the author introduces a rich collection of sample solutions of constrained optimization problems. One of the most widely circulated books in the domain.

Methodology

7.1 Introduction

We will certainly deceive those readers who were patient enough to read the present book until now and who would know which metaheuristic they should try first for solving the problem under their consideration. Indeed, this question is a perfectly legitimate one, but we must confess that it is not possible to recommend one specific technique or the other. It has been seen that the weak theoretical results known about the metaheuristics are almost of no use in practice. Indeed, in a sense these theorems state that, to ensure that the optimum is correctly determined, it is required to examine a number of solutions that is higher than the total number of solutions of the problem. In other words, they (trivially!) recommend to use an exact method if the optimum is needed to be determined absolutely correctly. However, the present chapter will make an attempt to draw some guidelines for elaborating a heuristic method based on metaprinciples discussed before. Following the same principles as we adopted in the chapter on tabu search, this illustration will be presented with the help of a given optimization problem. The vehicle routing problem has been chosen for this specific purpose. In order to make the illustrations as clear as possible, we limit ourselves to the simplest version of the problem, popularly known as *Capacitated Vehicle Routing Problem* (CVRP) in the literature. However, the proposed methodology is a general one and should be applicable for complex problems as well.

Academic vehicle routing problem

An academic problem, which is a simplification of practical vehicle routing problems can be described as follows. An unlimited set of vehicles, each one capable of carrying a volume V of goods, is required to deliver n orders to customers, starting from a unique depot, in such a fashion that the total distance traveled by the vehicles is minimum. Each order (or, commonly saying, customer) i has a volume v_i $(i = 1, \ldots, n)$. The direct distances d_{ij} between the customers i and j $(i, j = 0, \ldots, n)$, are known, with 0 representing the depot. The vehicles execute tours T_k, $(k = 1, 2, \ldots)$, that start from the depot and finish on their return to the depot. A variant of the problem imposes an additional constraint that the lengths of the tours must be upper bounded by a given value L. Figure 7.1 illustrates the shape of a solution obtained for a Euclidean problem instance considered in the literature [Christofides et al., 1979] with 75 customers (marked as circles, whose surface is proportional to the volume ordered) and a depot (marked as a black disk, whose surface is proportional to the volume of the vehicles).

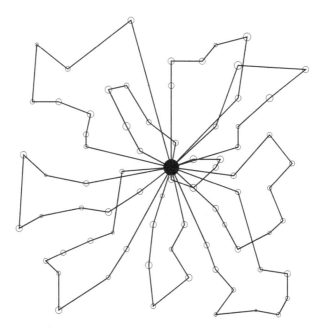

Fig. 7.1. Best solution known for a small academic vehicle routing problem with 75 customers. It is not yet proved that this solution is an optimal one.

A solution of this problem can be viewed as a partition of the customer set into a number of ordered subsets, the order defining the sequence in which each vehicle has to visit the customers constituting a tour. The first question to be addressed is about the number of tours that is needed to be created. It must be mentioned here that for practical problems, the vehicle fleet available is not unlimited and it is not always evident to even find a feasible solution for the problem. Indeed, the partitioning of the customers into a specified number of subsets of given weight is the well known NP-complete bin packing problem. Even if the number of vehicles is unlimited, it may not be useful to consider all feasible solutions of the problem. This is because this set may be composed of a very large number of poor quality solutions that can easily be shown to be inferior, for those instances containing a majority of tours with a single customer.

An initial guideline may be to try and limit the combinatorial explosion of solutions. This can be achieved with the knowledge about the problem and eventually employing a heuristic method. In other words, the size of the set of feasible solutions must be limited. A typical example for the CVRP is to impose the constraint that all the orders placed by the same customer must be placed on the same vehicle tour, provided that the later has sufficient capacity. Another idea consists in considering only those solutions that use at most one or two vehicles more than the lower bound of the number of vehicles required to deliver all customers (this lower bound can be easily determined by dividing the total volume of the orders by the volume of a vehicle). However, proceeding like this, it may be difficult to find a feasible solution or to define an adequate neighborhood structure.

7.2 Problem modeling

These considerations naturally lead us to the discussions about the definition of S, the set of feasible solutions. In fact it may happen that the shape of this set is very complicated, i.e. without the definition of a very large neighborhood, it is impossible to generate all feasible solutions, or, more precisely, it is not possible to reach an optimal solution starting from any feasible solution. In this case, to avoid the definition of an unmanageably large neighborhood (and therefore the computational effort required to perform one iteration of local search to be prohibitive), the set of feasible solutions is extended, while penalizing solutions violating constraints of the initial problem. Therefore, the problem is modified as follows:

$$\min_{s \in S^{\text{extended}}} f(s) + p(s)$$

where $S \subset S^{\text{extended}}$, $p(s) = 0$ for $s \in S$, and $p(s) > 0$ if $s \notin S$. This penalization technique, inspired from Lagrangean relaxation, is very useful for applications where finding a feasible solution is already difficult. For example, this is the case for school timetables, where the variety of constraints is plenty.

In the CVRP, the number of vehicles can be chosen a priori and solutions where some customers are not delivered can be accepted with some penalty. In this way, creating a feasible (but not operational) solution is a trivial job. The value of the penalty for not delivering an order can simply be the cost of a return trip between the depot and the customer.

The penalties can be modified during the search: If, during the last iterations, a constraint was systematically violated, then the penalty associated with the violation of this constraint can be increased. Conversely, if a constraint was never violated, then the penalty associated with this constraint can be decreased. This technique was used in the context of CVRP [Gendreau et al., 1994]. This technique is a well adapted one if only one constraint is relaxed. If several constraints are simultaneously introduced in the objective, then it may happen that only non feasible solutions are visited. This is due to the fact that the different penalties associated with different constraints could vary in phase opposition in such a way that at least one constraint is violated, the violated constraint changing during the search.

It is not always easy to model a problem, especially when the (natural) objective is to minimize a maximum. The choice of the function to minimize and the penalty function can be difficult. These functions must take a number of different values as large as possible over their definition domain, in such a way that the search can be efficiently directed. How can the choice of a suitable move to be decided upon, when a large number of solutions with the same cost exist in the neighborhood?

This last remark a priori assumes that a local search will be used. However, evolutionary algorithms or artificial ants do not refer to local searches, at least in their most elementary versions. But now, almost all efficient implementations inspired by these metaheuristics embed a local search, at least a simple improving method.

7.3 Neighborhood choice

It is required to proceed for the choice of the neighborhood structure(s), in conformation with the definition of the solution space. This problem is also far from being trivial, since, for the same generic problem, a well adapted neighborhood for a given instance could be bad for other instances. Typically, this is an empirical choice, even if the problem characteristics may provide some directions.

7.3.1 "Simple" neighborhoods

The model utilized, for the vehicle routing problem, used as an example for our illustrations, is very simple. Feasible solutions are extended by considering that it is not mandatory that all orders must be delivered. For this, a dummy tour T_0 with infinite capacity is added. The delivery mode of this tour consist of successive return trips: depot-customer-depot. With this model, finding a feasible solution is trivial: all orders can be placed on tour T_0.

For vehicle routing problems, many different neighborhoods have been proposed. The simplest one is moving an order from one tour to another. Then two orders belonging to two different tours can be exchanged. A relatively general neighborhood, known as CROSS [Taillard et al., 1997], consists in exchanging two paths belonging to two different tours. Figure 7.2 illustrates these three neighborhoods. The moves are evaluated by a simple insertion heuristic: a group of commands is placed on a tour where it implies the least detour; the delivery order of customers remaining in the tour are not modified. The vehicle directly travels from the previous customer to the next customer of those removed from the tour.

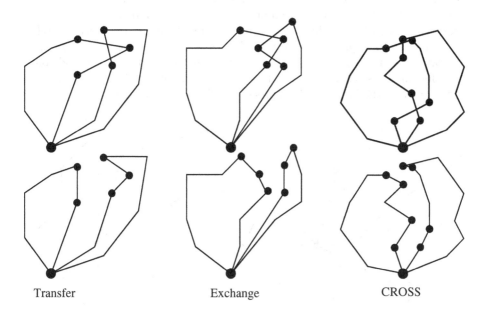

Transfer   Exchange   CROSS

Fig. 7.2. Three types of moves frequently used in vehicle routing literature.

The use of a simple deletion and insertion technique implies that the quality of the tours, i.e. the order in which the customers are visited, deteriorates as the local search proceeds. Theoretically, an NP-hard traveling salesman problem should be solved for each solution attempted. In order to be more efficient, the tours are periodically (but not very frequently) re-optimized. They are also re-optimized when a relatively good solution is found. These optimizations can be carried out either by an exact method (in case the number of customers in each tour is limited) or by a convenient heuristic.

In order to improve the efficiency of the neighborhood evaluation, it can be remarked that only two tours are modified from one solution to a neighboring one. Therefore by storing, for all moves, the modifications brought for both the tours concerned, only the update of the last tours must be computed to evaluate the whole neighborhood. Such a technique can significantly accelerate the search, at the cost of a very reasonable increase in memory consumption. In order to apply this technique, it is necessary to know the maximum number of vehicles m required to deliver the orders. If this number is unknown, generally it can be easily determined.

7.3.2 Ejection chains

An *ejection chain* is a technique to create potentially interesting neighborhood that facilitates to perform, in a single move, a significant modification of a solution. This technique can be well illustrated specially for the CVRP. For this problem, the simplest neighborhood is to transfer one customer from one tour to another. If this neighborhood is implemented in a local search, it is difficult to implement a "rotation" over a set or a subset of tours, i.e. to transfer a customer from the first tour to the second one, the second tour transferring a customer to the third tour, and so on until the last tour transfers a customer to the first one. This process is illustrated in Figure 7.3.

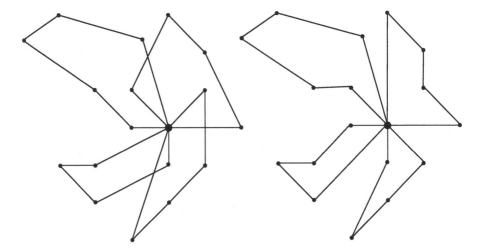

Fig. 7.3. Example of an ejection chain, consisting in transferring simultaneously many customers from one tour to another

A neighborhood can be completely scanned only for very limited subsets of tours, typically 2 or 3. However, an approximation can be performed for any number of tours by solving an auxiliary problem. Such an approach was proposed in [Xu and Kelly, 1996]. In this reference, the auxiliary problem is a minimum cost flow problem in a two layer network. Here, the first layer corresponds to the customers, and the second one corresponds to the tours. The network is built as follows: directed arcs with cost 0 and capacity 1 are created between a source-node and each customer-node and between each tour-node and a sink-node. If it is possible (and authorized, in case of tabu search) to move customer i in tour T_j, then an arc with capacity 1 is added between the customer-node i and the tour-node T_j. The cost of this arc is that of the insertion of the customer in the new tour diminished by the saving due to the removal of the customer from the old tour. The computation of all the minimum cost flows between a value of 1 and the number m of vehicle tours in this network provides an approximation of the global cost corresponding to the move of $1, 2, \ldots, m$ customers.

[Rego and Roucairol, 1996] proposed another mechanism for implementing an ejection chain. The idea is to perform a very partial enumeration of the solutions that can be obtained after a given number of ejections.

7.3.3 Decomposition into subproblems: POPMUSIC

When solving large size problem instances, a natural tendency is to proceed by decomposing into independent subproblems. These subproblems can then be solved employing an appropriate procedure. In this way, large size problem instances can be efficiently approached, since the global complexity of the method grows very slowly, typically as $O(n)$ or $O(nlog(n))$, where n is the problem size.

However, implementing an a priori decomposition of a problem may induce low quality solutions, since the subproblems have been created more or less arbitrarily and without considering the shape of the solution. Indeed, it is not easy to decompose a problem conveniently without having an intuition about the structure of good solutions. The idea behind POPMUSIC is to locally optimize parts of a solution *a posteriori*, once a global solution is known.

These local optimization procedures can be repeated until a local optimum — relatively to a very special neighborhood — is obtained. POP-MUSIC is the acronym for*Partial Optimization Meta-heuristic Under Special Intensification Conditions* [Taillard and Voß, 2002]. Several authors have proposed techniques that are slightly different from POPMUSIC. These techniques are sometimes less general and are given different names like LOPT (*Local OPTimizations* [Taillard, 2003a]), LNS (*Large Scale Neighborhood* [Shaw, 1998]), shuffle, MIMAUSA [Mautor and Michelon, 1997], VNDS [Hansen and Mladenović, 1999], hybrid branch & bound tabu search, etc.

For many combinatorial optimization problems, a solution S can be represented by a set of parts $s_1, \ldots, s_p$. For the vehicle routing problem, a part can be a tour, for example. The relations existing between each pair of parts may vary. To elaborate, two tours containing customers that are close each others will have a stronger interaction than tours located in opposite directions, relative to the depot.

The central idea of POPMUSIC is to build a subproblem with a *seed-part*, s_i, and a given number $r < p$ of parts $s_{i_1}, \ldots, s_{i_r}$ which are specially related to the seed-part s_i. These r parts build a subproblem R_i, smaller than the initial problem, that can be solved by an *ad hoc* procedure. In case if each improvement in subproblem R_i implies an improvement of the complete solution, then the frame for a local search can be defined. This local search is relative to a neighbourhood that consists in optimizing subproblems. So, by storing a set O of those parts that have been used as seeds for building a subproblem which are unable to improve the complete solution, the search can be stopped as soon as all p parts constituting the complete solution are contained in O. So, a special local search has been designed. This local search is parametrized by r, the number of parts constituting a subproblem (see algorithm 7.1).

1. Input: Solution S composed of parts $s_1, \ldots, s_p$
2. Set $O = \emptyset$
3. While $O \neq \{s_1, \ldots, s_p\}$ repeat
   a) Select $s_i \notin O$
   b) Create subproblem R_i composed of the r parts $s_{i_1}, \ldots, s_{i_r}$ that are the most in relation with s_i
   c) Optimize R_i
   d) If R_i has been improved, set $O \leftarrow O \backslash \{s_{i_1}, \ldots, s_{i_r}\}$, update S (as well as the set of parts).
      Else, set $O \leftarrow O \cup \{s_i\}$

Algorithm 7.1: POPMUSIC(r).

This technique corresponds exactly to an improving method which, starting from an initial solution, stops as soon as a local optimum, relative to a very large neighborhood, is obtained. Hence, the method was named LOPT (Local optimizations) in [Taillard, 2003a] and LNS (large neighborhood search) in [Shaw, 1998].

Indeed, the structure of the neighborhood so built contains all solutions s' that differ from s only by subproblem $R_i, i = 1, \ldots, p$. This means that the size of the neighborhood is defined by the number of solutions contained in the subproblems. This number is naturally very large and grows exponentially with parameter r (the subproblem created for $r = p$ is the whole problem).

The optimization of a subproblem is a hard problem which can only be exactly solved in a very few cases. Thus, a heuristic solution is frequently the only practical option.

Parts.

When a POPMUSIC-based intensification scheme is desired to be implemented, the first requirement is to define the meaning of a part of a solution. For vehicle routing problems, a tour (i.e. the set of orders delivered by a same vehicle) is perfectly convenient to define a part. This approach was used in [Taillard, 1993, Rochat and Semet, 1994, Rochat and Taillard, 1995]. It is also possible to consider each customer as a part. This approach was used in [Shaw, 1998]. If the problem instances are large enough and contain a relatively large number of tours, then considering a tour as a part has the advantage that the subproblems so defined are also vehicle routing problems. They can be solved completely independently.

Seed-part.

The second point not precisely specified in the pseudo code of POPMUSIC is the way the seed-part is selected. The simplest policy can be to systematically choose it at random. Another possibility can be to take those seed-parts into consideration that have been previously selected. In the case of parallel optimization of subproblems, the seed-parts are advantageously chosen as far as possible, so that the interactions between the subproblems are minimized. Other authors [Rochat and Semet, 1994] have suggested that successive seed-parts can be chosen as close as possible. In this reference, the authors have also suggested to change the value of the parameter r of POPMUSIC during the search, thus implementing what is now called a variable neighbourhood search.

Relations between parts.

The definition of the relations existing between different parts is the third point of discussion in POPMUSIC frame. Sometimes, this relation is naturally defined. For example, in case the parts are chosen as the customers of a vehicle routing problem, the distance between customers is a natural measure of the relation between parts. In case the parts are defined as the tour of a vehicle routing problem, the notion of proximity is not so easy to define. In [Taillard, 1993, Rochat and Taillard, 1995], who have treated Euclidean problems, the proximity is measured by the center of gravity of the tours. The quantity ordered by each client is interpreted as a mass. Figure 7.4 illustrates the principle of the creation of a subproblem from a seed-tour.

Optimization procedure.

Finally, the last point that is not specified in POPMUSIC frame is the procedure used to optimize subproblems. In [Rochat and Taillard, 1995] and [Taillard, 1993], the procedure employed is a basic tabu search. Shwa uses an exact method based on constraint programming [Shaw, 1998].

7.3.4 Conclusions on modeling and neighborhood

In conclusion, we particularly insist on the modeling aspect of a problem and on the choice of a neighborhood because in our opinion this is one of the most important phases to successfully design an efficient heuristic. Indeed, if even one of these points is poorly analyzed, the addition of another level (simulated annealing, tabu search, etc.) to obtain better solutions than those produced by a simple improving method can become deceptive.

For the purpose of illustration, we can cite the following example. Even if you choose the best jet or helicopter pilot (i.e. the best way to direct a local search), you will not be able to secure an alpinist in difficulty or cross the

Atlantic Ocean if you do not choose the right engine (i.e. the right model and the right neighborhood).

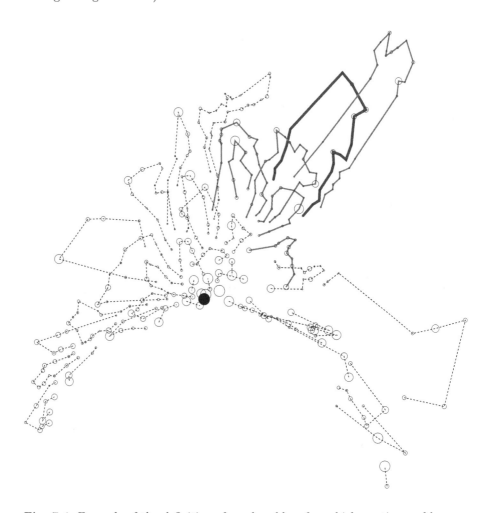

Fig. 7.4. Example of the definition of a subproblem for vehicle routing problems. The seed-part (tour) is drawn with a thick line, the tours most in relation with the seed-tour by normal lines and tours that are not considered in the optimization of a subproblem by dashed lines. The routes from or to the depot are not drawn, so that the figure is not overloaded

The vehicle routing also presents a good illustration of the importance of this phase: The method described in [Taillard, 1993], which was able to find many of the best solutions known for a set of benchmark problem instances [Christofides et al., 1979], is based on a very simple tabu search. (The moves considered are limited to the swapping of two customers or the move of one customer form one tour to another. The only component added to a very basic tabu search is the penalization of moves frequently used.) The entire method contains only few parameters, compared to much more sophisticated methods, like those of [Gendreau et al., 1994, Xu and Kelly, 1996], which incorporate various advanced mechanisms and parameters, and do not prove to be significantly superior.

7.4 Improving method, simulated annealing, taboo search. . . ?

To choose an appropriate model or a neighborhood, the knowledge of the type of metaheuristic to be implemented is essential. Now, it is not possible to decide which technique should be used on a theoretical basis. A simple improving method, with a convenient neighborhood, might reveal to be better than a simulated annealing or a tabu search. Indeed, an attempt to incorporate this neighborhood in one of these metaheuristics could lead us to an algorithm that requires too much computational effort to be effective. However the following methodology can be suggested: First a simple neighborhood can be designed and implemented. This neighborhood should be chosen in such a way that a neighboring solution can be evaluated fast. With such a base, the implementation of an improving method or a simulated annealing should be easy to design. Sometimes, like for the quadratic assignment problem, a complete evaluation of the neighborhood can be efficiently implemented. In such a case, the design of a tabu search seems to be a promising option (see Figure 7.8).

If possible, it is very useful to graphically represent solutions visited during an iterative search. In such a way, it is much easier to find, to analyze and to remedy problems than with a "black box" approach. In the later approach only the results obtained with given parameter settings are observed, without the exact know-how about what happens during the search.

Nevertheless, if the method seems to be incapable of exploring solutions with various structures, or if the parameters have to be set to extreme values such that the method behaves like a random search, then more global approaches should be attempted. Evolutionary algorithms or ant colonies are among these more global approaches. These methods require to embed a local search to be efficient. So the importance and the burden of implementing a local search is not lost!

7.5 Adaptive Memory Programming

A minute observation of recent implementations of evolutionary algorithms, scatter search or artificial ant colonies reveals that all these techniques seem to evolve toward a common framework, called *Adaptive Memory Programming*

(AMP) [Taillard, 1998, Taillard et al., 1998]. This framework can be the following.

Adaptive Memory Programming

1. Initialize memory
2. Repeat, until a termination criterion is satisfied:
   a) Build a new solution with the help of the memory
   b) Improve the solution with a local search
   c) Update memory with information carried by the new solution

Now, let us justify why various metaheuristics follow the same framework.

7.5.1 Ant colonies

In the spirit of Adaptive Memory Programming, trails of pheromone of ant colonies can be considered as a form of memory. This memory is utilized for building new solutions, following the specific rules of simulated ants, or, expressed in other terms, by following the magic formula, the belief in precepts of the designers of ant colony optimization. Initially, the process did not embed a local search. However, simulation experiments very soon revealed that the quality of the process was more efficient when a local search was incorporated. Unfortunately, the designer of ant colonies used to hide this component in the pseudo-code of the metaheuristic under the form of a "daemon action" which may consist, potentially, of anything!

7.5.2 Evolutionary or memetic algorithms

In the case of evolutionary algorithms, the population of solutions can be considered as a form of memory. Indeed, some characteristics of the solutions — hopefully the best ones — are transmitted and improved, from one generation to the next one. Recent implementations of evolutionary algorithms have replaced the "random mutation" metaphor by a more elaborated operator.

Instead of performing several local and random modifications to the solution obtained after crossover operation, a search for a local optimum is initiated. Naturally a more elaborated search can be executed, e.g. a tabu search or a simulated annealing. In the literature, this type of methods is called "hybrid genetic algorithm" or "memetic algorithm" [Moscato, 1999].

7.5.3 Scatter Search

Scatter Search is almost as old as genetic algorithms as the technique was originally proposed, completely independently, in 1977 [Glover, 1977]. However the technique started to gain prominence among the academic communities only by the end of the 90's. In contrary to the evolutionary algorithms, simulated annealing and tabu search, this method has almost not been used in the industrial world yet. Scatter search can be viewed as an evolutionary algorithm with the following specific characteristics:

1. Binary vectors are replaced by integer vectors.
2. The selection operator for reproduction may select more than two parent solutions.
3. The crossover operator is replaced by a convex or non convex linear combination.
4. The mutation operator is replaced by a repair operator that projects the newly created solution in the feasible solution space.

These characteristics may also be considered as generalizations of evolutionary algorithms which have been proposed and exploited later by various authors, especially [Mühlenbein et al., 1988]:

1. The use of crossover operators is different from the exchange of bits or subchains;
2. A local search is applied for improving the quality of solutions produced by the crossover operator;
3. More than two parents are used to create the child;
4. The population is partitioned with the help of classification methods instead of an elementary survival operator.

In scatter search, the production of new individuals from solutions of the population is a generalization of crossover in evolutionary algorithms. In "pure" genetic algorithms, solutions of a problem are only considered in the form of a fixed chain length of bits. For many problems, it is not natural to code a solution using a binary vector and, depending on the coding scheme chosen, a genetic algorithm may produce results of varying quality. In the initial versions of genetic algorithms, the main point was to choose an appropriate coding scheme, the other operators belonging to a standard set. On the contrary, scatter search advocates for a natural coding of solutions, implying the design of "crossover" operators (generation of new solutions from those of the populations) strongly dependent on the problem to be solved.

In the first reference [Glover, 1977], scatter search was applied to integer linear programming. Here, it was suggested to create a new solution by a linear combination of the solutions of the population. In the general case, e.g. for permutation problems, it is not possible to make a linear combination of the solutions. In this case, a specialized "crossover" operator must be designed, which must eventually be followed by a repair operator, in case the new solution is not feasible. Often, this repair operator consists of an elementary local search. One of the main drawbacks of evolutionary algorithms is the population convergence (genetic drift). Indeed, once the population has converged (i.e. all solutions of the population are clones), if the final solution is not satisfactory, there is no other option but to initiate a fresh search with a new population. To avoid such a situation, scatter search suggests to manage the population with automatic classification techniques. Here, the population is periodically classified into several groups, taking the similarity existing between solutions as the homogeneity criterion. The size of the population is then reduced by conserving only a few of the best solutions from each cluster. This new solution set is called the reference set or the elite solutions. In the construction phase, the solutions belonging to different clusters are used to build new solutions, in order to preserve the diversity of the solutions generated during the search.

7.5.4 Vocabulary building

Vocabulary building is also a concept introduced by [Glover and Laguna, 1997] in the context of tabu search, but the principles of this concept have certainly been used, under different names, by different authors. Vocabulary building can be conceived as a special scatter search (or evolutionary algorithm!). Here, instead of storing complete solutions in the memory, only fragments (or *words*) are memorized. These words are employed in building a *vocabulary*. A new solution (i.e. a *sentence*) is obtained by combining different fragments. In the context of the vehicle routing, a fragment — or part of a solution, conforming to the terminology of POPMUSIC — can be defined as a tour. Then the following procedure can be applied to build a new solution s, where M is a set of tours, each tour T being composed of a set of customers.

Building a new solution

1. $s = \emptyset$
2. Repeat, while $M \neq \emptyset$:
   a) Choose $T \in M$
   b) Set $s \leftarrow s \cup T$
   c) Set $M \leftarrow M \backslash T'$ $\forall T' \in M$ such that $T' \cap T \neq \emptyset$
3. If the solution s does not contain all customers, then complete it.
4. Improve the solution with a local search.

Thus, the idea is to build a solution by choosing tours successively belonging to a set of memorized tours. The chosen tours must not contain those customers which are already contained in the partially built solution.

For vehicle routing problems, this technique was first applied in [Rochat and Taillard, 1995]. It succeeded in obtaining several best solutions of benchmark problem instances of the literature. This method shows significant performance particularly for the following reasons: An elementary tabu search, embedded within the framework of POPMUSIC, is capable of finding very rapidly few of the tour of the best solutions known. Therefore, a lot of computational effort can be spared by collecting already existing tours without having to build them from scratch.

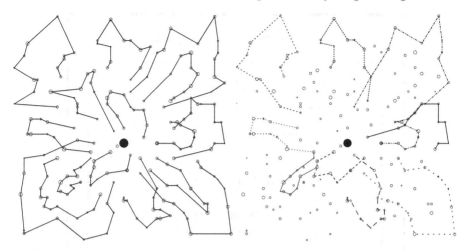

Fig. 7.5. An example of the utility of creating *words* (= tours) in vocabulary building. On the left, one of the best solutions known for a CVRP instance. On the right, few of the tours found within a few seconds with a POPMUSIC based search.

7.5.5 Path relinking

Path relinking [Glover and Laguna, 1997] is another technique that utilizes a population of solutions, also proposed by Glover in the context of tabu search. This technique can be viewed as a special scatter search. Here the objective is not to create one solution from many others, but to create a set of solutions, neighbors of each other, connecting two solutions of a population, thus creating a path between these two solutions. Each solution of the path may eventually be improved with a local search. The technique is also very similar to that of a tabu search, that is not guided by an objective function and tabu conditions, but by a target-solution to reach.

Hence, there can be infinite number of possibilities to extend a technique. The bottom-up methodology proposed here seems relatively logical to follow (see Figure 7.6). Indeed, the addition of a level that increases the complexity of a method is not very difficult to implement. For example, modifying an improving method, that will terminate in the first local optimum for creation of a simulated annealing, takes few minutes or few hours to code, if the first version is developed without implementing any algebraic or software optimization. Even, nowadays the users have several libraries to their disposal that allows them to embed a basic method into a more complex framework (see, for example, the articles published in [Voß and Woodruff, 2002]).

So, it can be inferred that coding a heuristic based on metaheuristic principles is relatively easy. However, it is much more problematic to find suitable parameters (e.g. annealing scheme, type and duration of tabu conditions, penalty factors, intensification and diversification mechanisms in a tabu

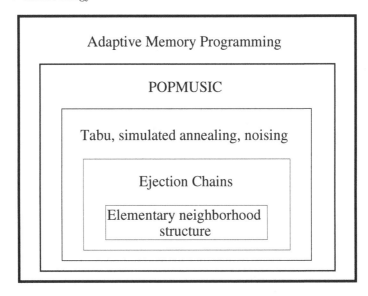

Fig. 7.6. Methodology suggested for designing a complex heuristic: The design can be initiated with the implementation of an improving method with a simple neighbourhood. If needed, the later may then be extended with the incorporation of the ejection chains technique, and then be used for a more elaborated local search, e.g. simulated annealing, noising methods or tabu search. If the problem is adapted to the technique and if the instances to solve are of large size, the local search can be embedded in a POPMUSIC frame. Finally, the entire design can be embedded in an Adaptive Memory Procedure that exploits statistical informations, as in artificial ant colonies, or a population of solutions, as in evolutionary algorithms or scatter search, for building new solutions.

search, coding scheme, crossover operators, population size in an evolutionary algorithm...)

In order to find good tuning of parameters, without performing elaborate numerical experiments, it is important to utilize of statistical tests, sometimes relatively specific. This leads us directly to a point that was quite neglected in the metaheuristic literature: this is the comparison of iterative heuristics.

7.6 Iterative heuristics comparison

The implementation of a heuristic method for solving a complicated combinatorial problem necessitates that the designer consider several choices. Some of them may be relatively easy to justify, but others, like the numerical tuning of parameters or the choice of a neighborhood may be much more hazardous. When theory or intuition cannot support the choice of the researcher, the later must justify his decision with the help of several numerical experiments. However, it is observed too often that the choices are not supported by scientific bases. The present section discusses few techniques for comparing improving heuristic techniques.

7.6.1 Comparing proportion

The first question that should be clarified concerns the comparison of success rates of two methods $\mathcal{A}$ and $\mathcal{B}$. Practically, the experiments are as follows: Method $\mathcal{A}$ is run n_a times and succeeds in solving the problem a times. Similarly when method $\mathcal{B}$ is executed n_b times and succeeds in solving the problem b times. So, the following question arises: is a success rate of a/n_a significantly higher than a success rate of b/n_b? A researcher, who is a perfectionist, should carry out a large number of experiments and work on a sufficiently large number of runs to conduct a standard statistical test based on the central limit theorem. Conversely, a less careful researcher will not conduct the 15 or 20 runs theoretically needed to validate his choice of method among several options, but will assume, for instance, that if $\mathcal{A}$ has 5 positive results over 5 runs, it will certainly be better than $\mathcal{B}$ that has only 1 positive run over 4. Is the above conclusion correct or not? A nonparametric statistical test developed in [Taillard, 2003b] shows that a success rate of 5/5 is significantly higher — with a confidence level of 95% — than a success rate of 1/4. The contents of table 7.1, which were originally proposed in [Taillard, 2003b], provide, for a confidence level of 95%, the couples (a, b) for which a success rate higher than or equal to a/n_a is significantly better than a success rate lower than or equal to b/n_b.

This table can be particularly useful to find good parameters for a technique, both quickly and in a rigorous manner. A suitable procedure can be to fix two different parameter sets (thus defining two different methods $\mathcal{A}$ and $\mathcal{B}$) and to compare the results obtained with both methods. In order to make proper use of the Table 7.1 it is required to define what a success is (for instance, the fact that the optimal solution or a solution of a given quality has been found for a given problem instance) and, naturally, it is assumed that the runs are conducted independently each other. It can be assumed that this will work with a given problem instance and nondeterministic methods $\mathcal{A}$ and $\mathcal{B}$ (like a simulated annealing) or will work with problem instances randomly chosen in a set (for example, randomly generated instances of a given size).

Table 7.1. Couples (a,b) for which a success rate $\geq a/n_a$ is significantly higher than a success rate $\leq b/n_b$, for a confidence level of 95%.

n_b	n_a								
	2	3	4	5	6	7	8	9	10
2	—	(3,0)	(4,0)	(5,0)	(5,0)	(6,0)	(7,0)	(7,0)	(8,0)
3	(2,0)	(3,0)	(3,0)	(4,0)	(4,0)	(5,0)	(5,0)	(6,0)	(6,0)
				(5,1)	(6,1)	(7,1)	(8,1)	(8,1)	(9,1)
4	(2,0)	(3,0)	(3,0)	(3,0)	(4,0)	(4,0)	(5,0)	(5,0)	(5,0)
		(3,1)	(4,1)	(5,1)	(5,1)	(6,1)	(7,1)	(7,1)	(8,1)
				(6,2)	(7,2)	(8,2)	(9,2)	(10,2)	
5	(2,0)	(2,0)	(3,0)	(3,0)	(3,0)	(4,0)	(4,0)	(4,0)	(5,0)
		(3,1)	(4,1)	(4,1)	(5,1)	(5,1)	(6,1)	(6,1)	(7,1)
			(4,2)	(5,2)	(6,2)	(7,2)	(7,2)	(8,2)	(9,2)
							(8,3)	(9,3)	(10,3)
6	(2,0)	(2,0)	(2,0)	(3,0)	(3,0)	(3,0)	(4,0)	(4,0)	(4,0)
	(2,1)	(3,1)	(3,1)	(4,1)	(4,1)	(5,1)	(5,1)	(6,1)	(6,1)
		(3,2)	(4,2)	(5,2)	(5,2)	(6,2)	(7,2)	(7,2)	(8,2)
				(5,3)	(6,3)	(7,3)	(8,3)	(9,3)	(9,3)
									(10,4)
7	(2,0)	(2,0)	(2,0)	(3,0)	(3,0)	(3,0)	(3,0)	(4,0)	(4,0)
	(2,1)	(3,1)	(3,1)	(4,1)	(4,1)	(4,1)	(5,1)	(5,1)	(6,1)
		(3,2)	(4,2)	(4,2)	(5,2)	(6,2)	(6,2)	(7,2)	(7,2)
			(4,3)	(5,3)	(6,3)	(6,3)	(7,3)	(8,3)	(9,3)
					(6,4)	(7,4)	(8,4)	(9,4)	(10,4)
8	(2,0)	(2,0)	(2,0)	(2,0)	(3,0)	(3,0)	(3,0)	(3,0)	(4,0)
	(2,1)	(3,1)	(3,1)	(3,1)	(4,1)	(4,1)	(5,1)	(5,1)	(5,1)
		(3,2)	(4,2)	(4,2)	(5,2)	(5,2)	(6,2)	(6,2)	(7,2)
		(3,3)	(4,3)	(5,3)	(5,3)	(6,3)	(7,3)	(7,3)	(8,3)
				(5,4)	(6,4)	(7,4)	(8,4)	(8,4)	(9,4)
						(7,5)	(8,5)	(9,5)	(10,5)
9	(2,0)	(2,0)	(2,0)	(2,0)	(3,0)	(3,0)	(3,0)	(3,0)	(3,0)
	(2,1)	(2,1)	(3,1)	(3,1)	(4,1)	(4,1)	(4,1)	(5,1)	(5,1)
	(2,2)	(3,2)	(3,2)	(4,2)	(4,2)	(5,2)	(5,2)	(6,2)	(6,2)
		(3,3)	(4,3)	(4,3)	(5,3)	(6,3)	(6,3)	(7,3)	(7,3)
			(4,4)	(5,4)	(6,4)	(6,4)	(7,4)	(8,4)	(8,4)
				(5,5)	(6,5)	(7,5)	(8,5)	(9,5)	(9,5)
							(8,6)	(9,6)	(10,6)
10	(2,0)	(2,0)	(2,0)	(2,0)	(2,0)	(3,0)	(3,0)	(3,0)	(3,0)
	(2,1)	(2,1)	(3,1)	(3,1)	(3,1)	(4,1)	(4,1)	(4,1)	(5,1)
	(2,2)	(3,2)	(3,2)	(4,2)	(4,2)	(5,2)	(5,2)	(5,2)	(6,2)
		(3,3)	(4,3)	(4,3)	(5,3)	(5,3)	(6,3)	(6,3)	(7,3)
		(3,4)	(4,4)	(5,4)	(5,4)	(6,4)	(7,4)	(7,4)	(8,4)
			(4,5)	(5,5)	(6,5)	(7,5)	(7,5)	(8,5)	(9,5)
					(6,6)	(7,6)	(8,6)	(9,6)	(10,6)
								(9,7)	(10,7)

7.6.2 Comparing iterative optimization methods

In the context of metaheuristics presented in this book, it is often difficult to define success rates, since there is no clear goal to reach. More precisely, the processes we are interested in have two objectives: in addition to improving solution quality, it is required to minimize the computational burden. But this last parameter can be freely chosen by the user, for instance by modifying the number of iterations of a simulated annealing or a tabu search, or the number of generations of an evolutionary algorithm or the number of solutions built with an ant colony. Moreover, most of these methods are nondeterministic, which indicates that two successive runs on the same problem instance generally produce two different solutions.

Traditionally, the measure of quality of a method is the average solution value it produces. The computational burden is measured in terms of computer CPU time consumed in seconds. Both of these measures are not satisfying. If the computational burden is fixed for two nondeterministic methods $\mathcal{A}$ and $\mathcal{B}$, and if it is desired to rigorously compare the quality of the solutions produced by these methods, both methods must be executed several times and a statistical test comparing two methods must be conducted. Unfortunately, the distribution function of the solutions quality produced by a method is unknown and generally not Gaussian. Therefore it is not possible to use a standard statistical test unless large samples are available. This means that the numerical experiments are repeated a large number of times — practically, this may correspond to many hundreds of times, contrary to the common belief that a sample size of 20 or 30 is large enough.

In case quality may be measured by some other method than the average solution values obtained, interesting comparisons can be performed with very few runs. One of these nonparametric methods may consist in ranking the set of all solutions obtained with methods $\mathcal{A}$ and $\mathcal{B}$ and to compute the sum of the ranks obtained by one method. If this sum is lower than a value — that depends on the level of significance, which can be read in numerical tables —, then one cannot exclude the fact that a run of this method has a probability significantly higher than $1/2$ to obtain a better solution than a run of the other method. In the literature [Conover, 1999], this test is known as the *Mann-Whitney* test.

Naturally, if iterative methods must be compared using such a test, the test must be repeated each time with a fixed computational effort. In practice, as mentioned before, computational time in a given computer is used for measuring the computational effort. This is a relative measure as it depends on the hardware used, on the operating system, on programming language, on compiler, etc. To have a more rigorous comparison, an absolute computational burden must be used. Typically, the evaluation of the neighboring solutions is the most demanding part of a metaheuristic-based method, such as simulated annealing, tabu search, evolutionary algorithms or ant colonies (provided that the last two techniques are hybridized with a local search).

Thus, it is often possible to express the computational burden not in seconds but in iteration numbers and to specify the theoretical complexity of one iteration. For instance, one iteration of tabu search proposed in Chapter 2 for the quadratic assignment problem has a complexity of $O(n^2)$. By making the code of this method available in public domain, everyone can now express the computational burden of his/her own method for a given problem example in terms of the equivalent number of tabu search iterations. So, there is no necessity to provide a reference to a computational time relative to a given machine — which will very soon become obsolete in near future.

By applying these principles, it is possible to generate two diagrams on the same figure, indicating the functional evolution of the computational effort: the first providing the average solution value and the second providing the probability that a method will be better than the other, this probability being computed on the base of a Mann-Withney test. This can be realized by means of the *STAMP* [Taillard, 2002] software. An example of comparison is given in Figure 7.7.

Finally, let us extract the information that is of real interest — is Method $\mathcal{A}$ significantly better than Method $\mathcal{B}$? —, where the first diagram providing the evolution of average solution values can be removed and only the second diagram providing the probability that a method will be better than the other can be drawn. If the problem is approached in such a way, the surface needed to provide the essential information is reduced in large proportions. So, it is possible to draw many probability diagrams on the same figure, for example to compare many methods with each other for the same problem instance or to compare two methods solving different problem instances. This possibility is illustrated in Figure 7.8, where not less than 5 heuristic methods are pairwise compared when they are run on a problem instance, taken from the literature.

These diagrams provide, at a first glance, much more information than a table comprising traditional numerical results. The two main advantages are that they show comparisons, in a continuous manner of computational burden and they provide exactly the requisite information (is this method better than the other one?). For instance, it can be seen in Figure 7.8 that the simulated annealing method is significantly worse than the other ones (where long FANT runs are excepted), followed by the FANT. The three other methods cannot be really ordered, as the order depends on the computational burden. Finally, let us mention that this order is not the same for other problem instances.

7.7 Conclusion

It is sincerely hoped that this chapter will partially guide those researchers who are engaged in the design of a heuristic based on techniques presented in the previous chapters. We are well aware of the fact that each practical problem will be a specific case and that our advices sometimes may not be

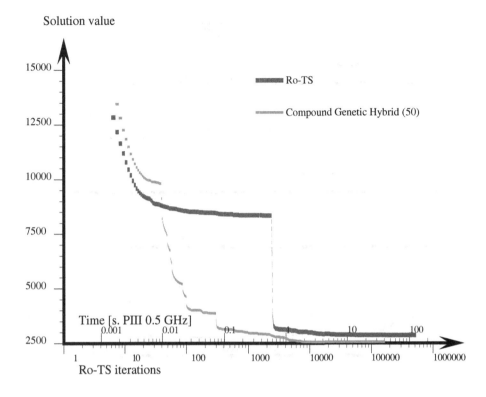

Solution value

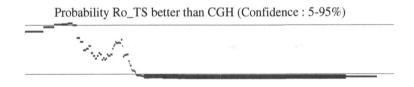

Fig. 7.7. Comparison of a tabu search and an hybrid genetic algorithm for the quadratic problem instance *tai27e01*. The upper diagram provides the average solution values obtained by both methods over 20 runs. The lower diagram provides the probability that a tabu search run gives a better solution than a hybrid genetic algorithm run. A value below the lower horizontal line, indicating a confidence level of 5 %, means that tabu search is significantly worse than the genetic hybrid. A value above the upper line, indicating a confidence level of 95 %, means that tabu search is significantly better than the genetic hybrid. The STAMP software indicates by a bold line a statistical difference that remains significant even if one of the two methods runs 2 times faster or slower.

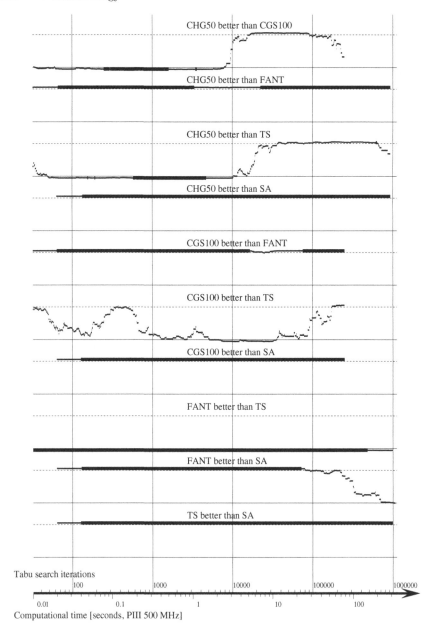

Fig. 7.8. Comparison of a simulated annealing (SA), a tabu search (TS), an algorithm based on the artificial ant colonies (FANT) and two versions of a hybrid genetic algorithm (CHG50 and CGS100) for the quadratic assignment problem *dre30*. Each of the 10 diagrams gives the probability that one of these methods is better than another. The threshold of confidence of 5 % is indicated by fine horizontal lines whereas the threshold of 95 % is in dotted lines.

judicious. For example, for the travelling salesman problem, one of the best heuristic methods available till now is a simple improving method, which is based on the appropriate neighborhood. For the p-median problem, one of the best methods is based on a POPMUSIC that does not embed other metaheuristic principles such as tabu search or simulated annealing. Finally, let us mention that the evolutionary algorithms or scatter search implementations do not necessarily embed ejection chains, partial optimization or other metaheuristic principles.

However, in our opinion, the researchers should be more careful concerning the methodology for comparing iterative heuristics. Indeed, in the literature, tables that formally contain no reliable information are too often presented, and their authors draw conclusions that are not supported by the experiments performed. This is why we hope that the last part of this chapter, where the comparison of improving heuristics is presented, should lead to research topics that will gain an increasing importance in the near future.

7.8 Annotated bibliography

In the literature there are relatively few works available that deal with design methodology of heuristic methods. For example, the *Journal of Heuristics* has very recently devoted a specific scope for this topic. Perhaps, this situation is due to the fact that the metaheuristics are a relatively new domain, in which each author who proposes a new paradigm desires to make publicity for his own work. Indeed, our experience shows that the statistical tests induce less affirmative conclusions than those which are sometimes expressed on purely instinctive basis, leading to the choice of a computational effort specially in favour of a method. Among works that try to unify the presentation of some metaheuristics, let us mention:

[Hertz and Kobler, 2000]: This reference proposes a framework for the description of various methods that qualify as evolutionary, where this term is not to be understood in the usual meaning used in the present book, but in a less specific meaning, depending on the fact that the methods embed data structures that evolve. In this chapter, we prefer to describe these method as adaptive. Among several methods considered in this reference are genetic algorithms, scatter search, ant colonies, multipopulation genetic algorithms and vocabulary building (strangely presented under the name of adaptive memory programming).

[Taillard, 1998, Taillard et al., 1998]: These references present, under a unified framework, evolutionary algorithms hybridized with a local search, artificial ant colonies, scatter search, GRASP-type procedures as well as some tabu search principles.

[Taillard, 2002]: This reference presents, under a unified framework, various techniques for decomposing problems into subproblems and optimizing them.

As far as the comparison of iterative heuristic methods is concerned, references are virtually non existent, even if many works opined that the researchers should use statistical tests in presenting their numerical results [Barr et al., 1995, Hooker, 1995, Coffin and Saltzman, 2000]. Moreover, these few references only consider the quality of the solution produced by the heuristic methods and there are almost no suggestions for comparisons with varying computational effort. One of the only suggestions is to superpose vertical bars, indicating the standard deviation of the measures, with the curve (average quality/computational time). Such a diagram can be sometimes difficult to read, and does not provide the reader with enough meaningful information about the significance of a curve to be above the others...

Part III

Case Studies

Optimization of UMTS Radio Access Networks with Genetic Algorithms

Sana Ben Jamaa, Zwi Altman, Jean-Marc Picard and Benoît Fourestié

France Télécom R&D, Département Interface Radio et Ingénierie pour les Réseaux Mobiles, 38, rue du Général Leclerc, 92794, Issy les Moulineaux
`zwi.altman@francetelecom.com`

8.1 Introduction

Third generation (*3G*) mobile network, the *UMTS* ("Universal Mobile Telecommunications System"), will provide, in addition to the voice services, new services and applications: Internet, transfer of data, video telephony, etc. Real time (*RT*) services, with guaranteed bit rates, such as voice and multi-media services and non-real-time (*NRT*) services such as downloading of files or electronic messaging will both be supported. The deployment of the *UMTS* networks in Europe involves a huge investment for the operators due to prohibitive costs of infrastructures and licenses. The installation of the radio access network represents approximately 80% of the total investments in infrastructures. In this context, the optimization of the radio access networks becomes, for an operator, a fundamental stake enabling to save investments, to reduce the number of sites to be deployed, and to guarantee good quality of service for the users.

The majority of the operators make use of manual optimization, based on empirical and nonsystematic methods. The main objective of this case study is to automate network optimization, so as to test a range of configurations much larger than the range permitted by manual optimization, to improve the performances of the manually optimized networks and to save the time of radio experts. The significant number of parameters to be optimized on one hand,

and the complexity of the evaluation of the network on the other hand, make a combinatorial optimization approach almost mandatory. The choice of the Genetic Algorithm (*GA*) as the technique for optimization proved to be effective and powerful for this problem [Altman et al., 2002, Ben Jamaa et al., 2003]. This network optimization is known as *Automatic Cell Planning* and the tool performing this optimization as the *Automatic Cell Planner* (*ACP*).

This chapter is organized in the following manner: the first part briefly presents the operation of *UMTS* networks, and the quantities involved in the analysis of its performance. The second part defines the optimization problem for planning the *UMTS* network. Next, we present the application of the *GA* to the network planning problem, and finally, the results obtained for a realistic network are analyzed.

8.2 Introduction to mobile radio networks

The objective of a mobile radio network, fixed by the licenses of operation, is to provide a large number of users with the facility of reaching the network using a portable terminal (telephone, computer, etc), starting from any point within a wide territory. In general, the licenses define some specified objectives for accessing network, or *coverage*, in terms of the percentage of population covered or as the percentage of covered territory. The users must be able to move within this territory without their communication quality being excessively degraded.

8.2.1 Cellular network

In order to allow the users to be mobile, a radio channel connects the mobile terminal to the network. The radioelectric signal is attenuated during its propagation (diffraction, reflections etc). The received power at the handset level must be sufficiently high so that the information intended for it be correctly reconstituted. The powers of the various transmitters of the radiomobile system being limited, several access points for the network, called the base stations, are installed within the territory (figure 8.1). A mobile which moves within a territory is attached to the base station which provides it the best radio link. A cell associated with a given base station is defined as the zone where a mobile is attached to that base station. The transfer of the mobile from one cell to another (which is also called *handover*) must be carried out in a transparent manner for the user, without interruption of communication, or excessive degradation in the quality of service. In order to save installation and maintenance costs of the sites, the operators carry out trisectorisation (see figure 8.2). Three base stations with distinct antennas are installed on the same site; typically, directive antennas are utilized separated, in the horizontal plane, by angle of approximately $2\pi/3$ radians.

Fig. 8.1. A mobile network. The base stations make it possible to cover the territory.

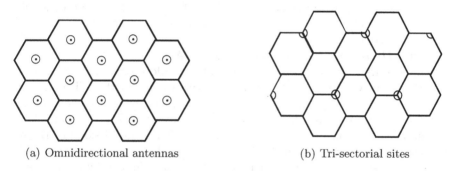

(a) Omnidirectional antennas   (b) Tri-sectorial sites

Fig. 8.2. Tri-sectorization: the number of sites is divided by three in the case of tri-sectorial sites (b) compared to the case where the antennas are omnidirectional (a).

8.2.2 Characteristic of the radio channel

The use of the radio connection in the communication poses certain problems, related to the nature of the transmission medium. This medium is by nature dispersive, it is common to all the users, and thus should be shared. The radio spectrum, and consequently the capacity available for the radio access, are generally limited by the regulations (imposed by the licenses). It thus represents an expensive resource, which should be essentially saved, as far as practicable. The radio channel is prone to multiple variations due to the mobility of the users and the changes of the environment characteristics, e.g. the shadowing or fast fading [Sizun, 2003]. The channel is disturbed by interferences, and the transmitted signals are propagated along multiple paths. These phenomena vary in space and time.

Spectrum sharing and access methods

In mobile radio systems, the frequency band available for communication is limited. This band must thus be judiciously used to achieve the maximum number of communications. It is divided into several channels which are allocated at the request of the mobiles or the base stations. The nature of the transmission channels depends on the access method used. The three principal techniques of multiple access, i.e. of radio resource sharing, are as follows:

Frequency Division Multiple Access (**FDMA**):

The oldest multiple access method is the FDMA method. Here, the frequency bandwidth is divided into narrow bands around a carrier frequency. For example, in the French radio-mobile system of the first generation, Radiocom 2000, each channel occupied $25 kHz$. Each carrier (or channel) is used to convey a single call, in one direction at a time (from the station to the mobile or vice-versa). The capacity of the system is thus limited by the number of carriers available.

Time Division Multiple Access (**TDMA**):

The TDMA technique is used in the radio-mobile systems of second generation like the GSM, where it is combined with FDMA method. The carrier (radio frequency) is divided into N time intervals (TI) or *time slots* and can thus be used by N terminals, each one using a particular TI. There are thus several users, transmitted on the same frequency. The transmission in TDMA is discontinuous: a mobile which transmits on the TI i must wait for the TI $N + i$ to transmit again. The capacity of the system is limited by the number of carriers and the number of TI for each carrier.

Code Division Multiple Access (**CDMA**):

The principle of the CDMA access method is to modulate the useful signal by a pseudo-random digital code with a frequency bandwidth much larger than that utilized by the useful signal. This modulation results in the spreading of the frequency spectrum of the signal. In reception, the signal is correlated with a synchronized counterpart of the spreading code, which enables to restore the initial information. The simultaneous access over a single carrier of several users is made possible by assigning to each one an independent pseudo-random code (orthogonal codes). The reconstruction of the original signal for a user is carried out by applying the specific code assigned to this user, which results in canceling the signals intended for the other users. This technique of spreading the spectrum distributes the electromagnetic power on a frequency band much broader than that necessary for the transmission of the useful signal. The spectral density of power per user is low, the signal is almost undetectable and not very sensitive to interference. Hence, the importance of Wideband-CDMA or WCDMA systems, which belong to third generation systems, can be clearly understood. Their capacity is limited by the interference between the signals and the number of codes.

The three access methods are presented in the figure 8.3.

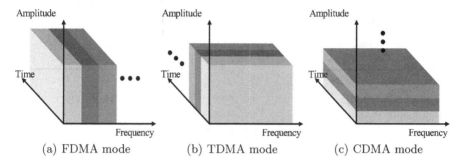

(a) FDMA mode   (b) TDMA mode   (c) CDMA mode

Fig. 8.3. Access Methods.

Modes of duplexing

Duplexing is the way in which the two directions of connection share the radio resources. The upward link (*reverse link* or *uplink*, noted as UL) describes the transmission of the mobile towards the base station. The downward link (*forward link* or *downlink*, noted as DL) describes the transmission of the base station towards the mobile.

In *frequency duplexing* or FDD (*Frequency Division Duplexing*), the base station and the mobile terminal use different frequencies of transmission and they transmit simultaneously. This mode is of particular interest in the macro-cellular systems, because it does not require synchronization.

In *temporal duplexing* or TDD (*Time Division Duplexing*), the base station and the mobile terminal use the same carrier, but transmit at different moments. These two modes of duplexing are presented in the figure 8.4.

8.2.3 Radio interface of the UMTS

The third generation mobile radio systems aim at providing a large variety of services such as voice, video, Internet, and data transfer, with bit rates capable of reaching $2\,Mb/s$ in downlink. Two access modes for the radio interface of the UMTS were specified by the organizations of standardization: the first is associated with the FDD duplexing mode (FDD/WCDMA) and the second with a TDD (TDD/WCDMA) mode [Holma and Toskala, 2000].

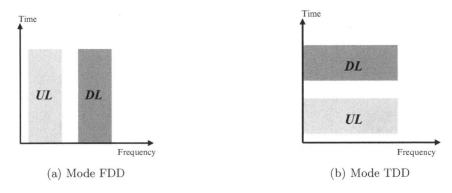

(a) Mode FDD   (b) Mode TDD

Fig. 8.4. Duplexing modes.

In this chapter, our interest is focused on the FDD/WCDMA mode, which is used in the first version of the UMTS deployed in Europe. In frequency division duplexing (FDD), the uplink and downlink signals cannot interfere. For the same direction of connection, the users share the same frequency band and interfere mutually, which limits the capacity. The performances of the network, in terms of coverage, capacity and quality of service, and its optimization are thus closely related to the management of the interferences. For this reason, the following paragraph will describe the various types of interferences in the system, followed by the descriptions of the concepts of capacity, coverage and macro-diversity.

Evaluation of the interferences

Let us consider the communication of a mobile with a given base station. This communication is interfered by the communications of the other mobiles. In each direction of connection, the interferences caused by the communications in the same cell are distinguished, and denoted as the "intra cellular" interference, and interferences caused by the communications of the mobiles situated outside the cell, are denoted as the "inter cellular" interference. To better understand the different types of interferences, a simple example of a two-cell system is described below.

Intracellular interferences

The figure 8.5 illustrates the intracellular interference experienced by the communication of a mobile m_i. In the downlink, the mobile m_i receives a useful signal coming from the base station, but it also receives a collection of signals transmitted to the other mobiles of the cell. At the transmission antenna, these various signals are orthogonal (synchronous multiplexing by orthogonal codes). Because of the multipath propagation, the orthogonality is partially preserved on reception. A factor of orthogonality can be defined which measures the degree of orthogonality. The expression of the intracellular interference received by a mobile m_i (in the downlink) is as follows:

$$I_{intra}^{DL} = \frac{\alpha_{orth} \left(\sum_{j \neq i} P_j^{TCh} + P_{CCh} - P_{SCh} \right) + P_{SCh}}{Aff\left(B, m_i\right)}$$

In this equation:

α_{orth} is the factor of orthogonality, ranging between 0 and 1. For $\alpha_{orth} \to 0$, the orthogonality is perfect; on the contrary, if $\alpha_{orth} \to 1$, the orthogonality is null.

P_j^{TCh} is the power of transmission of the base station towards the mobile m_j with $j \neq i$.

P_{CCh} is the power of transmission of the common channels. These common channels, or beacons, are used to provide information or signalling in the cell. They are transmitted at constant power.

P_{SCh} is the power of the synchronization channel. It is a particular channel which is not orthogonal to the other channels.

$Aff\left(B, m_i\right)$ is the path loss of the signal between the base station B and the mobile m_i.

In the uplink (from mobile to base station), the signal transmitted by the mobile m_i is interfered by the signals coming from all the mobiles belonging to the same cell B. These signals are not synchronized, and the codes of Gold (scrambling codes that differentiate the users) are used. The signals of the other users are thus perceived as white noise [Holma and Toskala, 2000].

$$I_{intra}^{UL} = \sum_{j \neq i} \frac{P_j}{Aff\left(B, m_j\right)}$$

In this equation:

P_j is the transmitted power of the mobile of index m_j with $j \neq i$.

$Aff\left(B, m_j\right)$ is the path loss between the base station and the mobile m_j.

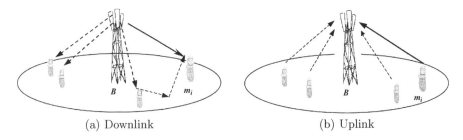

(a) Downlink   (b) Uplink

Fig. 8.5. Intracellular interferences related to mobile m_i.

Intercellular interference.

The figures 8.6 and 8.7 illustrate the intercellular interferences that degrade the communication of the mobile m_i in a system with two cells. In the general case of a network with N stations, the mobile m_i attached to a station of index B is interfered by the signals originating from all the other base stations of the network in the downlink:

$$I_{inter}^{DL} = \sum_{b \neq B} \frac{P_b^{tot}}{Aff(b, m_i)}$$

In this equation:

P_b^{tot} is the total transmission power of the base station of index b.
$Aff(b, m_i)$ is the path loss of the signal between the base station of index b
               and the mobile m_i.

In the uplink, the intercellular interference is due to the transmission of all the mobiles in the other cells. The transmission of a mobile attached to the base station of *index* B is perturbed by the following signal:

$$I_{inter}^{UL} = \sum_{j \notin B} \frac{P_j}{Aff(B, m_j)}$$

In this equation:

P_j is the transmitted power of a mobile of index j, attached to a
               base station other than B.
$Aff(B, m_j)$ is the path loss of the signal between the base station B and the
               mobile of index j.

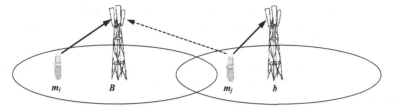

Fig. 8.6. Downlink intercellular interference.

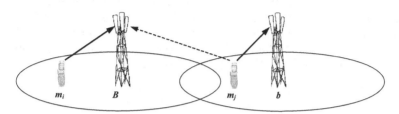

Fig. 8.7. Uplink intercellular interference.

Coverage and capacity

Capacity.

The profitability of a network is closely related to its capacity, i.e. with the quantity of information that can be exchanged simultaneously. In a mono-service context, the number of users defines the capacity. In a UMTS system, where several services are offered and where consumption in radio resources differs from one service to another, rather than relating to the number of mobiles, the capacity can be defined as the aggregate bit rate transmitted in the network for example. The maximum number of communications does not depend solely on the "hard" resources, namely the number of codes available, but also on the interferences, and therefore on the distribution of traffic in the network and its characteristics. Hence the concept of *"soft capacity"* can then be introduced.

Coverage.

A mobile is covered by the network if the three following conditions are satisfied:

- It can decode the information from the network. The mobile must receive at least a pilot signal with a sufficient quality. Hence, the term "pilot" coverage is used.
- The power necessary for the transmission of the station towards this mobile should be lower than the maximum power of a traffic channel. It can then be said that the mobile is covered in the downlink.

• The necessary transmission power of this mobile towards the base station is lower than the maximum transmission power of the mobile. The mobile is then covered in the uplink.

In the three cases, the coverage of a mobile strongly depends on the interferences, and therefore on the distribution of the traffic in the network. Thus, a base station that serves many mobiles can see that its zone of coverage is reduced. For example, in the figure 8.8, the station b is much more loaded than its neighbor c (continuous line) and its coverage is thus lower. This phenomenon is denoted as cell breathing. If overlapping with neighboring cells is not sufficient, then coverage holes will appear, and calls may be blocked or dropped. To avoid coverage holes related to the increase in traffic, admission control algorithms should be implemented. Hence coverage and capacity are strongly interdependent in the WCDMA networks.

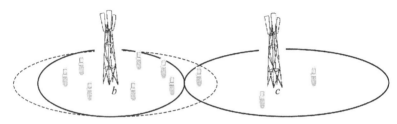

Fig. 8.8. Cell breathing. In continuous line, a situation in which the station b is much more loaded than the station c.

Macro-diversity.

A mobile connects to the station which offers the best quality of radio link for the pilot channel. When a mobile passes from one cell to another, the pilot of the first station weakens, and that of the second gradually grows. When the intensity of the two pilots is rather close ($3\,dB$ difference for example), a radio link is established with each of the two cells (see figure 8.9). The mobile is then attached to the two base stations at the same time, making it possible to combat, for example, the effects of fading and shadowing [Durrenbach et al., 2003], and to guarantee continuity of service, namely mobility between the two cells. The mobile recombines the two received signals, in order to extract the maximum information (the *Maximum Ratio Combining* algorithm). In the uplink, the mobile is received by the two base stations and the network reconstitutes the useful signal by evaluating the best signal received on the two links, at every moment (using a selection algorithm). In the particular case where the base stations belong to the same site, the recombination algorithms can be used in the uplink.

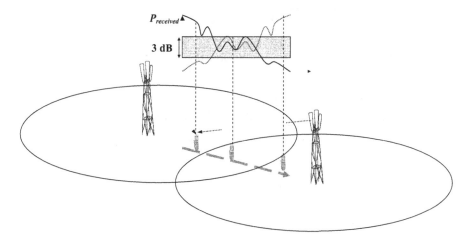

Fig. 8.9. Establishment of two radio links, or macro-diversity.

8.3 Definition of the optimization problem

In this section, we present the optimization of the parameter setting of the UMTS networks. In the first paragraph, the various phases of the process of planning are described, by specifying the role of an automatic tool for planning in this chain. In the second paragraph, the optimization problem is defined, by specifying the parameters and the objectives.

8.3.1 Radio planning of a UMTS network

The process of radio planning of UMTS networks can be divided into several stages presented in the figure 8.10. Initially, dimensioning consists in estimating the density of sites necessary for the coverage of the territory and choosing the sites among a set of possible sites. This choice is constrained by the infrastructure costs and the authorizations of installation of the antennas. For the 3G networks, the operators thus seek to maximize the reuse of the existing GSM sites.

Once the first phase is completed, the network is parameterized to serve the users as well as possible. The term *design* is often associated with this stage. In order to satisfy a significant traffic demand with a good quality of service, the operator must parameterize the network in order to minimize interferences. This optimization is a difficult problem that involves hundreds of parameters. Hence, within this framework, the use of an automatic tool for planning is relevant.

The automatic tool for cell planning, the ACP, involves the search for an optimal set of parameters that allows the network to adapt itself as well as possible to satisfy the traffic demand. The evolutions of the network such

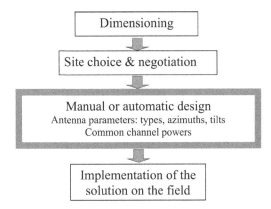

Fig. 8.10. Stages of the planning of a mobile network.

as the introduction of a new service, or the appearance of zones of strong traffic (*hot spots*), can make the existing parameter setting nonoptimal. It is thus necessary to adapt the parameter setting by using the ACP. In addition, it is possible that the optimization problem does not have a solution with acceptable quality. It is then essential to add base stations to ensure coverage, which is known as network densification. If the optimization of the parameter setting is more localized (i.e. modifications centered around a particular site), one can also use the automatic tool for planning. Thus, as the figure 8.11 shows, the ACP receives an initial configuration of the network as an entry and then determines an optimized network. The initial network can result from an initial deployment or *roll out*, from a process of manual dimensioning or from an operation of densification. The solution suggested by the ACP is then evaluated, before being implemented on the terrain.

8.3.2 Definition of the optimization problem

The optimization problem is defined by the parameters to be adjusted and the objectives to be optimized.

The optimization parameters

Two types of parameters can be distinguished: the antenna parameters and the system parameters.

Antenna parameters.

Three parameters characterize the antenna of a tri-sectoral site: the type of antenna, its tilt and its azimuth.

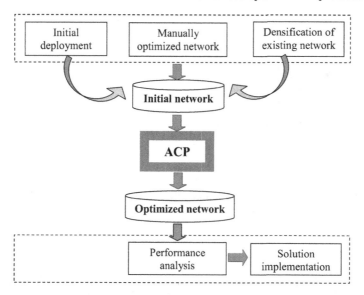

Fig. 8.11. Utilization scenarios of a cell planning tool in the design process.

Type of antenna: The antenna type influences the extent and the form of the zone covered by the base station. It also has an effect on the interference created on the nearby cells. For sectorial sites for example, one uses reflector antennas, which are characterized by the width of the principal lobe (beam) of their radiation pattern in two perpendicular planes. A pool of various types of candidate antennas is defined from which the genetic algorithm chooses the antennas.

Tilt: The tilt of an antenna corresponds to its angle of inclination in a vertical plane. By "tilting" the antenna downwards (corresponding to an increase in the angle of the tilt), the zone covered by the antenna decreases and the intensity of the average power received in the cell increases. The influence of the tilt on the surface covered by the antenna is illustrated in the figure 8.12. The tilt can be modified mechanically (the angle of the antenna changes physically), or electrically, by modifying the radiation pattern of the antenna, without changing its inclination. Tilting an antenna requires human intervention on the site, which introduces a cost related to this parameter modification. In the near future, a remote electrical tilt will be possible.

Azimuth: The azimuth angle corresponds to the orientation of the principal lobe of the antenna in the horizontal plane. Ideally, in a tri-sectorial site, the orientation of the antennas is separated by $2\pi/3$. A modification of the azimuth angle can be useful following a masking effect due to the landscape or buildings, which can cause undesirable reflections and interferences.

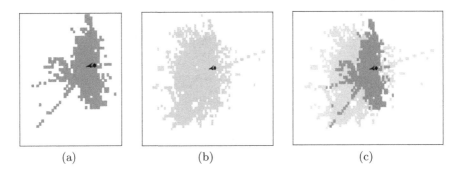

(a)   (b)   (c)

Fig. 8.12. Zone covered by an antenna tilted by 10 (a), by 6 (b), and the super-position of both coverages (c).

System parameters.

Pilot powers: The power of the pilot channel is the only system parameter considered in the optimization. The pilot power indicates to the mobile the cell to which it should be attached, thus defining the area of the cell. According to the received pilot powers, the mobile decides how many stations it will be attached to. This renders, on one hand, the possible mobility throughout the network, and, on the other hand, reinforces the radio link.

Optimization objectives

The objectives of the optimization are the following:

- to satisfy the objectives of coverage for the various services and the various situations (*outdoor, in car, indoor...*);
- to maximize the capacity of the network;
- to dimension the zones of macro-diversity, so as to ensure the continuity of service;
- to improve the quality of service (QoS);
- to minimize the cost related to the implementation of the solution. An economic cost is associated with each type of parameter modification.

A cost function incorporating the relevant objectives (criteria) is used. The choice of the criteria and the way to aggregate them depend on the strategy implemented by the operator. For example, for the criterion of coverage, the priorities granted to the various services influence the choice of the services to optimize. In a similar way, the compromise desired between coverage and capacity intervenes in the choice of the criteria and their weightings.

8.4 Application of the genetic algorithm to automatic planning

The general principle of the automatic tool for planning is described in figure 8.13. A genetic algorithm guides a generation of networks towards a global solution corresponding to a global optimum or a good local optimum. At each iteration, a fast UMTS network evaluator calculates the various quality criteria which allows "to give a mark" to the suggested network, namely to assign a fitness to the parameter settings. Then, the GA uses these "notes" and proposes a new set of parameters for the network.

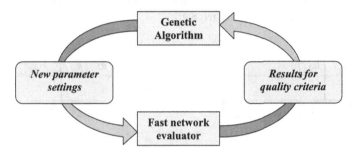

Fig. 8.13. General diagram of the automatic tool for planning.

8.4.1 Coding

To implement the GA, the coding of information is carried out in the following manner: a chromosome corresponds to a particular set of parameters of the network. Each chromosome consists of several genes. A gene corresponds to the parameter setting of a particular base station. It is thus a quadruplet consisting of the type of antenna used, its tilt, its azimuth and the power of the pilot channel (figure 8.14).

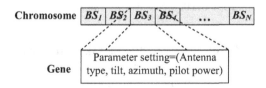

Fig. 8.14. Coding of the various parameter settings of the network.

8.4.2 Genetic operators

Selection

The traditional selection operator, the *roulette wheel selection* is used, guaranteeing the increase in the number of high quality individuals from generation to generation.

Crossover

The traditional crossover consists in choosing, at random, an index of station in the network, and exchanging the parameters of the stations on both sides of this point, as shown in the figure 8.15.

Fig. 8.15. Crossover at a point.

This crossover operation can be carried out at two points of the chromosome, which consists in exchanging the configurations (parameters) of the stations ranging between two indices chosen at random (shown in figure 8.16).

Fig. 8.16. Crossover at two points.

These two examples of crossover do not take into account any information on the geographical proximity of the various stations in consideration, since the indices of the stations in the chromosome do not reveal their relative positions. It would be more judicious, for our problem, to exchange the configurations between two chromosomes for the same zone of the network. A method of geographical crossover is proposed in [Meunier et al., 2001]. It suggests to choose a site at random, and to exchange the configurations of all the stations contained within a circle with this chosen site at the center, for a given radius, which can also be modified in a random manner (see figure 8.17).

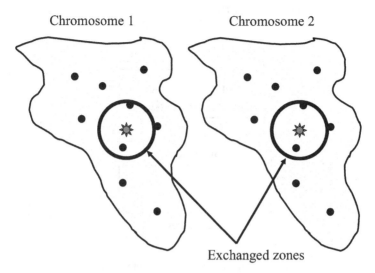

Fig. 8.17. Geographical crossover.

Mutation

The mutation operator acts in two phases (figure 8.18). It initially chooses the gene to be modified, i.e. the base station, then a parameter of this base station.

8.4.3 Evaluation of the individuals

The evaluation of the individuals is carried out by a *statistical evaluation* module of the quality of UMTS networks. This module enables us to carry out a large number of simulations in a short time, making it possible to compare the quality of the networks for a large number of different configurations.

During the evaluation, the basic quantities of a UMTS network, such as the transmitted powers of the mobiles and base stations, the loads and the interferences in both uplink (UL) and downlink (DL), are calculated taking into account traffic distribution in the network, macro-diversity, load control, etc. These quantities influence the calculation of the cost function.

8.5 Results

In this section we describe an example of application of the GA to the optimization of a UMTS network. The studied network consists of 172 base

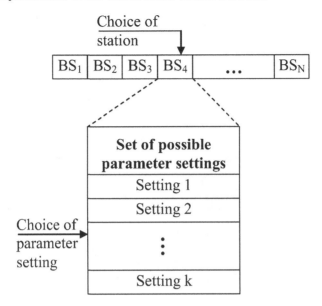

Fig. 8.18. The mutation operator acts in two steps.

stations, on tri-sectoral sites (figure 8.19). It is characterized by a heterogeneous, urban and dense urban environments, and an inhomogeneous traffic, which is denser in the zone of strong density of sectors (shown by the zone on the right of the figure). An initial network, manually optimized, is introduced as the input to the ACP. The hilly landscape shown in the figure renders the task of manual optimization particularly difficult.

For each base station, three parameters are optimized: the type of antenna, selected from six candidate antennas, the tilt and the powers of the pilot channels. A total of 516 parameters are set for possible modification by the GA. In this example, we assume that the azimuths of the antennas are fixed, the values being imposed by the constraints of cohabitation with an existing GSM network. For a problem of this size, the GA converges in approximately 3−4 hours on a UNIX workstation. This relatively short optimization time can be attributed to the ultra-rapid evaluation algorithms of the network, which ensure that the evaluation time of the cost function is particularly short. It can also be attributed to the landscape of the of the solution space.

The GA modifies the parameters of the majority of the base stations. The impact of optimization on the tilts of the network is presented in the figure 8.20. For the initial network (shown in black), the tilts of approximately half of the antennas have an identical value, which result in a significant peak in the center of the histogram. High tilts make it possible to ensure the network coverage. The results of the optimized network show that the tilts of the majority of the antennas have been adjusted. Indeed, the large number of

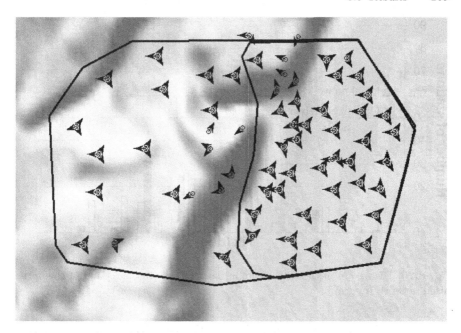

Fig. 8.19. UMTS network with 172 base stations in a heterogeneous environment, urban and dense urban. The majority of the sites are tri-sectoral (three base stations are installed on the same site).

configurations tested by the GA produces a wide spread histogram. Similar results are obtained for the powers of the pilot channels: spreading out of the histogram and reduction of the powers of the majority of the pilot channels.

The adjustments of the tilts, of the powers of the pilot channels and the choice of best alternative for the antennas lead to a reduction of the interferences and an optimization of the traffic allocation to the base stations of the network. The improved usage of the radio resources results in a significant gain of capacity and quality of service. Next, the quality of the optimized network is evaluated and the improvement brought about by the optimization is analyzed.

8.5.1 Optimization of the capacity

The capacity is a central indicator of the profitability of a network, and should be evaluated carefully. For a single service, the capacity is proportional to the number of served mobiles. For a multi-service traffic, one can use either the aggregate bit-rate, or the rate of successful transmissions. For packet mode, erroneous bits are retransmitted.

The figures 8.21 and 8.22 present the evolution of the traffic demand as a function of the satisfaction rate for $64\,kb/s$ and voice services respectively.

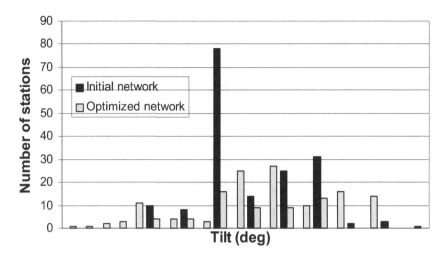

Fig. 8.20. Histogram of the tilts of the initial network (in black) and of the optimized network (in gray).

The product of the traffic demand and the satisfaction rate indicates the capacity of the network. For a given satisfaction rate, the capacity of the optimized network increases by approximately 30 % compared to the initial network for the 64 kb/s service, and by 10 % for the voice service. The optimization has been carried out for a target (dimensioning) service with high bit rate, which explains the more significant gain obtained for the 64 kb/s service. The optimized network leads to comparable capacity gains for various levels of traffic demand. This result is important because it shows the robustness of the solution obtained by the GA with respect to the traffic.

8.5.2 Optimization of the loads

The improved usage of radio resources for the optimized network is translated into a lower average load value of the base stations of the network. Conversely, for a given average load value, the optimized network can serve more traffic. The variation of the served traffic as a function of the average network load is presented in the figures 8.23 and 8.24. The improvement in DL is particularly significant due to the optimization of the pilot channel powers.

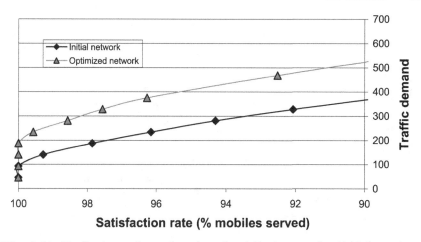

Fig. 8.21. Traffic demand as a function of satisfaction rate for 64 kb/s service.

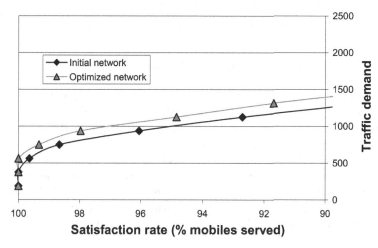

Fig. 8.22. Traffic demand as a function of satisfaction rate for the voice service.

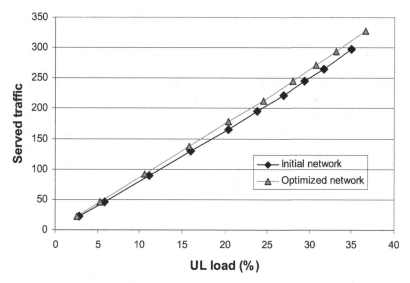

Fig. 8.23. Served traffic as a function of the average UL load.

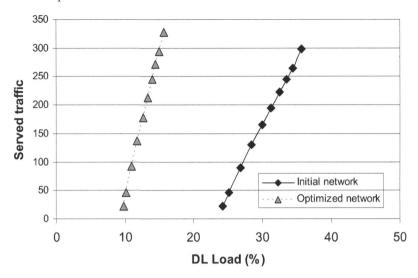

Fig. 8.24. Served traffic as a function of the average DL load.

8.5.3 Optimization of the intercellular interferences

The optimization of the network and the gains in capacity are closely related to the reduction of the interferences. It is thus particularly interesting to observe the interference maps in the network. Figures 8.25a and 8.25b present the intercellular interferences of the initial and the optimized networks respectively. The gray levels correspond to the received powers of interference (figure 8.25c). The clearer interference map of the optimized network highlights the significant reduction of the intercellular interferences. Figure 8.25d illustrates that the proportion of highly interfered surfaces is considerably reduced in the optimized network.

8.5.4 Optimization of the coverage

The coverage of the network is satisfied for the initial and the optimized networks for both UL and DL and for the three services studied: voice, 64 and $144kb/s$. The analysis of the results shows that the network is downlink limited, namely part of surface is on the limit of coverage. The DL coverage for the $144\,kb/s$ service for the initial and optimized networks are presented in figures 8.26a and 8.26b respectively. The gray level at a given point represents the power of the traffic channel necessary to cover a mobile located at this point (see figure 8.26c). The shift towards clear colors for the optimized network illustrates the improved usage of the radio resources. The proportion of surface on the limit of coverage is presented in figure 8.26d. The surface on the limit of coverage has practically disappeared for the optimized network.

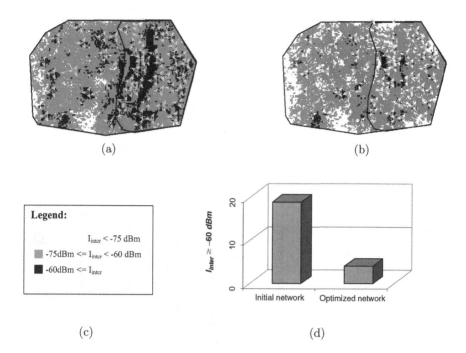

Fig. 8.25. Intercellular interferences of the initial network (a) and the optimized network (b). The legend is presented on (c) and the comparison between the highly interfered surfaces with $I_{inter} \geq -60\,dBm$ on (d).

8.5.5 Optimization of the probability of access

The probability of access to the network is a significant indicator of the quality of service offered. A mobile can establish a communication if it is covered in UL and DL by a traffic channel and by the common channels. Figure 8.27 presents the probability of access for the voice service with a penetration margin of $15\,dB$ added to the signal path-loss. This margin takes into account an additional attenuation corresponding to the penetration inside buildings. The gray levels of the meshes represent the probability of access defined by the legend (figure 8.27c). The proportion of surface having a probability of access below 98 % has decreased considerably for the optimized network. The comparison between the surfaces with probability of access below 98 % is presented in figure 8.27d.

The study of this example has highlighted the fundamental contribution of optimization for the improvement of the network performance. Automatic planning makes it possible to improve usage of the available radio resources,

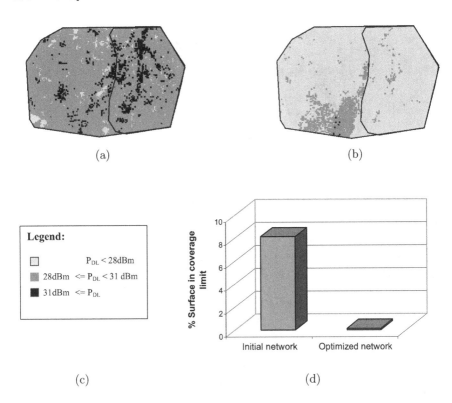

Fig. 8.26. DL coverage maps of the traffic channels for the initial network (a) and the optimized network (b). The legend is presented on (c) and the comparison of surfaces on the limit of coverage on (d).

to decrease the level of interference in the network, to increase its capacity and to improve the quality of service.

8.6 Conclusion

This chapter has described the use of the Genetic Algorithm for the optimization of UMTS networks. The Genetic Algorithm adjusts the antenna parameters and the powers of the pilot channels of the network base stations. The number of parameters to be optimized varies typically between a few tens and several hundreds. An efficient evaluator has been developed to allow very fast calculation of the quality criteria of the network from which the cost function is computed. The large number of parameters to be optimized on the one hand and their interdependence on the other hand make the development

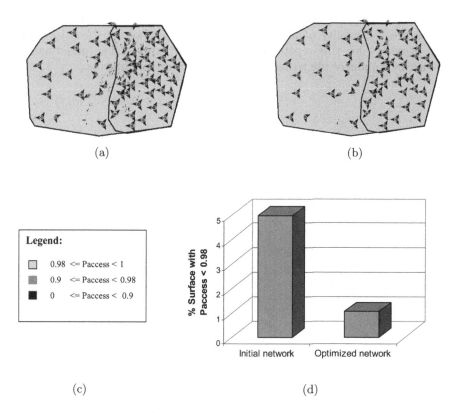

Fig. 8.27. Probabilities of access for the initial network (a) and the optimized network (b). The legend is presented in (c) and the percentages of surface with a probability of access lower than 98% in (d).

of optimisation solutions based on deterministic heuristics particularly difficult and render the Genetic Algorithm a natural tool, well adapted for this problem.

The optimization of the network allows to considerably improve its performances in terms of capacity, coverage and quality of service. For an inhomogeneous network, the gain obtained by automatic optimization will be more obvious. Furthermore, for networks with inhomogeneous traffic, heterogeneous environments or with hilly landscape, manual optimization is particularly difficult. In these cases, the optimal solutions diverge from a homogeneous parameter setting and the optimization gains, obtained by the Genetic Algorithm, are considerable.

Finally, the optimization of the network by the Genetic Algorithm allows to improve the quality of manually optimized networks and to reduce the time necessary for this repetitive task by radio experts. The performance enhancement induces the reduction of infrastructure investments.

9

Genetic Algorithms Applied to Air Traffic Management

Nicolas Durand and Jean-Baptiste Gotteland

Laboratoire d'Optimisation Globale, Centre d'Études de la Navigation Aérienne, École Nationale de l'Aviation Civile
`durand@recherche.enac.fr`

The constant increase in air traffic, since the beginning of the commercial aviation, has led to problems of saturation on airports, approaching areas, or higher airspace. Whereas the aircraft are largely optimized and automated, the air traffic control is still essentially relying on human experience.

The present case study details two problems of air traffic management (ATM) for which a genetic algorithm based solution has been proposed. The first application deals with the *en route* conflict resolution problem. The second application deals with the traffic management problem in an airport platform.

9.1 *En route* conflict resolution

An air traffic control (ATC) can be represented by a set of filters, where each filter has a specific objective and manages distinct spatial and temporal horizons. One can coarsely distinguish five levels:

In long term (more than 6 months), the traffic is organized in a macroscopic manner. For example, here people are concerned with the traffic orientation diagrams, the measures of the committee regarding the hourly schedule or the inter-centre agreements and the agreements with the military who allow their own air zones to be used for civil aviation during peak periods e.g. in the Friday afternoon.

In shorter term, pre-regulation is often talked about: it consists in organizing a day of traffic, on the day before or two days before. At this stage, one has a clear idea about most of the flight plans, the control capacity of each center [1] is well known. The maximum flow of aircraft that can penetrate in one sector[2] is called the sector capacity. This job is performed by the CFMU [3].

The very same day, adjustments are carried out according to the last events. The transatlantic traffic, for example, is taken into account at this stage. Airways, taking off hours are adjusted, unused time slots are re-allocated, weather conditions are taken into account. Generally, this job is performed by the FMP [4] in each center.

The last filter is the tactical filter: it deals with the control inside a sector. The average time spent by an aircraft in a sector is about fifteen minutes. Here, the visibility of the controller is a little higher as it receives the flight plans a few minutes before the entry of the aircraft into the sector. The controller ensures the task of monitoring, resolves conflicts and performs coordination with the neighbouring sectors. In this context, it is desirable to specify the definition of a conflict: two aircraft are known to be in conflict when the horizontal separation distance between them is lower than 5 nautical miles[5] and their difference in altitude is lower than 1000 feet[6]. The methods used by the controllers to resolve conflicts are mostly based on previous experience and very rarely require any creative knowledge. When several such couples of aircraft interact in the same conflict, they start by simplifying the problem in order to have only elementary conflicts to solve.

The emergency filter is not supposed to intervene except when the control system is found missing or is weakened: for the controller, the *safety net* predicts the trajectory of each aircraft with a temporal horizon of a few minutes, using the last radar positions and the continuing algorithms, and sets off an alarm in the event of a conflict. It does not propose a solution for the detected conflicts. On board, the TCAS[7] is supposed to avoid such a collision. The temporal prediction is less than a minute (between 25 and 40 seconds). It is then too late for the controller to maneuver the aircraft

---

[1] In France the air traffic control is de-centralized into five centers (Paris, Reims, Brest, Bordeaux and Aix en Provence), each center manages a part of the french airspace.

[2] Each control center manages between 15 and 20 elementary sectors which can be gathered according to the traffic density and the available controller teams.

[3] "Central Flow Management Unit", located in Brussells.

[4] "Flow Management Position"

[5] 1 nautical mile is equivalent to 1852 meters

[6] one foot is equivalent to 30.48 cm.

[7] "Traffic alert and Collision Avoidance System": an embarked system for aircraft which was made mandatory by the United States for all aircraft carrying more than 30 passengers.

as it is estimated that it requires a minimum duration of 1 to 2 minutes to analyze a situation, to find a solution and to communicate it to the aircraft. Currently, the TCAS detects the surrounding aircraft and delivers an advise for solving the conflict to the pilot (for the moment in the vertical plane). This filter must solve the non-foreseeable conflicts like, for example, a plane exceeding its assigned flight level, or a technical problem which would significantly degrade the performances of the aircraft.

The application proposed in this section deals with the tactical filtering: knowing the positions of the aircraft at a given moment and their future positions (with a given precision), which are the maneuvers to be ordered to these aircraft so that the trajectories do not generate any conflict and minimize the generated delay.

The solution is based on a certain number of assumptions.

- An aircraft cannot modify its speed (or can, but slightly), except during the period of descent.
- It cannot be considered that an aircraft flies at a constant speed, except possibly when it is leveled and when there is no wind. Moreover during climb and descent, its trajectory is not rectilinear. Hence it is almost impossible to analytically describe it. The evaluation of the future positions of an aircraft requires the use of a simulator.
- Aircraft are constrained in their turning rates, generally pilots prefer lateral maneuvers than vertical maneuvers, except during climb or descent.
- Although, nowadays autopilots are largely more powerful than human pilots (in normal flight situations), for the moment it does not appear realistic to consider those trajectories which are not achievable by a human pilot.
- Uncertainties in climbing and descending rates are very large (between 10% and 50% of the vertical speed). During cruising, uncertainty in speed is reduced (in the vicinity of 5%). Laterally, uncertainty does not grow with time, as an aircraft, in general, holds its altitude quite well during cruising.

As trajectory prediction can only be done by a simulator, it is almost impossible to search for the analytical solutions for the conflict resolution problem and the implementation of traditional optimization methods employing gradient based techniques or the Hessian criterion is impossible. However, the principal difficulty arises essentially from the complexity of the problem itself.

The first part of this chapter is devoted to the introduction of some definitions which facilitate to understand the complexity of the conflict resolution problem. The second part is devoted to a short history of the algorithms tested for this problem and their limitations. The third part details the modeling of the problem. The development of the genetic algorithm for the problem is detailed in the fourth part, which is followed by the numerical results obtained.

9.1.1 Complexity of the conflict resolution problem

A conflict can be defined as follows:

> A potential conflict is defined as a conflict between two aircraft during a given time window of trajectory forecast, taking the uncertainties in trajectories into account.

The relation "is in conflict with", or "is in potential conflict with", defines an equivalence relation. The equivalence classes associated will be called "clusters" of conflicting aircraft or simply "clusters".

A cluster of size n can involve up to $\frac{n\,(n-1)}{2}$ potential conflicts. Considering only the horizontal plane, it was shown [Durand, 1996] that the entire set of the acceptable solutions contains $2^{\frac{n\,(n-1)}{2}}$ connected components, under the assumption that a local optimization method (continuous deformation of trajectories) has been used which requires as many executions of the search algorithm. Thus, for a 6 aircraft cluster, it represents 32768 connected components. In practice, if the performances of the aircraft are taken into account, all the connected components do not need to be explored. Nevertheless, the theoretical presence of as many disjoined sets and the remote possibility of knowing a priori which set contains the optimal solution make the problem strongly combinatorial. By relaxing the separation constraint, the problem becomes similar to a global optimization problem comprising at least as many local optima as the connected components.

The addition of the vertical dimension does not reduce the combinatorial character of the problem as one does not simultaneously propose a maneuver in the vertical and the horizontal planes.

9.1.2 Existing resolution methods

Operational approaches

The first air traffic control automation project (AERAIII[8] [Niedringhaus, 1989]) was American and appeared in the beginning of the 80s, but it was unable to solve clusters of size 3 or more. The European project ARC2000[9] [Fron et al., 1993] proposed a method of continuous deformation of four-dimensional tubes to optimize the trajectory of the $n + 1^{th}$ aircraft in an environment of n previously computed trajectories. This modeling did not take uncertainties into account and was unable to deal with high traffic densities. It could not find the global optimum for large clusters and only used an iterative method (the trajectory of the first aircraft is defined first, then that of the second, by considering the trajectory of the first plane as a constraint and so on...). Finally the Experimental European project FREER[10] [Duong and Faure, 1998], implemented in 1995, proposed to offset

---

[8] "Automated In-Road Air Traffic Control"

[9] "Automatic Radar Control for the 21st Century"

[10] "Free-Road Encounter Resolution"

the task of solving the conflict aboard aircraft. The problem of coordination between aircraft is managed by employing priority rules, which is like using an iterative method as in ARC2000. It was unable to deal with large clusters [Granger, 2002].

Theoretical approaches

Among the theoretical approaches employed for the problem, we can first mention the reactive techniques of Zeghal [Zeghal, 1994]. According to this method, aircraft are "attracted" by their objective and pushed back by the close aircraft. The method seems to perform satisfactorily when the density is low, but becomes chaotic when the traffic is dense. In addition, the model assumes that flights are completely automated, as trajectories can be continuously modified. Similar approaches using the potential fields are tested by the aeronautics department of Berkeley [Gosh and Tomlin, 2000], but for the moment they are unable to solve more than 3 aircraft clusters. This was very similar to the performance shown by the neural networks based methods tested in LOG(CENA-ENAC) [11] [Durand et al., 1996] which could not be extended to the cases of complex clusters. Lastly, among the global approaches for complex clusters (involving more than five aircraft), the first major work was done by Feron [Frazzoli et al., 1999] . He used semi-definite programming to determine the direction of resolution for each pair of conflicting aircraft. Then, a convex optimization method involving convex constraints is implemented to calculate the maneuvers. However, this method does not give an acceptable solution in all cases. The addition of a random noise helps to improve the success rate. The simplified framework of the selected model (assuming constant speed, horizontal maneuvers, no uncertainty...) leaves little scope for its successful application in complex situations. Lastly, the LOG also tested an Interval "Branch and Bound" type method [Médioni, 1998], which could optimally solve the problem for small clusters (four aircraft), but was not able to extend it successfully to more significant clusters, the size of the search space becoming too significant.

Till date, only a genetic algorithm based algorithm could solve large clusters (up to thirty aircraft) in a reasonable time.

9.1.3 Modeling of the problem

Taking uncertainty into account

First of all, a detection time window T_w is defined and a simulator evaluates the future positions of the aircraft in the time window. The simulator takes into account uncertainties in the horizontal and vertical speeds of the aircraft,

---

[11]Global Optimization Laboratory of the Centre d'Etudes de la Navigation Arienne Navigation and the Ecole nationale de l'Aviation Civile

as shown in figure 9.1. Time is discretized in practice employing 15 seconds of sampling steps. In the horizontal plane, the aircraft is represented by a point at the initial moment. In due course of time, this point becomes a segment whose length keeps on increasing. When direction is changed (at $t = 4$), the segment gets deformed while following the new speed vector . The aircraft is then represented by a parallelogram. Implementing a new change of heading (at $t = 7$) transforms the parallelogram into a hexagon and, more generally speaking, into a "convex". In the vertical plane, a cylinder can be defined whose height grows with time. When the plane reaches its requested flight level (at $t = 8$), the top of the cylinder does not change its altitude any more and the bottom of the cylinder continues to go up until the flight level is reached.

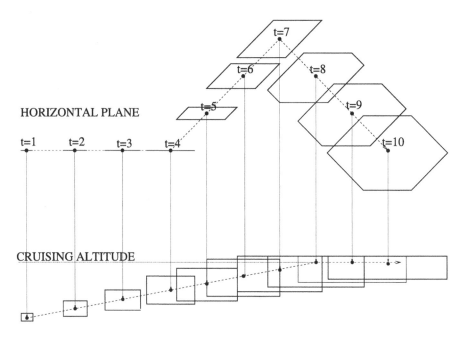

Fig. 9.1. Modeling the uncertainty.

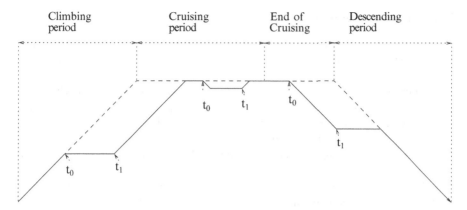

Fig. 9.2. Maneuvering in the vertical plane.

Conflict detection

To detect potential conflicts between aircraft, we need to measure, at each time step, the horizontal distance between the convexes and the vertical distance between the cylinders representing the two aircraft. A conflict occurs when the vertical and the horizontal standards are simultaneously violated.

Modeling the maneuvers for avoidance

In order to respect both pilots and aircraft performances we define simple maneuvers: in the horizontal plane, a maneuver is a heading change of 10, 20 or 30 degrees to the right or to the left. The maneuver begins at time t_0 and ends at time t_1. In the vertical plane, the maneuvers proposed depend on the phase of flight in which the airplane is. Thus, as shown in figure 9.2, when the aircraft is climbing, it can stop its climb at t_0 and resume its climb at t_1. In the cruising phase, it can descend to the nearest lower flight level (1000 feet down) at t_0 and join the initial flight level at t_1. When the aircraft is less than 50 nautical from the beginning of its descent, it can anticipate its descent at t_0 and stop descending at t_1 to join its trajectory of descent. In order to make the maneuver achievable, only one maneuver is given to the pilot at a time. A new maneuver could be proposed to him only when the first maneuver is finished.

A maneuver is thus modeled by three variables. The first is a discrete variable indicating the type of maneuver (10, 20, 30, −10, −20, −30 degrees, or vertical maneuver), the two others, t_0 and t_1, are integer variables indicating the beginning and the end of the maneuver. A resolution of a n aircraft cluster is thus modeled by $3\,n$ variables.

Real time management

The resolution is operated on the forecast time window T_w (a fixed value, chosen between 10 and 15 minutes) and the situation is updated each δ minutes (2 or 3 minutes in practice). The figure 9.3 details the real time modeling. Three periods are distinguished in the time window. The first one is the time duration of δ minutes which is called the locked period. No modification of trajectory can be effected during this period. Indeed, during the time necessary for evaluation of the situation, the resolution of possible conflicts and the transmission of the orders of maneuvers, the aircraft continue to fly. It is consequently not possible to modify their trajectories. The following period is called the final period, because the orders of maneuvers given for this period could not be modified during the next iteration. The last period is the period of predicted maneuvers. These maneuvers will be reconsidered during the next iteration. Because of uncertainty, certain conflicts can disappear when one approaches the point of conflict.

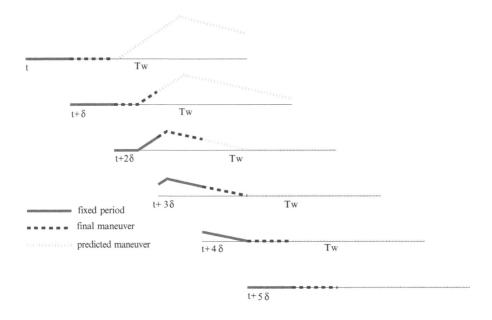

Fig. 9.3. Modeling in real time.

The traffic simulator

The traffic simulator controlled CATS[12] is an arithmetic simulator which uses a tabulated model to make aircraft fly. It takes into account flight plans for

---

[12]Complete Air Traffic Solver

one day of traffic. Every δ minutes (2 or 3 minutes in practice), the simulator forecasts the trajectories for the next T_w minutes. It detects the conflicts for each pair and then builds the clusters of aircraft in conflict. Each cluster is solved by a solver using a genetic algorithm which proposes maneuvers for the aircraft. A new forecast for the trajectories, taking the modified trajectories into account, is then carried out in order to detect possible conflicts between two aircraft, not belonging to the same cluster. When two aircraft of two different clusters are in conflict, the two clusters are joined together and a new resolution is operated. If a non conflicting aircraft interferes with an aircraft of a cluster, it is integrated into the cluster and a new resolution is operated. The process is repeated as many times conflicts remain between aircraft not belonging to the same cluster.

9.1.4 Implementation of the genetic algorithm

The function to be optimized for each cluster takes several different criteria into account in order to:

- ensure all separations between aircraft.
- minimize delays.
- minimize the number of maneuvers and the number of aircraft undergoing maneuvers.
- minimize the duration of maneuvers so that the aircraft are freed as soon as possible.

General description

The genetic algorithm implemented is a simple algorithm as described in [Goldberg, 1994].

An initial population of $3n$ variables is randomly created (the size of the population being proportional to the number of aircraft with a maximum of 200 individuals). Then the fitness of each individual (representing a configuration of maneuvers) is evaluated. The best individuals are then reproduced and selected according to their adaptation (the selection technique used is the "Stochastic Reminder Without Replacement"). A part of the population (50 %) is then crossed: from two "parents", two "children" are created; they replace the parents in the population. Then a certain number of individuals undergo mutation (15 %). The mutation generally consists in modifying the maneuver of an aircraft in the cluster. The distance used to distinguish two configurations for the "sharing" operator is simple. Two maneuvers are considered equal if they are both in vertical or in horizontal direction and, in the later case, if they are carried out to the same side. To measure the distance between the two configurations, the number of different maneuvers are computed. An elitism process is used: at each generation, the best individuals of the population are preserved so that they do not disappear during a crossover or a mutation.

Taking the temporal requirements imposed by the real time traffic management into account, the termination criterion used consists in stopping the optimization procedure at the end of a certain number of generations (generally twenty). However this number keeps increasing if the algorithm is unable to find a solution without conflict (the maximum number of generations is kept limited to forty).

The horizon effect

The solver has only one short-term vision of the aircraft trajectories. With the cost function simply consisting in limiting the delay generated by a maneuver, the solver is sometimes tempted to defer a conflict beyond the temporal window without solving it. In order to counter this "horizon effect", one can measure the effectiveness of the resolution of a conflict and modify the fitness function of the algorithm for resolving the conflict.

For any pair of aircraft under consideration in a cluster:

- If the aircraft are not in conflict, it is not necessary to penalize the cost function.
- Else, if the trajectories between the current positions of the aircraft and their destinations cross, the cost function is penalized when the aircraft are still not crossing each other at the end of the time window.

The fitness function

For each configuration, a matrix F of size $(n \times n)$ is used to store the following information:

- The diagonal term $F_{i,i}$ measures the lengthening of aircraft i's trajectory. It is zero if no maneuver is given to aircraft i.
- The term $F_{i,j}$ with $i < j$ measures the violation of separation between aircraft i and aircraft j. It is zero when the two aircraft are not conflicting.
- The term $F_{i,j}$ with $i > j$ measures the effectiveness of the conflict resolution between aircraft i and aircraft j.

The fitness function chosen is:

$$\exists (i,j),\ i \neq j,\ F_{i,j} \neq 0 \Rightarrow F = \frac{1}{2 + \sum_{i \neq j} F_{i,j}}$$

$$\forall (i,j),\ i \neq j,\ F_{i,j} = 0 \Rightarrow F = \frac{1}{2} + \frac{1}{1 + \sum_i F_{i,i}}$$

It guarantees that a configuration without conflict has a better fitness than a configuration with one or more conflicts remaining.

As described above, the genetic algorithm could hardly solve large clusters, but it was shown that the use of the partially separable structure of the fitness function makes it possible to define crossover and mutation operators adapted for the problem.

Use of the partial separability

Let us consider the minimization problem of a function F of n variables $x_1, x_2, \ldots, x_n$, sum of m terms F_i, each of which depends only on a subset of the variables of the problem.

Such a function (that is denoted as partially separable) can be expressed as:

$$F(x_1, x_2, \ldots, x_n) = \sum_{i=1}^{m} F_i(x_{j_1}, x_{j_2}, \ldots, x_{j_{n_i}})$$

The adapted crossover operator

The intuitive idea is the following: for a completely separable problem, the global minimum is obtained when the function is separately minimized for each variable. In this case, the function to be minimized can be written as:

$$F(x_1, x_2, \ldots, x_n) = \sum_{i=1}^{n} F_i(x_i)$$

Minimizing each function F_i leads to the global minimum of the function.

A crossover operator which chooses, for each variable x_i, among the two parents, the variable which minimizes the function F_i, creates an individual which is better than the two parents (or at least equal).

This strategy can be adapted for partially separable functions. To create a child starting from two parents, the idea is to choose, for each variable, the one among the two parents which minimizes the sum of the partial functions F_i in which it intervenes.

First, we define a *local fitness* $G_k(x_1, x_2, .., x_n)$ for variable x_k as follows:

$$G_k(x_1, x_2, .., x_n) = \sum_{i \in S_k} \frac{F_i(x_{j_1}, x_{j_2}, .., x_{j_{n_i}})}{n_i}$$

where S_k is the set of i such that x_k is a variable of F_i and n_i the number of variables of F_i.

The local fitness associated with a variable isolates the contribution of this variable in the global fitness.

When minimizing F, if:

$$G_k(parent_1) < G_k(parent_2) - \Delta$$

then child 1 will contain variable x_k of parent 1. Else, if:

$$G_k(parent_1) > G_k(parent_2) + \Delta$$

then child 1 will contain variable x_k of parent 2. If:

$$|G_k(parent_1) - G_k(parent_2)| \leq \Delta$$

then variable x_k of child 1 will be randomly chosen, or can be a random linear combination of the kth variable of each parent when dealing with real variables. If the same strategy is applied to child 1 and to child 2, children may be identical, especially if Δ is small. This problem can be avoided by taking a new pair of parents for each child.

Let us consider the following completely separable function:

$$F(x_1, x_2, x_3) = x_1 + x_2 + x_3$$

for x_1, x_2 and x_3 integers include in $[0, 2]$. Variable k's local fitness is: $G_k(x_1, x_2, x_3) = x_k$. Let us cross parents $(1, 0, 2)$ and $(2, 1, 0)$ which have the same fitness $F = 3$. With $\Delta = 0$, child 1 will be $(1, 0, 0)$: $F = 1$. With $\Delta = 1$, child 2 may be $(2, 1, 0)$, $(2, 0, 0)$, $(1, 1, 0)$, or $(1, 0, 0)$. The children's fitness are always better than the parents' fitness when $\Delta = 0$ which is not the case with a classical crossover operator.

As it is completely separable, this function is obviously too simple to show the interest of the adapted crossover operator. In the next paragraph, a simple partially separable function is introduced and the improvement achieved is theoretically measured.

9.1.5 Theoretical study of a simple example

Let us define the following function:

$$F(x_1, x_2, .., x_n) = \sum_{0 < i \neq j \leq n} \delta(x_i, x_j) \tag{9.1}$$

$(x_1, x_2, .., x_n)$ is a bit string and $\delta(x_i, x_j) = 1$ if $x_i \neq x_j$ and 0 if $x_i = x_j$. It must be noticed that the function is only partially separable and has 2 global minima, $(1, 1, 1, .., 1)$ and $(0, 0, 0, .., 0)$.

For $x = (x_1, x_2, ..x_n)$, we define the local fitness $G_k(x)$ by:

$$G_k(x) = \frac{1}{2} \sum_{i=1}^{n} \delta(x_k, x_i)$$

We define $I(x)$ as the number of bits equal to 1 in x. Then, it is easy to establish that:

$$F(x) = I(x)(n - I(x))$$

$$G_k(x) = \frac{I(x)}{2} \quad if \ x_k = 0$$

$$= \frac{n - I(x)}{2} \quad if \ x_k = 1$$

In the following discussion,we use a classical n point crossover operator; A_1 and A_2 represent 2 parents randomly chosen in a population and C represents their child.

In paragraph 9.1.5, the probabilities of increase of the fitness with the adapted or the classical crossover operator are compared. The interested reader will find a detailed study for this example in [Durand and Alliot, 1998].

Probability of improvement

For function (9.1), the probabilities of increase of the fitness with the classical or the adapted operator can be mathematically computed for every possible couple of parents.

Let us define $P_{1-1}(i, j, k)$ as the probability to find k bits equal to 1 at the same position in both parents A_1 and A_2, with $I(A_1) = i$ and $I(A_2) = j$. As $P_{1-1}(i, j, k) = P_{1-1}(j, i, k)$, we will assume in the following discussion that $i \leq j$. It can be shown that:

- if $k > i$, then:

$$P_{1-1}(i, j, k) = 0$$

- if $k \leq i$, then:

$$P_{1-1}(i, j, k) = C_i^k \prod_{l=0}^{k-1} \frac{j - l}{n - l} \prod_{l=k}^{i-1} \frac{(n - l) - (j - k)}{n - l}$$

The classical crossover used is the n point crossover that randomly chooses bits from A_1 or A_2 (the order of the bit string has no influence on the fitness).

For the adapted crossover (respectively for the classical crossover), let us define $P_a(i, j, k)$ (resp. $P_c(i, j, k)$) as the probability that if $I(A_1) = i$ and $I(A_2) = j$ then $I(C) = k$. As $P_a(i, j, k) = P_a(j, i, k)$ and $P_c(i, j, k) = P_c(j, i, k)$, we will assume int the following discussion that $i \leq j$. Then, it can be shown that for the classical crossover:

$$P_c(i, j, k) = \sum_{l=\max(0, \frac{i+j+1-n}{2})}^{\min(k, i+j-k)} P_{1-1}(i, j, l) \frac{C_{i+j-2l}^{k-l}}{2^{i+j-2l}}$$

For the adapted crossover(with $m = \min(k, n - k)$):

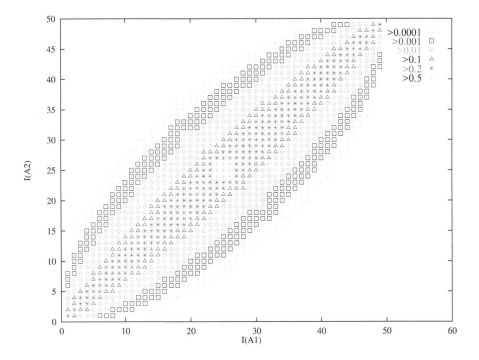

Fig. 9.4. $Prob(F(C) > \max[F(A_1), F(A_2)])$ according to $[I(A_1), I(A_2)]$ — traditional crossover — $n = 50$.

$$i + j < n : P_a(i, j, k) = P_{1-1}(i, j, k)$$
$$i + j > n : P_a(i, j, k) = P_{1-1}(n - i, n - j, n - k)$$
$$i + j = n : P_a(i, j, k) = \sum_{l=0}^{m} P_{1-1}(i, j, l) \frac{C_{i+j-2l}^{k-l}}{2^{i+j-2l}}$$

As $P_{1-1}(i, j, k) = 0$ if $k > \min(i, j)$, then:

- if $i + j < n$ and $k > \min(i, j)$, then $P_a(i, j, k) = 0$
- if $i + j > n$ and $k < \max(i, j)$, then $P_a(i, j, k) = 0$

Consequently:

- if $i + j < n$ and $P_a(i, j, k) > 0$,
  then $k < \min(i, j, n - i, n - j)$
- if $i + j > n$ and $P_a(i, j, k) > 0$,
  then $k > \max(i, j, n - i, n - j)$

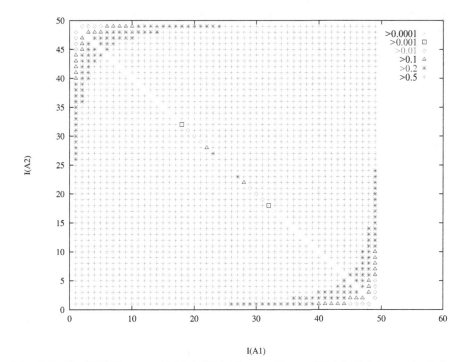

I(A1)

Fig. 9.5. $Prob(F(C) > \max[F(A_1), F(A_2)])$ according to $[I(A_1), I(A_2)]$ — adapted crossover — $n = 50$.

Thus, if $i + j \neq n$, then $F(C) \geq \max[F(A_1), F(A_2)]$. If $i + j = n$, local fitness of variables of each parent are equal and the adapted crossover behaves like a classical n points crossover.

Figures 9.4 and 9.5 present the probability for a child to have a better fitness than its parents (for all the possible combinations of the parents). On this example, the adapted crossover widely improves crossover efficiency. The small square in the center of the figure 9.4 represents a probability of improvement larger than 0.5. It becomes a very large square in figure 9.5.

Application for the problem of resolution of conflicts

For the conflict resolution problem, the "local fitness" associated with each aircraft is defined as follows:

$$F_i = \sum_{j=1}^{n} (F_{i,j})$$

The adapted crossover operator is described in figure 9.6. For each aircraft i, if the local fitness of aircraft i of parent A is definitely lower than that of parent B, then the maneuver of aircraft i of parent A is chosen for both children. In the opposite case (for example for aircraft 3), the maneuver of aircraft i of parent B is chosen for both children. When the local fitness are close, a combination of both maneuvers is used.

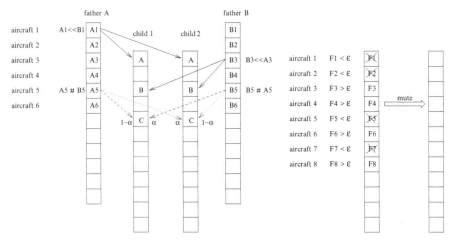

Fig. 9.6. The adaptive crossover and mutation operators.

An adaptive mutation operator is also used (figure 9.6). An aircraft is chosen among those whose local fitness are higher than a given threshold (for example, the aircraft which are still in conflict).

9.1.6 Numerical application

Example of complex conflicts

In this example obtained from a simulation, at 10h42 (figure 9.7), 5 aircraft are cruising at flight level 350 (35,000 feet). 4 conflicts are detected between aircraft A and B, B and C, C and D, D and E.

The genetic algorithm uses a population of 100 individuals. A solution without conflict is obtained after 5 to 10 generations (without the use of the adapted crossover operator, about sixty generations would be required)[13]. The algorithm is terminated 20 generations after the achievement of a solution without conflict (between the 25th and the 30th generation).

The best individual proposes to resolve the conflict with only two maneuvers: it proposes a descent of 1000 feet for aircraft B and D. However, as the

---

[13]These numbers were obtained by repeating the genetic algorithm a hundred times, each time starting from a different initial random population.

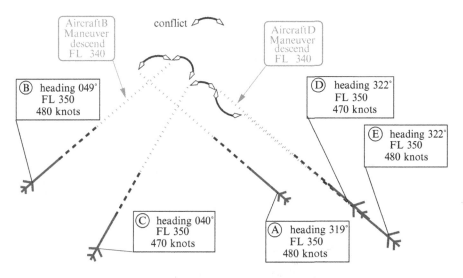

Fig. 9.7. Situation at 10h42.

maneuvers begin during the prediction period, they are not communicated to the pilots, because they can be modified 3 minutes later, at the next update. At 10h45 (figure 9.8), 5 aircraft are detected with 5 conflicts (4 preceding conflicts and a new additional conflict between aircraft C and E). The maneuvers previously calculated do not solve the conflict between the aircraft C and E. The resolution algorithm proposes, from now on, 3 maneuvers of which one takes effect during the final period (turn to the left for the aircraft D) and two become effective during the prediction period. Three minutes later, because of the reduction of uncertainty, the conflicts disappeared. Finally, only aircraft D underwent a turn to the left during one minute. A complete simulation over one day of traffic is proposed in the following paragraph.

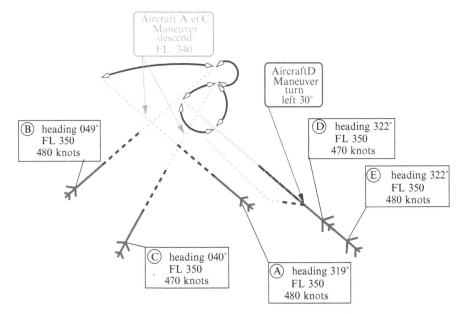

Fig. 9.8. Situation at 10h45 (3 minutes later).

Statistics over one day of traffic

The results obtained with the simulator over one day of traffic in the French airspace (Friday May 21 1999: 7540 flights carried out) are demonstrated in this paragraph. More complete results can be found in [Granger, 2002]. The simulation was carried out with three levels of uncertainties:

- 2 % in the horizontal plane and 5 % in the vertical plane;
- 5 % in the horizontal plane and 15 % in the vertical plane;
- 10 % in the horizontal plane and 30 % in the vertical plane.

2140 real conflicts are observed during the day above flight level 100 (10000 feet) when the resolution process is not used.

For each level of uncertainty, the simulator is able to solve all the conflicts. It is significant to note that the simulator adds a random noise to the real trajectories of the aircraft so that they do not maintain the nominal trajectories exactly. The table 9.1 shows the number of times the solver was called, the number of maneuvers, the average duration of the maneuvers, the proportion of the flight constrained by the maneuvers and the execution time of the simulation[14] for the various levels of uncertainties. It is observed that with a weak uncertainty, the number of maneuvers carried out (2461) is slightly

---

[14]The simulations were realized using 12 PCs of which the most powerful one was Pentium IV 2.53 MHz machine, the resolutions were carried out in parallel.

higher than the number of real conflicts (2140). Hence it can be assumed that the uncertainty causes some useless maneuvers. With 10 % and 30 % of uncertainties, the number of maneuvers is almost three times more significant than with 2 % and 5 % of uncertainties, and the number of times the solver is called is more than twice.

Table 9.1. Numerical results.

Uncertainty	Number of clusters	Number of maneuvers	Monthly duration by plane	Proportion of constrained flight	Duration of the simulation
2 % and 5 %	8539	2461	34 s	1, 27 %	26 mn
5 % et 15 %	12831	3881	78 s	2, 85 %	35 mn
10 % et 30 %	19390	6819	236 s	8, 43 %	55 mn

The table 9.2 shows the influence of uncertainty on the size of the clusters. It is observed that the increase in uncertainty plays a significant role in determining the size of the clusters to be solved and thus the difficulty in solving problems grows significantly.

Table 9.2. Influence of the uncertainty on the size of the clusters.

cut	2	3	4	5	6	7	8	9	10	11-17	18-37
2 %–5 %	7205	1021	224	56	23	6	3	1			
5 %–15 %	9970	1855	586	218	100	42	24	14	11	11	
10 %–30 %	12859	3326	1317	741	388	245	153	81	77	157	46

9.1.7 Remarks

The automation of the air traffic control is certainly not for tomorrow. Whereas the various phases of a flight can be completely automated, the air traffic management remains a complex problem for which no classical optimization method can propose a satisfactory solution. The genetic algorithms make it possible to take account of the operational constraints of the problem: necessity to simulate the trajectories, to take the uncertainties into account, to model the maneuvers with discrete variables etc. In addition, the partially separable structure of the problem enabled us to develop effective crossover and mutation operators to increase the size of the problems dealt with to about thirty aircraft. Till date, no other method tested in our laboratory or by other teams all over the world could solve clusters of this size. It is thus difficult to compare the performances of the genetic algorithm with those of another algorithm. For the moment, the developed tool has only a vocation of simulation. A model was recently adapted in order to take account of the current structure of the air routes. It should be well understood that this tool does not control in the same manner as the human controllers do. In the detailed example with 5 aircraft, a human operator would prefer to divide the problem into two smaller clusters. Nevertheless, this tool for simulation makes it possible to make a certain number of measurements on the complexity of the traffic and to compare various airspace (European space and American space). Its speed of execution enables us to develop a statistical tool that is capable of absorbing complete days of traffic over Europe.

9.2 Ground Traffic optimization

Traffic delay due to airport congestion and ground operations becomes more and more penalizing in the total gate-to-gate flight cycle. This phenomenon can be largely attributed to recent development of the hubs, as all departures and arrivals are tending to be scheduled at the same time.

Moreover, the uncertainty on departure and arrival times is largely increased by ground delays and can easily reach several tens of minutes during the peak of traffic, which is extremely damaging for all the actors of air traffic flow management.

In this application, an airport simulation tool is used to compare the ability of different optimization methods to solve efficiently some ground traffic situations: these methods use genetic and graph exploration algorithms to find the best path and/or the best holding positions for each aircraft. The efficiency of each method is measured by the correlation between the number of taxiing aircraft and the resulting total delay.

9.2.1 Modeling

The problem is to find an optimal set of acceptable trajectories for all the aircraft, where a trajectory is defined by a beginning time, a path and some holding positions in this path.

An *optimal* set of trajectories can have various definitions and will be globally considered as the one minimizing a cost function, which will be defined below.

The trajectories are *acceptable* when the path of each aircraft is compatible with the airport exploitation constraints (see 9.2.1) and when aircraft separation rules (detailed below) are ensured.

Cost function

The global criterion to minimize can be defined as a function of several factors: for example, the length of the paths assigned to the aircraft or the total taxiing time can appear relevant. However, holding on a taxiway can be interpreted more or less penalizing than increasing the length of the path or holding at the gate position...

In the current version, the cost function is defined as the total taxiing time (including queuing for runway delay), added to the time spent in lengthened trajectory:

$$f_c = \sum_{i=1}^{N} f_{c_i} \text{ where } f_{c_i} = r_i + d_i$$

With this definition, lengthening trajectory is twice more penalizing than holding position.

The airport

In order to assign to each aircraft a set of realistic alternative paths, the airport is described by a graph linking its gates, taxiways and runways.

The cost from a taxiway node to its connected nodes is the time spent to proceed via this taxiway, taking a speed limitation due to its turning rate into account. The cost from other nodes (gates and runway positions) to their connected nodes is zero.

Some taxiways can be described as "one-way": in this case, the cost of the opposite direction is balanced by a multiplicative coefficient representing the inconvenience for an aircraft to follow this way, according to operational controllers procedures.

In this context, classical graph algorithms can be used to compute a set of alternative paths for aircraft.

The Dijkstra algorithm [Ahuja et al., 1993] can compute all the best paths and the corresponding minimal taxiing times from a given node to every other node. This information is then useful to find the k_0 best paths linking two given points of the airport, using a Recursive Enumeration algorithm [Jimenez and Marzal, 1999]. By performing m iterations of this process while increasing the cost of selected taxiways, we can obtain up to mk_0 "different enough" paths.

Each aircraft is thus assigned a set of k possible paths ($k \leq mk_0$) between its gate to its runway entry points or between its runway exit points to its gate.

Figures 9.9 and 9.10 represent the graphs of Roissy and Orly and show an example of a set of paths in these graphs.

The traffic

Aircraft intentions are described by their *flight-plan*, containing their departure or arrival time, the type of the aircraft used, the gate position, the requested runway and eventually a CFMU slot. The wake turbulence category

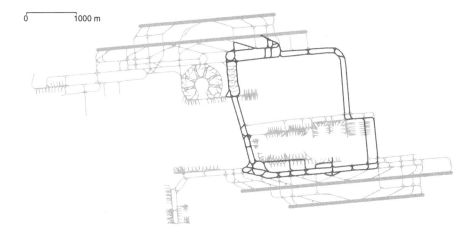

Fig. 9.9. Roissy airport graph

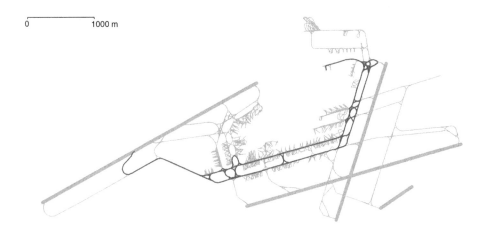

Fig. 9.10. Orly airport graph

(low, medium or high) and the takeoff or landing distance (restricting the choice of the runway exit or entry points) can be deduced from the type of the aircraft.

Aircraft separation rules

In order to detect the problems to solve in each traffic situation, a model for aircraft separation is defined. This model takes into account runways area, 90 meters away from each side of the runway (or 150 meters away in bad weather conditions). In these area, aircraft are considered on the runway even if they are not taking off or landing.

Aircraft separation is then defined as follows :

- Aircraft in gate position are separated from all other aircraft.
- The distance between two taxiing aircraft must never be lower than 60 meters.
- No more than one aircraft at a time can take off or land on a given runway.
- A time separation of 1, 2 or 3 minutes (depending on the aircraft category) is necessary after a take off to clear next takeoff or landing from wake turbulence.
- When an aircraft is proceeding for takeoff or landing on a given runway, other aircraft can be taxiing on the same runway area only if they are behind the proceeding one.

When one of these rules is not ensured in the traffic prediction, there is a *conflict* between the two concerned aircraft.

Speed uncertainty

In the simulation, the traffic prediction takes the uncertainty relative to aircraft speeds into account: this uncertainty is modeled as a fixed percentage of the initial defined speed (which is a function of procedures and turning rate). Therefore, an aircraft is considered to occupy multiple potential positions at a given time.

Separation rules are ensured if all of the possible aircraft positions are separated with other aircraft positions, as defined before. However, two special cases must be treated differently:

- When an aircraft is following another one, its speed will trivially depend on the speed of the first aircraft, so that the two aircraft are assumed separated, even if some of their uncertainty positions are not.
- When an aircraft must hold position, the uncertainty decreases, as the position and the time until which the aircraft must hold is rigorously fixed.

Simulations

Like in the first application, the simulation works with a shifting windows model: at each simulation step (every Δ minutes), traffic prediction is performed for the next T_w minutes (T_w is called the time window for traffic prediction). The pairs of conflicting aircraft are extracted from this prediction. At this simulation step, the problem consists in choosing a path and some holding positions for each aircraft, in order to ensure separations between them. These paths and holding positions are then used to build the new situation, *Delta* minutes later.

As a consequence of the limited time window for traffic prediction, some negative effects could appear and must be explicitly by-passed:

- Two aircraft can be brought one in front of the other, which will definitively freeze the future traffic situations.
- An aircraft can be stopped in a runway area while an arrival (that was not predicted in the last situation) is coming...

For these reasons, a special analysis of aircraft positions at the end of the time window is necessary: two aircraft must not be one in front of the other and the runway area must absolutely be cleared.

These new rules complete the separation rules and correspond to a predictive conflict detection: they will be applied for each resolution method detailed in the next sections.

9.2.2 BB: the 1-against-n resolution method

In this resolution method, aircraft are sorted and considered one after another.

The optimization problem is therefore reduced to one aircraft: the algorithm must find the best path and the best holding positions for the aircraft, avoiding the other already considered aircraft. In this point of view, the initial aircraft considered have higher priorities than the latter ones.

Graph definition

Given one particular path for the aircraft, the solution of the 1-to-n problem for this aircraft can be found with a graph exploration:

- A node of the graph is a timed position of the aircraft.
- The root node of the graph is the initial position of the aircraft, at the beginning of the time window.
- The terminal nodes are made of solution nodes: all the non conflicting positions of the aircraft at the end of the time window or at the end of the path and of no solution nodes: all the conflicting positions of the aircraft at any time.

- Each non terminal node has two sons, representing the two possibilities for the aircraft at each time step: moving forward or holding position. If the first possibility can reach to a solution node, then it is the best solution for the aircraft.

The best solution for the aircraft could be found by iterating this best-first search on each of its paths.

However, each node of the graph relative to a particular path can be linked to the current delay of the aircraft. This consideration allows to bound the graph exploration with the minimum value of delay found in the already explored paths: when the current delay is greater than this bound, the exploration can be aborted.

Thus, the graph exploration for the complete set of paths of the aircraft becomes a *Branch & Bound* algorithm [Horst and Tuy, 1995] with a best-first exploration strategy.

Aircraft classification

As the later considered aircraft are extremely penalized (they must avoid all earlier considered aircraft) the way to sort aircraft is a determining factor.

A simple way to assign priority levels is to consider the flight-plan transmission time to the ground controllers.

This option seems the most realistic one as ground controllers can hardly take an aircraft without its flight-plan into account. In the simulation context, this is equivalent to sorting aircraft by their departure or arrival time.

However, this option must be refined :

- As landing aircraft can not hold position before exiting runway, their priority level must be higher than all taking off aircraft.
- Queuing for runway aircraft should be sorted in their queue order.

In order to satisfy these principles, a time T_a is imposed on each aircraft as a function of its beginning time T_0 and its remaining time t_r: $T_a = T_0 + t_r$ for departures, and $T_a = T_0 - 1 hour$ for arrivals. Aircraft are sorted by increasing values of T_a.

9.2.3 GA and GA+BB : genetic algorithms

Two resolution methods using classical Genetic Algorithms and Evolutionary Computation principles such as those described in the literature [Goldberg, 1989, Michalewicz, 1996], are developed.

In the first method, the algorithm finds a path and an optional holding position for each aircraft. In the second one, the algorithm finds a path and a priority level for each aircraft, and uses the BB algorithm (see 9.2.2) to build the resulting trajectories.

Data structure

In the first method, the trajectory of an aircraft a is described by 3 parameters (n_a, p_a, t_a): n_a is the index of the path to follow and p_a is the position where the aircraft must wait until time t_a (if p_a is reached after t_a, the aircraft does not stop).

The second method needs 2 parameters (n_a, k_a) for each aircraft: n_a is the number of the path to follow and k_a its priority level. The detailed trajectory of the aircraft is the result of the BB algorithm applied with the classification given by (k_a) and restricted to one path per aircraft. The case of an aircraft for which the BB algorithm can not find any solution is interpreted as a conflict involving this aircraft.

Fitness function

For the two methods, the fitness function must ensure that a solution without any conflict is always better than a solution with a conflict: the fitness of conflicting solutions is always less than $\frac{1}{2}$ while the fitness of acceptable solutions is greater than $\frac{1}{2}$.

Thus, for a solution with n_c remaining conflicts,

$$F = \frac{1}{1 + n_c}$$

For a solution without any conflict,

$$F = \frac{1}{2} + \frac{1}{2 + \sum_{i=1}^{N} d_a + l_a}$$

where d_a is the delay of aircraft a and l_a the time spent by aircraft a in lengthened trajectory.

Crossover and mutation operators

The partially separable property of the conflict resolution problem is exploited one more time.

A local fitness F_a is computed for each aircraft a, as a function of the number of conflicts n_{c_a} for this aircraft and the cost function (see 9.2.1):

$$\text{if } n_{c_a} > 0, \ F_a = K n_{c_a} \text{ else, } F_a = f_{c_a}$$

(where K is a constant parameter such that $K \gg f_{c_a}$)

Sharing

The problem is very combinatorial in nature and may have several local optima. In order to prevent the algorithm from a premature convergence, the sharing process introduced by Yin and Germay [Yin and Germay, 1993a] is developed.

To implement this sharing process, a distance between two chromosomes must be defined, in order to separate different clusters in the population. In the experiments, the following distance is introduced:

$$D(A, B) = \frac{\sum_{i=1}^{N} |l_{A_i} - l_{B_i}|}{N}$$

where l_{A_i} (respectively l_{B_i}) is the length of the path of aircraft i^{th} in the chromosome A (respectively B).

Termination criteria

As the time to solve each problem would be limited in a real time application, the number of generations is limited: as long as no available solution is found, the number of generation is limited to 50 and the algorithm is stopped 20 generations after the first acceptable solution (with no remaining conflict) is found.

Clusters of conflicting aircraft

In order to lower the complexity of the problem as often as possible, a transitive closure is applied on conflicting aircraft pairs and gives the different clusters of conflicting aircraft. The different clusters will be solved independently at first. When the resolution of two clusters creates new conflicting positions between them, the two clusters are unified and the resultant one is then solved.

9.2.4 Experimental results

Simulations

Simulations are carried out with a real flight-plan sample at Roissy Charles De Gaulle and Orly airports in a complete day (May 18^{nth} 1999).

The three resolution methods are compared with the following parameters:

- Paths per aircraft : $k = 30$
- One-way : applied
- Time window for traffic prediction : $T_w = 5mn$
- Resolution step : $\Delta = 2mn$
- Speed uncertainty : $\delta = 10\%$

Comparing the three methods

Figure 9.11 gives the mean value of the generated delay as a function of the number of taxiing aircraft for the different methods.

As far as light traffic situations are concerned, the GA method provides the best results: the aircraft are not sorted so that the solutions found can approach the global optimum.

When the number of aircraft increases, the GA+BB method generates less delay than the two other ones: the fact that aircraft are sorted becomes less and less penalizing when the traffic density increases, and, perhaps, the size of the problem becomes too large for the GA method.

Generally, the results obtained with the deterministic BB method are less interesting because of the fixed priority levels of aircraft, which are never modified during the simulation.

Thus, assigning priority levels to the aircraft seems to be an efficient way to solve the ground traffic situations, under the condition that these priority levels are regularly adapted to each new situation.

Figure 9.12 gives, for the three methods, the number of aircraft simultaneously moving at each period of the day. It appears that the GA+BB method always keeps a lower number of moving aircraft during heavy time periods.

This result puts focus on an important phenomenon concerning the airport traffic: a good resolution of a situation allows to decrease the delay in the short term but also leads to better situations (with less moving aircraft) in the long term.

Remarks

This preliminary work has shown that the ground delay at such busy airports as Roissy Charles De Gaulle and Orly is very sensible to the resolution method used, which means the way the traffic is dispatched in the airport. This first conclusion gives an idea of the potential benefits that could be obtained with the development of some decision support tools for the airport controllers.

Once again, genetic algorithms seem well adapted to treat this kind of combinatorial problems as they can produce some unexpected solutions most often close to the global optimum, while deterministic algorithms (1-against-n method) must get limited in finding a local optimum, relative to a simplified problem.

It can also be noticed that the modeling was easily improved with the new runways of Roissy Charles De Gaulle, different speeds and uncertainties, specific one-ways... without changing the genetic algorithm itself. In this sense, simulations can be useful to evaluate some new operational procedures or future infrastructures for the airport.

Of course, a lot of improvements are still to be incorporated: future work will concentrate in refining the global criteria for genetic algorithms, taking,

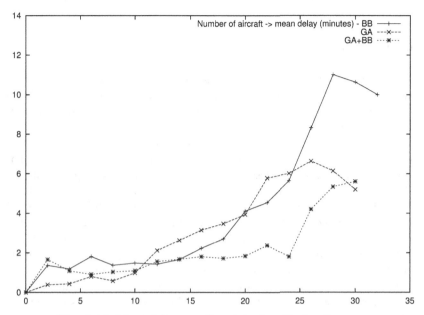

Fig. 9.11. Mean delay as a function of traffic density

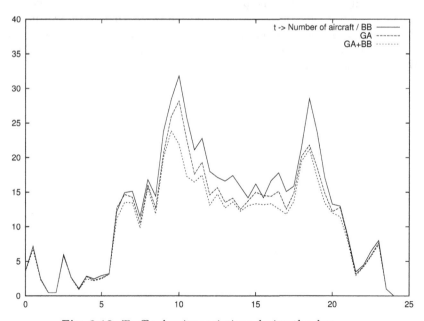

Fig. 9.12. Traffic density variations during the day

for example, takeoff sequencing needs of approach sectors or priority levels for departures which are constrained by a takeoff slot, into account.

9.3 Conclusion

The two applications presented in this chapter show that the air traffic gives rise to combinatorial problems which are very difficult to solve. The constraints related to the operational environment (uncertainties, human operators etc.) do not facilitate to define simple models. The functions to be optimized do not have analytical expressions but are obtained from the result of simulations. It seems that the genetic algorithms are quite efficient for these two applications. Nevertheless, the use of such algorithms requires rather finer adjustments and a good knowledge about the problem. The introduction of an adapted crossover operator makes it possible to deal with large size problems without losing the efficiency of the algorithm. The combination of genetic algorithms with local or deterministic methods is also sometimes very useful. From practical point of view, nowadays, the first application today makes it possible to make comparative statistical studies for the air traffic structures in various European countries or in the United States. It would be difficult to develop decision-making tools that could be very useful as the human capabilities to analyze large clusters are limited. Currently the controller never finds itself in a situation where it must solve a conflict with five aircraft like that presented in the paragraph 9.1.5. The problems are regulated upstream, to leave only the elementary problems to be solved by the controller. Using a conflict solver, its role would become purely as that of an executive. On the other hand, the second application has more operational future, even if the current modeling procedure does not take into account all the constraints related to the activities on an airport.

Constraint Programming and Ant Colonies Applied to Vehicle Routing Problems

Sana Ghariani and Vincent Furnon

ILOG S.A. 9, rue de Verdun, 94253 Gentilly Cedex
{sghariani,vfurnon}@ilog.fr

The ant colony algorithm is inspired by the behavior of real ants. It provides an original approach to solve combinatorial optimization problems. We present an industrial application of this method, in the context of constraint programming, focused on solving vehicle routing problems.

10.1 Introduction

One of the main concerns in the industry is to improve the effectiveness of their logistic chain, to be able to organize a better service at a lower cost and to maintain the flow of their goods. Thus a fundamental component of any logistic system is the planning of the distribution networks by fleets of vehicles. In recent times, significant research efforts have been devoted to the modeling of such problems and the implementation of suitable algorithms to solve them. A fruitful collaboration between the specialists in the area of mathematical programming and combinatorial optimization on one side and the transport managers on the other side, resulted in a great number of successful implementations of software for vehicle routing optimization. The interest in the implementation of the quantitative methods, for the optimization of the transport activities, becomes obvious as the importance of the distribution cost is paramount. The practical applications of this type of problems include: public transportation, newspaper distribution, garbage collecting, fuel delivery, distribution of products to department stores, mail delivery or preventive supervision of road inspections. However vehicle routing problems can be extended by various constraints [Osman and Laporte, 1995] that metaheuristic algorithms cannot usually manage.

The ant colony algorithm is a metaheuristic which is inspired by the behavior of real ant colonies [Colorni et al., 1992] in their search for food. It belongs to the family of evolutionary algorithms and is characterized by the combination of a constructive approach and a memory based learning mechanism. In this chapter, we will present an industrial case study based on this method, solving vehicle routing problems including both pickup and delivery orders. Constraint programming will be used for modeling the problem and will provide various services such as constraint propagation. Thereafter, we will present a resolution algorithm based on ant colonies, supported by experimental results.

10.2 Vehicle routing problems and constraint programming

10.2.1 Vehicle routing problems

Vehicle routing problems are defined by a set of vehicles which must be used to satisfy a set of demands corresponding to visits at various sites. The vehicles are, for example, trucks, boats, planes, etc. The sites are factories, hospitals, banks, post offices, schools, etc. The demands specify one or more destinations, which must be visited by only one vehicle. These problems can be constrained, thus the possible paths for the vehicles which satisfy these demands are limited.

The routing problem can be solved by assigning the visits to the vehicles first and then building a tour (i.e. placing the visits in order) for each one of these vehicles. Each tour must satisfy an entire set of constraints such as capacity constraints and time window constraints. The goal is to minimize the total cost of the tours.

Capacity constraints

They denote the fact that a vehicle cannot transport more than the limit of its capacity (in weight, volume. . .) i.e. the quantity of goods at each point should not exceed a certain value.

Time window constraints

They become effective when the service can only take place in a specified interval of time or when the vehicles are available only during certain periods of time. The problem is in this case a vehicle routing problem with time windows. Several other constraints corresponding to real situations can also be taken into account in vehicle routing models such as visit precedence constraints, or working time and rest constraints.

The cost function

The total cost of the tours is calculated as the sum of the costs of each vehicle. Generally, the cost of a vehicle is the linear combination of the total distance traveled on the tour and the duration of the tour. A fixed cost can be added when the objective is also to minimize the number of vehicles.

In this chapter, we will focus our attention on the "Pickup and Delivery Problem" (PDP) which is a vehicle routing problem with loading and unloading visits. It deals with the problem of picking up goods at a customer site (or at a depot) and delivering it to another customer using the same vehicle. The pickup must naturally be performed before the delivery.

The method implemented to solve this problem uses ILOG Dispatcher [ILOG, 2002a], a C++ library dedicated to solving routing problems, based on the constraint programming engine and the search algorithms provided by ILOG Solver [ILOG, 2002b]. The next paragraph details the concepts of constraint programming as well as the modeling of the problem in this context.

10.2.2 Constraint programming

In this section, we present concepts of constraint programming useful for the rest of the chapter.

Constraint programming [Rossi, 2000, Tsang, 1993, Hentenyrck, 1989] is a technique which is becoming increasingly significant in the field of optimization. Its main advantage, compared to other techniques, is that it facilitates the description and the resolution of general models, while keeping both notions separate.

A problem modeled using constraint programming is defined by a set of variables and a set of constraints. Each variable is associated to a finite set of possible values (called a domain), and each constraint relates to a subset of variables. Each one of these constraints indicates which partial assignments, involving the variables appearing in the constraint, satisfy (or violate) the constraint. The problem of constraint satisfaction is then to find an assignment (one value in the domain) for each variable such that no constraint is violated. This problem is NP-hard [Garey and Johnson, 1979]. A variant to this problem — optimization — seeks an assignment for each variable, while minimizing the value of one of the variables.

Tree search

Constraint programming is usually based on *tree search*, and in the simplest case on depth-first search. Depth-first search is both complete and light in memory consumption: it guarantees that a solution will be found if there is one, or proves there is no solution, if the problem is infeasible. In the worst case, tree search examines all the combinations of values for all the variables, but this worst case is seldom reached. Generally, significant parts of the search tree are pruned thanks to constraint propagation.

Let us take an example to show how the *depth-first search* works. Let us consider a problem with three variables a, b and c which have the following respective domains $\{0,2\}$, $\{0,1\}$ and $\{0,1\}$. Let us also consider the three constraints $a \neq b$, $a \geq c$ and $b \leq c$. The solutions to this problem are:

- $a = 2$, $b = 0$, $c = 0$;
- $a = 2$, $b = 0$, $c = 1$;
- $a = 2$, $b = 1$, $c = 1$.

Five other possible combinations of values for a, b and c violate at least one constraint and are thus not acceptable. The tree search presented here considers the variables in the lexicographical order: initially a, then b, and finally c. The values of the variables are considered in increasing order. The description of the order in which the variables and the values are chosen, and more generally the description of the implementation of the search tree, is called a *goal* in constraint programming. The search tree for this goal is presented in figure 10.1. The nodes of the tree represent the states of the assignment of the variables. When a node is marked by $\otimes$, it indicates that at least one constraint is violated at this stage. The arcs of the tree are transitions between the states and are labeled with the assignments which take place during the transition. Finally, the leaves of the tree not marked by $\otimes$ are *solution states*, where all the variables are assigned and no constraint is violated.

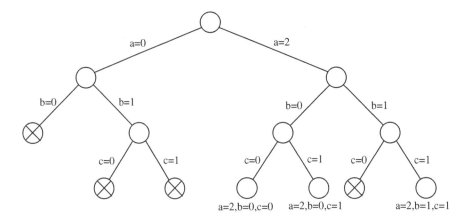

Fig. 10.1. Depth-first search.

Tree search is carried out as follows. Initially, the variable a is considered and the smallest value of its domain, 0, is assigned to it, then b is assigned the smallest value of its domain, 0. Although we do not have a complete solution yet, we can check the validity of the constraint $a \neq b$ since all the variables under consideration in this constraint have a fixed value. In other words, no other assignment will be able to change the state of the constraint — satisfied or violated. In this case, it can be seen that the constraint $a \neq b$ is violated. This state is called *failure* (marked by $\otimes$). One of the assignments $a = 0$ or $b = 0$ must be changed. In "Depth-first search", the assignments are backtracked by undoing the most recent assignment. The search then returns to the state immediately before the assignment $b = 0$ — the movement is known as *backtracking* — and b is assigned to the next value which is 1. This time, the constraint $a \neq b$ is satisfied, and one can thus continue the search by assigning a value to c. In fact, no value for c can satisfy the constraints as 0 violates the constraint $b \leq c$ and 1 violates the constraint $a \geq c$. Hence it is proved that there is no solution for $a = 0$ since one implicitly explored all the combinations for b and c with $a = 0$. One can now return to the top of the tree by undoing all the previous assignments and assign $a = 2$. In this case if we assign $b = 0$ and $c = 0$, the constraints are satisfied. Hence, there is a solution. If we were interested in only one solution, we could have stopped there. However, one can also continue to find other solutions by backtracking. Thus, by changing the assignment of c with $c = 1$, another solution is found. The assignment $b = 1$ following that of $c = 0$ leads to a failure (($b \leq c$ is violated), but the assignment $c = 1$ after backtracking produces the last solution.

Constraint propagation

The technique for testing the constraints described above is known as *backward checking* since the constraints are tested after all the variables involved in the constraint have been assigned to values. It is an improvement over the *generate and test* algorithm which postpones the tests of the constraints after all variables have been instantiated. This improvement is however in general too weak to solve anything but very simple problems. Normally, constraint programming prunes branches of the search tree using a much more effective method, known as *constraint propagation*. Constraint propagation is a technique much more active than backward checking. Instead of checking the validity of a constraint, domains are filtered using a dedicated algorithm. This procedure very often results in a reduction of the number of failures — i.e., dead ends in the search are noticed higher up in the tree, which avoids explicit sub-tree exploration.

In the case of backward checking, each variable already has an assigned value, or no value at all. On the contrary, constraint propagation maintains the *current domains* of each variable. The current domain is initially equal to the initial domain, and is filtered as the search progresses in the tree. If it is found that a value is impractical for a variable (because it would violate at least one constraint), it is removed from the current domain. These removed values are not considered any more and the search tree is thus more effectively pruned than with backward checking alone. In general, filtering algorithms are usually kept polynomial. General methods [Bessiere and Regin, 1997, Mackworth, 1977] and specific algorithms for certain types of constraints exist [Belideanu and Contjean, 1994, Regin, 1994, Regin, 1996].

Domains filtering is carried out *independently* for each constraint. Communication between constraints is only based on the changes in the current domains of the variables. In general, constraint propagation algorithm works by having each constraint filter the domain of the variables related to them. This process is carried out either until a variable has an empty domain (no more possible value), or until no more domain reduction occurs. In the first case, a *failure* and a backtrack occur. In the second case, a fixed point is reached. If all the variables have only one possible value, a solution is found. If not, it is necessary to branch on the values of the remaining variables to either find a solution or to prove that there is none. The fixed point found by constraint propagation should not depend on the order in which the constraints filter the values of the domains.

The figure 10.2 shows how constraint propagation interacts with depth-first search on our small problem. As it can be seen, this method is more effective than backward checking in terms of number of failures . There is only one failure, all the other branches leading to a solution. At the top of the tree, before branching, a propagation is carried out. In this case however, no filtering can be done by considering constraints individually. A branch is thus created, and a is assigned to 0. The constraint $a \neq b$ deduces that $b \neq 0$ and thus $b = 1$. Constraint $a \geq c$ deduces that $c \neq 1$ and thus $c = 0$. Constraint $b \leq c$ deduces that $c \neq 0$ which causes a failure as $c = 0$, and that $b \neq 1$ (that also causes a failure as $b = 1$). It can be noticed that only one failure will occur whether one or the other deduction is carried out first, with finally either b, or c having an empty domain. When the failure occurs, the constraint solver backtracks to the root node and domains are restored. The right-hand side branch is then taken which assigns a to 2. Similarly to the root node, no constraint can reduce the domains, and branching is still needed, $b = 0$. No filtering is done and a final branching $c = 0$ leads to the first solution. After backtracking and assigning c to 1, the second solution is obtained. Then we backtrack higher up in the tree, just before the assignment of b, and the branch to the right-hand side of $b = 1$ is taken. In this case, the constraint $b \leq c$ filters, which removes the value 0 from the domain of c. It is the only filtering which can be done at this node and it results in a solution . It is not necessary to branch on c as the variable does not have more than one possible value. This is the last solution to the problem.

Note on optimization

Until now, we explained how to solve a decision problem which satisfies all the constraints. Sometimes however the problem at hand is an optimization prob-

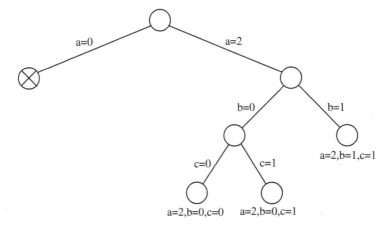

Fig. 10.2. Depth-first search with propagation.

lem. Such problems are solved like a succession of decision problems. When a solution of cost v is found, a constraint is added to the problem specifying that any solution of cost v or more is not valid. This process is repeated until one cannot find a better solution, and in this case the last solution is the optimal solution. Each time a — better — solution is found, it is not required to start the search again from the beginning. One can simply continue the search with the new upper bound on v. This upper bound becomes more constraining as the search progresses.

A powerful modelling and a powerful search.

In the preceding discussion, for pedagogical reasons, we only considered simple constraints with simple propagation rules. When complex industrial problems are solved, the models are more complex and constraint programming provides more advanced constraints to take this complexity into account. These constraints are often based on powerful propagation algorithms which perform strong filtering on domains, for example [Regin, 1994, Regin, 1996]. Here are examples of more complex constraints usually available:

- $y = a[x]$: y is constrained to be equal to the xth element of a, where a is a table of constants or variables.
- $all - diff(a)$: all the variables of the table a must take different values.
- $c = card(a, v)$: the variable c is forced to be equal to the number of occurrences of the value v in the table of variables a.
- $min - dist(a, d)$: all the pairs of variables in the table a must take values which differ by at least d.

Some open constraint programming frameworks let the users write their own constraints, by writing the corresponding propagation rules. This is often very useful as a framework cannot provide all possible constraints.

All constraint programming frameworks give the option of writing custom search goals, i.e. a description of how to assign values to variables during the search. This can be useful because the efficiency of the constraint propagation can depend on the order in which variables are considered. For example, in our example with three variables, no failure occurs if one considers the variables in the order b, a, c instead of a, b, c. For vehicle routing problems, it is usually more interesting to extend a tour rather than to build many chains which will finally merge. Such variable selection heuristics can often improve the performance by an order of magnitude. The extensibility and the flexibility of constraint programming make it possible for the users to write effective optimization software.

10.2.3 Constraint programming applied to PDP: ILOG Dispatcher

Problem formulation in constraint programming

The formulation of the PDP in constraint programming is facilitated by the clear separation between the description of the problem using constraints and decision variables, and the resolution of the problem.

Decision variables.

In the usual linear model, there are $O(m\,n^2)$ Boolean decision variables where m is the number of vehicles and n the number of customers to be visited. But it is possible to describe the PDP with a linear number of variables. Each visit i is associated with two variables with finite domains $next_i$ and veh_i representing, respectively, the visit which follows i and the vehicle that serves the visit i. The decision variables are $next_i$ whose domain is the finite set of visits. This set of visits consists of the visits to the customers and the visits to the depot (2 visits to the depot by vehicle, starting and ending visits of the tour corresponding to only one depot, but having different indices for each vehicle). The value of veh_i can be obtained by propagation of the following constraints: $next_i = j$ if the visit j immediately follows the visit i in the tour and $veh_i = veh_j$ if $next_i = j$:

- If N is the set of the indices corresponding to the visits other than the ones at the depot, S the set of the visit indices corresponding to the departures from the depot and E the complete set of visit indices corresponding to the returns to the depot, $next_i = j$ is possible if $i \in N \cup S$ and $j \in N \cup E$, which ensures the continuity of a tour;
- The constraints $\{next_i = j$ is possible if $i \in N \cup S$ and $j \in N \cup E\}$ associated with the constraints $\{(next_i = j \wedge veh_i = k) => veh_j = k\}$, are equivalent to the collection of the linear constraints:

$$\forall i \in V, \qquad \forall k \in M: \qquad \sum_{j \in V} x_{ij}^k = \sum_{j \in V} x_{ji}^k = 1$$

with $V = N \cup E \cup S$.

Path Constraints.

This concept allows us to take into account dimension constraints, such as capacity and time constraints. Path constraints propagate quantities accumulated along a tour. Formally these constraints can be described as follows:

$$\forall (i,j) \in ((S \cup N) \times (N \cup E)) \qquad next_i = j \Rightarrow \sigma_i + f(i,j) = \sigma_j$$

where σ_i is a variable representing the accumulation of the quantity. $f(i,j)$, called transit, can be the distance from i to j or simply the quantity of goods to be delivered at i.

Example:

- $f(i,j) = q_i$: demand of the customer i (weight, volume...). σ_i represents the load of the vehicle in i, thus the path constraint is simply:

$$\forall (i,j) \in ((S \cup N) \times (N \cup E)) \qquad next_i = j \Rightarrow \sigma_i + q_i = \sigma_j$$

The capacity of a vehicle can be expressed by the following constraint:

$$\forall i \in S \cup N \cup E \qquad 0 \le \sigma_i \le C$$

where C is the capacity of the vehicle;
- σ_i: represents the arrival time at i, $f(i,j) = s_i + t_{ij} + w_i$, $t_{i,j}$ being the travel time between i and j, s_i the service time in i and w_i the waiting time between i and j. Time window constraints can be expressed by $a_i \le \sigma_i \le b_i$.

Visit precedence and same vehicle constraints.

In a PDP, the two visits of the shipment (pickup and delivery visits) must be carried out by the same vehicle: if i is a pickup and j is the corresponding delivery, $veh_i = veh_j$.

Moreover a pickup must precede the corresponding delivery. Thus, if σ_i represents the rank of the visit i in the tour and σ_j the rank of j, then $\sigma_i < \sigma_j$.

The cost function.

The goal of the problem is to minimize the total cost of the tours:

$$\sum_{k \in E} c_k$$

c_k being the cost of the tour of the vehicle k, c_k is generally defined by:

$$\sum_{(i,\,j) \in \{(i,\,j) /\, next_i = j \text{ et } veh_i = k\}} f(i,j)$$

$f(i,j)$ being the traveling cost from i to j.

10.3 Ant colonies

The ant colony algorithm is a metaheuristic introduced by Marco Dorigo et al. [Colorni et al., 1992]. It was inspired by the studies on the behavior of real ants.

10.3.1 Behavior of the real ants

Real ants are able to find the shortest path joining their nest to a source of food without using visual indicators, by only exploiting information from a pheromone trail. Indeed, during a movement, an ant deposits a substance called pheromone on its way and follows, with a certain probability, the quantity of pheromone left by the other ants.

Assuming that a colony of ants moves towards a source of food and meets an obstacle, the problem is overcome by some ants choosing the shortest path and others the longest path. The choice is done at random. But, since the ants move almost at a constant speed, the ants which chose the shortest path, reach the source of food faster than the others and return faster to their nest. The rate of pheromone accumulated on this path is then more significant than on the longest path. Thus the ants will tend to follow the first path. The ant system (AS) and the ant colony system (ACS) are algorithms which are inspired by the behavior of the ants. Artificial ants cooperate to find the shortest path on a graph, by exchanging information through the quantity of pheromone deposited on the arcs of the graph. The ACS has been applied to the combinatorial optimization problems like the TSP (Travelling Salesman Problem) [Dorigo and Gambardella, 1997b] or the VRP [Bullnheimer et al., 1997, Bullnheimer et al., 1999].

10.3.2 Ant colonies, vehicle routing problem and constraint programming

The AS was applied in [Bullnheimer et al., 1997, Bullnheimer et al., 1999] to solve the VRP with capacity constraints. In the first version, the proposed algorithm is very similar to [Dorigo and Gambardella, 1997b], dedicated to the TSP. The only difference is that each ant builds a complete tour for each vehicle, at each iteration. In a second version, the algorithm is closer to ACS due to the way it updates the quantity of pheromone.

The ACS was applied in [Gambardella et al., 1999] to solve the VRP with time windows. The algorithm uses two ant colonies, one to minimize the number of vehicles used and the other to minimize the total cost of the tours. It should be noted that local search techniques were used in the two algorithms to improve the solution.

Constraint programming appears well adapted to the principles of search by ant colonies. Indeed, constraint propagation makes it possible to direct the movement of the ants so that the choice of a destination does not violate the constraints of the problem.

10.3.3 Ant colony algorithm with backtracking

Description

The proposed algorithm is inspired from [Gambardella et al., 1999]. It uses two ant colonies, one to minimize the cost of the solution and the other to minimize the number of vehicles (see algorithm 10.1). Each ant colony has its own trail of pheromone. In this algorithm, the minimization of the number of vehicles used is favored. Indeed, this criterion is often the most significant and remains difficult to optimize. Moreover, the minimization of the number of vehicles can sometimes help obtaining better global costs.

Solutions accepted during the iterations of this algorithm are not necessarily feasible. A solution is not feasible if it contains one or more unperformed visits. Let C_1 be the colony of ants minimizing the number of vehicles and C_2 be the one minimizing the cost. Let $trail_1$ denote the trail of pheromone of C_1 and $trail_2$ denote pheromone trails of C_2.

S^* : best solution found
S^* := solution found by the savings heuristic
Repeat NbIteration times
    v := number of vehicles used in S^*
    Initialization of $trail_1$
    Initialization of $trail_2$
    n := 1
    While (S_1 is not feasible) or ($n < n_1$)
        S_1 := Best solution found by C_1
        Update the quantity of pheromone on the arcs belonging to S_1
        n := n + 1
    If S_1 is feasible then
        (a) $S^* := S_1$
        (b) Update the quantity of pheromone on the arcs belonging to S^*
        (c) Proceed to the next iteration
    n := 1
    While (S_2 is not feasible) or ($\text{cost}(S_2) \geq \text{cost}(S^*)$) or ($n < n_2$)
        S_2 := Best solution found by C_2
        Update the quantity of pheromone on the arcs belonging to S_2
        n := n + 1
    If S_2 is feasible and cost of $S_2 < $ cost of S^* then
        (a) $S^* := S_2$
        (b) Update the quantity of pheromone on the arcs belonging to S^*
        (c) Proceed to the next iteration

Algorithm 10.1: A broad outline of an ant-based algorithm with backtracking.

Remarks:

n_1 and n_2 are parameters.

Initialization of pheromone trail.

The quantity of pheromone is represented by the same data structure in the two ant colonies. The initial value of pheromone for each couple of visits (i, j) is:

$$t_{ij} = \begin{cases} \frac{1}{Nc} & \text{if } i \neq j \\ 0 & \text{otherwise} \end{cases}$$

where N is the number of visits and c is the cost of the solution obtained by the "savings" heuristic [Clarke and Wright, 1964] for a given number of vehicles (at most v vehicles for the ant colony optimizing the cost and $v - 1$ vehicles for the one minimizing the number of vehicles).

Search procedures used by the two ant colonies.

The search procedures used by the two ant colonies are identical, with the following limitations:

- Each colony manages its own trail of pheromone.
- The ant colony which minimizes the number of vehicles seeks a solution with at most $v - 1$ vehicles whereas the other colony seeks a solution with v vehicles.
- The two colonies do not use the same local search procedure.

The algorithm used by the two colonies is given in algorithm 10.2.

At this stage, the quantity of pheromone is updated according to the formula: $\tau_{ij} = (1 - \rho) * \tau_{ij} + \rho * 1/L_{best}$, if $(i, j) \in$ the best solution found by the ants in this iteration. Here L_{best} is the cost of the best solution found by the ants in the current iteration and ρ a evaporation coefficient.

Repeat for each ant
    Repeat
        Stage 1 Choose a random unused vehicle
        Stage 2 Build a tour for this vehicle
    Until there is no more unused vehicles
    Apply local search to the solution found
Store the best solution found by the ants
Update the quantity of pheromone.

Algorithm 10.2: A broad outline of the algorithm used by the two ant colonies.

Construction of a tour for a vehicle.

An ant is initially positioned at the depot, corresponding to the starting position of the vehicle. The constraints of the problem are propagated and the domains of the variables are possibly reduced. This results in a filtering of the possible next visits (figures 10.3 and 10.4).

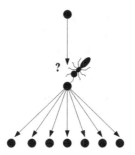

Fig. 10.3. Possible next visits before propagation.

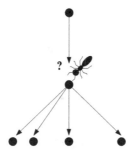

Fig. 10.4. Possible next visits after propagation.

Choice of the next visit.

The choice of a next visit is carried out according to the following rule:

$$j = \begin{cases} j \text{ corresponding to } max_{h \in \Omega}\{[\tau_{ij}]^{\alpha}[\eta_{ij}]^{\beta}\}, & \text{if } q \leq q_0 \quad (intensification) \\ j \text{ found applying the following, probability rule} & \text{otherwise } (diversification) \end{cases}$$

$$p_{ij} = \begin{cases} \dfrac{[\tau_{ij}]^{\alpha}[\eta_{ij}]^{\beta}}{\sum_{h \in \Omega}[\tau_{ih}]^{\alpha}[\eta_{ih}]^{\beta}} & \text{if } j \in \Omega \\ 0 & \text{otherwise} \end{cases}$$

where:

ω is the set of possible next visits for visit i;

q is a random number between 0 and 1;

q_0 is a parameter between 0 and 1;

η_{ij} $= 1/\max\left(1, c_{ij} - (\text{number of times visit } j \text{ has not been performed})\right)$;

c_{ij} is the travel cost from visit i to visit j;

α and β are of the parameters determining the relative importance of the pheromone compared to the cost ($\alpha > 0$ and $\beta > 0$).

The parameter q_0 determines the relative importance of diversification and intensification. When q_0 is close to 1, the selected arc corresponds to the choice of the other ants, which can lead to a premature trapping of the search in a local minimum. In order to favor diversification and to push the ants to move towards less explored arcs, the value of q_0 is gradually decreased until an ant succeeds in finding a better solution.

Once the next visit is chosen, the constraints are propagated again (figure 10.5).

- If the propagation fails then the search "backtracks" and the selection process is started over after removing from the domain of the next visit variable the visit which led to a failure (figure 10.6),
- Otherwise, start the process again from the new current visit.

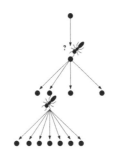

Fig. 10.5. No failure of the constraint propagation.

Fig. 10.6. Failure of the constraint propagation.

Local update of the pheromone quantity.

Each time a next visit is selected, the quantity of pheromone is updated for the arc created according to the formula:

$$\tau_{ij} = (1 - \rho) * \tau_{ij} + \rho * [\text{initial quantity of pheromone}]$$

where:

i corresponds to the current visit and j the next visit;
ρ is the coefficient of evaporation.

Each ant is thus influenced in its search by the solutions found by the preceding ants and also by the current best solution.

Global update of the pheromone quantity.

The quantity of pheromone is updated on the arcs belonging to the current best global solution, according to the formula:

$$\tau_{ij} = (1 - \rho) * \tau_{ij} + \rho * \frac{1}{L_{best}} \qquad if(i,j) \in (\text{the best current solution})$$

where:

L_{best} is the cost of the current best global solution;
ρ is the evaporation coefficient.

Improving the solution using local search.

Each time an ant finds a solution, it can be improved by a phase of local search. This phase uses the principle of local search in constraint programming proposed in [De Backer et al., 2002].

The ant colony which minimizes the number of vehicles uses the following move operators.

MakePerform: creates new solutions (neighboring solutions) by inserting an unperformed visit after a visit assigned to a vehicle.

MakeUnperform: creates new solutions (neighboring solutions) by making a visit unperformed.

SwapPerform: creates new solutions (neighboring solutions) by exchanging a visit assigned to a vehicle with an unperformed visit.

The ant colony which minimizes the cost uses the move operators described previously and TwoOpt, OrOpt, Cross, Relocate and Exchange described in figure 10.7.

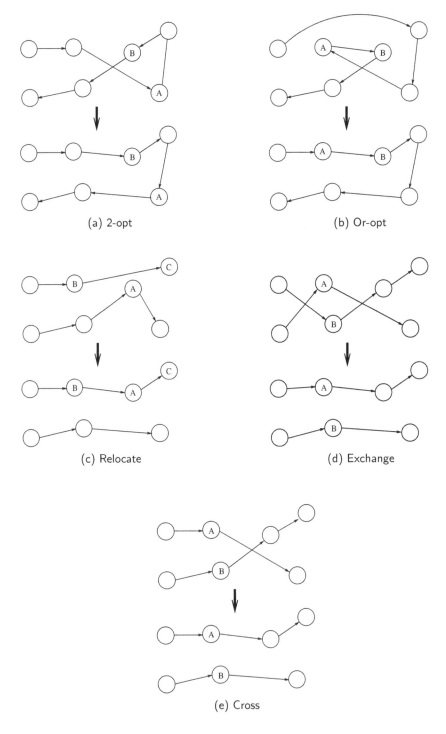

(a) 2-opt

(b) Or-opt

(c) Relocate

(d) Exchange

(e) Cross

Fig. 10.7. Move operators.

10.4 Experimental results

The data files which have been used as test bench are derived from those of Solomon [Solomon, 1987]. They were modified to make some visits pickups and others deliveries.

There exist two series of Solomon instances: series 1 is the series for which the time windows are tight and series 2 is the series for which the time windows are looser. Series 1 is easier to solve because there are more constraints: tight time windows limit the number of possible next visits, for a given current visit. Moreover, the test instances are divided into three classes: the class "C" in which the visits are divided into several compact groups from a geographical point of view (they are the easiest problems to solve), the class "R" in which the co-ordinates of the visits are distributed at random and "RC" for which part of the visits are grouped geographically and others are grouped at random. Each derived instance is a problem with 200 visits.

For the sake of simplicity, in our proposed algorithm the two ant colony algorithms are not run in parallel. The ant colony algorithm which minimizes the number of vehicles is run first. If this algorithm succeeds to find a feasible solution, using fewer vehicles than the best current solution, this solution becomes the best current solution. The quantity of pheromone is updated and the algorithm is started over. If the algorithm does not find a feasible solution after five iterations, this ant colony algorithm is stopped and the one minimizing the cost is started. If this algorithm succeeds in finding a better solution, then it is stopped and the trail of pheromone on the arcs belonging to the best solution is updated. The first algorithm is then re-started. If the second algorithm does not find a feasible solution after three iterations, the colony algorithm is stopped and the algorithm which minimizes the number of vehicles is re-started, without reinitializing the pheromone trail.

The results obtained with this algorithm are compared to the results obtained using guided local search (GLS) [De Backer et al., 2000] [Kilby et al., 1997], which is a popular metaheuristic that usually achieves satisfactory performance. The tests were carried out using the following parameter values:

- 10 ants;
- $\alpha = \beta = 1$;
- $q_0 = 0.9$;
- $\rho = 0.75$;
- $n_1 = 5$;
- $n_2 = 3$.

The main comparison criterion is the number of vehicles used in the solution. The computing time for each instance, for each of the two algorithms is two hours. To trace the evolution of the optimal solution, the value of the solution is recorded every 10 seconds during the first 300 seconds, and then recorded every 50 seconds up to 7200 seconds. Figures 10.8, 10.9 and 10.10 show the evolution of the number of vehicles used in the solution.

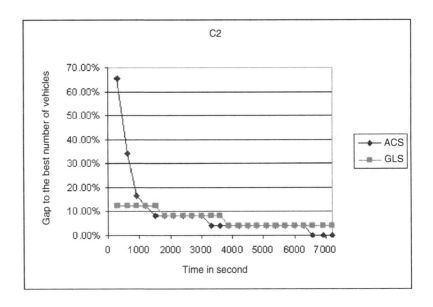

Fig. 10.8. C2 : problems of series 2 and class C.

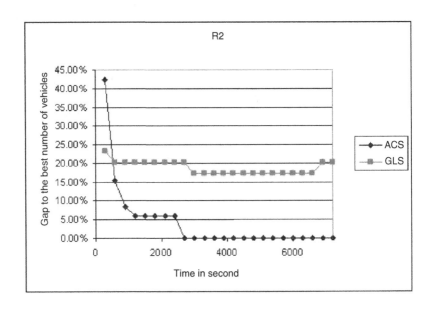

Fig. 10.9. R2 : problems of series 2 and class R.

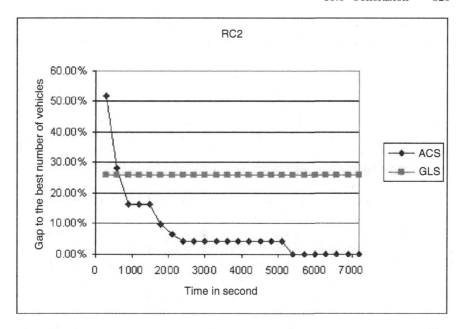

Fig. 10.10. RC2 : problems of series 2 and class RC.

If we restrict ourselves to the results obtained for the series 2, it is clearly understandable that the ACS scores highly above the GLS from the point of view of the number of vehicles used. It is also known that the problems R2 and RC2 are often regarded as more difficult to solve. Regarding the actual solution costs (which are not presented here), they remain comparable with those obtained with GLS. Although these results are still at a preliminary stage, it can be assumed that the ant colony algorithm possesses a certain robustness, since it succeeds in finding good solutions for difficult problems, which other metaheuristics cannot reach, e.g. GLS.

10.5 Conclusion

In this chapter, an integration of ant colony algorithms is presented in the context of constraint programming. This enabled us to combine the features of the two approaches.

Firstly, constraint programming provided us with a modelling procedure, making it possible to represent the problems in an expressive and concise way. From the solution point of view, the propagation of the constraints and the resulting reduction of the domains of the variables made the depth-first search approach effective. However, this effectiveness is often at the expense of good branching heuristics.

The "expertise" of the user can help designing effective custom heuristics. However, this may not always be possible, in particular when the users of such techniques are not experts.

The ant colony algorithms can significantly help in such cases. In this chapter, we described how one can make use of ant colonies to obtain heuristics which can be simultaneously robust and generic in nature. The robustness characterizes the ability to finding a *good* solution, whatever the nature of the data. This objective is primarily fulfilled by memorizing the movements of the ants (in fact, of the states of the variables) and the corresponding costs. Thus it is possible to reach good quality solutions, by alternating intensification and diversification phases.

Since the only knowledge ant colony algorithms have of the problem is the costs connecting variables to values (in the routing problems, that corresponds to the costs of the arcs between visits), it can be assumed that this approach is generic, maintaining a certain independence between the description of the problem and the way it is solved. Initial experimental results on vehicle routing problems with pickup and delivery visits confirm this assertion, especially when minimizing the number of vehicles. It would be very interesting to study the impact of replacing depth-first search by other more sophisticated procedures, such as "Slice-Based Search" (SBS) proposed in ILOG Solver [ILOG, 2002b].

Acknowledgement

We would like to thank Mr. Bruno De Backer and Mr. Paul Shaw for their assistance in the development of this chapter.

Conclusion

This work has shown multiple facets of metaheuristics proposed, in the last 20 years, for the approximate solution of the "difficult optimization" problems. The success of these procedures should not mask the principal difficulty that the user will encounter, in the presence of a concrete optimization problem: that of the choice of an "efficient" method capable of producing an "optimal" solution — or of acceptable quality — at the cost of a "reasonable" computing time. Compared to this pragmatic concern, the theory is not yet of a great help, because the convergence theorems are often non-existent, or applicable under very restrictive assumptions. Moreover, the "optimal" adjustment of the various parameters of a metaheuristic, which can be recommended by the theory, is often inapplicable in practice, because it induces a prohibitory cost of calculation. Consequently, the choice of a "good" method, and the adjustment of its parameters, generally require the knowledge and the "experience" of the user, rather than the faithful application of well laid down rules.

The efforts of research in progress aim at rectifying this situation, essential in the long term for the credibility of the metaheuristics. Considering the possible expansion of the field, it becomes essential to light the user in the choice of a metaheuristic or a hybrid method, and in the adjustment of its parameters.

Let us mention a first significant direction of the research tasks in course: systematic exploitation of hybridizations and co-operations between methods (emergence of the multi-agent or self-organized systems, development of a taxonomy of the hybrid methods...). The literature on this subject is abundant. The reader is directed, for example, to [Renders and Flasse, 1996], which describes several hybrid methods exploiting the evolutionary algorithms. In presence of these several possibilities, the need for a classification is imminent. In a recent work [Talbi, 2002], E.G. Talbi proposes a taxonomy of the hybrid metaheuristics: the author draws from them a common terminology and mechanisms for classification, at the same time. The proposal is illustrated through the classification of a large number of hybrid methods described in the literature.

It seems that the credibility of metaheuristics is reduced by an artificial bulk-heading of the various techniques. Indeed, to take only one example, what conceptually differentiates an ant colony from an approach of the GRASP type, if not some details, such as the origin of the inspiration? Ultimately, the two techniques rest on repeated construction, in a probabilistic and adaptive way, of new solutions. As the fact of finding the local optimum associated with a new solution does not conform to the metaphor of an ant colony, the inventors of this last technique do not insist on the need for a local search. However, the majority of the heuristics, established on the concept of the ant colonies, make use of it. On the other hand, the possibility of building several solutions in parallel is naturally based on the metaphor of the ant colonies, whereas it will be about an extension of GRASP.

In our point of view, we should attempt to go beyond these polarizations, which do not help the field of the metaheuristics to progress, and to think globally. This is what one of the authors of this book attempted by employing the "adaptive memory programming" or "POPMUSIC", which, under unified schemes, forms a broad group of techniques having various names and origins. To develop effective heuristics for a given problem, it seems more interesting to us to consider the set of principles contained in the metaheuristics, like the use of a neighborhood (simple neighborhood, extended or composite; list of candidates), a memory (population of solutions, trails of pheromone, tabu lists), noising effects (noising effects of data or solutions; penalization of the movements) and to choose, among this set of principles, those which seem most suitable for the problem to solve. Certain authors even suggest techniques, that they term as "hyper-heuristic", to carry out these choices in an automatic way.

The principal justification for the use of a method developed on the basis of metaheuristic being to produce solutions of high quality, one could observe a race for the good solutions, which did harm to the metaheuristics. To present tables of "demonstrative" results that a new method is effective, one generally tends to overload the heuristic with options, parameters and mechanisms which reduce the possibility that one can propose a heuristic of simple design. It is the reason which pushed us to encourage the use and the development of statistical tests to compare more scientifically the nondeterministic iterative methods, just as to publish in appendix the source of a complete program, based on simple principles, but not inevitably naive. Going back more than a decade — a great age, for this kind of techniques —, this code is still useful, if one looks at the quality of the solutions which it produces and its simplicity. It still remains the harsh reality that the theoretical analysis of metaheuristics is particularly difficult and the results obtained until now are extremely thin. The future, at the theoretical level, can perhaps encompass the definition of complex neighborhood and the analysis of the "energy landscapes" which result from these techniques. The measures of the efficiency of certain neighborhoods, such as for example the coefficient of roughness, were proposed, but these theoretical analyses have not yet led to really exploitable general results in practice.

Appendices

A

Modeling of Simulated Annealing Through the Markov Chain Formalism

Let R be the complete space of all the possible configurations of the system, and $r \in R$ be a "state vector", whose components entirely define a specified configuration (or "state"). Moreover, let us consider that the set I_R consists of the numbers assigned to each configuration of R:

$$I_R = (1, 2, \ldots, |R|)$$

where $|R|$ is the cardinality of R. Finally let us denote by $C(r_i)$ the value of the cost function (or "energy") in the state i, r_i the state vector, and $M_{ij}(T)$ the probability of transition from the state i to the state j at the "temperature" T. In case of the simulated annealing algorithm, the succession of the states forms a Markov chain, in the sense that the probability of transition from the state i to the state j depends only on these two states, and not on the states former to i. In other words, all the past information about the system is summarized in the current state. When the temperature T is maintained constant, the probability of transition $M_{ij}(T)$ is constant, and the corresponding Markov chain is known as *homogeneous*. The probability of transition $M_{ij}(T)$ from the state i to the state j can be expressed in the following form:

$$M_{ij}(T) = \begin{cases} P_{ij} \cdot A_{ij}(T) & \text{if } i \neq j \\ 1 - \Sigma_{k \neq i} P_{ik} \cdot A_{ik}(T) & \text{if } i = j \end{cases}$$

which includes the following notations:

P_{ij} the probability of perturbation, i.e. the probability of generating the state j when one is in the state i;

and:

$A_{ij}(T)$ the probability of acceptance, i.e. the probability of accepting the state j when one is in the state i, at the temperature T.

The first factor P_{ij} can be easily calculated. In fact, the system is generally perturbed by randomly choosing a movement from the allowed elementary movements. It results from it that:

$$P_{ij} = \begin{cases} |R_i|^{-1} & \text{if } j \in I_{R_i} \\ 0 & \text{if } j \notin I_{R_i} \end{cases}$$

where R_i denotes the subset of R comprising of all the configurations which can be obtained in only one movement starting from the state i, and I_{R_i} denotes the set of the numbers of these configurations. As for the second factor $A_{ij}(T)$, it is often defined by the Metropolis rule. Aarts and Van Laarhoven noted that, more generally, the simulated annealing method makes it possible to impose the following five conditions:

1. the configuration space is *connected*, i.e. two unspecified states i and j correspond by a completed number of elementary movements.
2. $\forall i, j \in I_R : P_{ij} = P_{ji}$ (reversibility).
3. $A_{ij}(T) = 1$, if $\Delta C_{ij} = C(r_j) - C(r_i) \leq 0$
   (the movements which result in a reduction of energy are systematically accepted).
4. if $\Delta C_{ij} > 0$ $\begin{cases} \lim\limits_{T \to \infty} A_{ij}(T) = 1 \\ \lim\limits_{T \to 0} A_{ij}(T) = 0 \end{cases}$
   (the movements which result in an increase in energy are all accepted at infinite temperature, and all refused at zero temperature).
5. $\forall i, j, k \in I_r \mid C(r_k) \geq C(r_j) \geq C(r_i) : A_{ik}(T) = A_{ij}(T) \cdot A_{jk}(T)$

Asymptotic behavior of the homogeneous Markov chains

By using the results obtained for the homogeneous Markov chains, one can establish the following properties.

Property 1.

Let us consider the Markov process generated by a mechanism of transition which observes the five conditions stated above. This mechanism is applied n times, at constant temperature, starting from a specified initial configuration, arbitrarily chosen. When n tends towards infinity, the Markov chain obtained has an equilibrium vector and only one, called $q(T)$, which is independent of the initial configuration. This vector, which consists of $|R|$ components, is called *distribution of static probability* of the Markov chain. Its ith component, i.e. $q_i(T)$, represents the probability that the system is in the configuration i when, after an infinity of transitions, the static condition is reached.

Property 2.

$q_i(T)$ is expressed by the following relation: $q_i(T) = \dfrac{A_{i_0 i}(T)}{\sum\limits_{i=1}^{|R|} A_{i_0 i}(T)}$,

where i_0 denotes the number of optimal configurations.

Property 3.

When the temperature tends towards infinity or zero, the limit values of $q_i(T)$
are given by : $\lim_{T \to \infty} q_i(T) = |R|^{-1}$ and $\lim_{T \to 0} q_i(T) = \begin{cases} |R_0|^{-1} & \text{if } i \in I_{R_0} \\ 0 & \text{if } i \notin I_{R_0} \end{cases}$
where R_0 denotes the set of the optimal configurations:

$$R_0 = \{r_i \in R \mid C(r_i) = C(r_{i_0})\}$$

Property 3 results immediately from property 2 when condition (4) is used.
Its interpretation is the following: for larger values of the temperature, all the
configurations can be obtained with the same probability. On the other hand,
when the temperature tends towards zero, the system reaches an optimal
configuration with a probability equal to the unit. In both cases, the result is
obtained at the end of a Markov chain of infinite length.

Remarks.

If one chooses the probability of acceptance $A_{ij}(T)$ recommended by Metropolis (see in the reference [Aarts and Van Laarhoven, 1985] a justification for
this choice independently of any analogy with physics):

$$A_{ij}(T) = \begin{cases} e^{\frac{-\Delta C_{ij}}{T}} & \text{if } \Delta C_{ij} > 0 \\ 1 & \text{if } \Delta C_{ij} \leq 0 \end{cases}$$

one finds, in property 2, the expression for the distribution of Boltzmann.

Choice of the annealing parameters

We saw in the preceding paragraph that the convergence of the simulated
annealing algorithm is assured when the temperature tends towards zero. A
Markov chain of infinite length undoubtedly ends in the optimal result if it is
built at a sufficiently low temperature (though nonzero). But this result is not
of any practical utility because, in this case, the balance is approached very
slowly. The Markov chain formalism makes it possible to theoretically examine the convergence speed of the algorithm. One can show that this speed is
improved when one starts from a high temperature and this temperature is
decreased in stages. This procedure requires the use of an annealing program,
which defines the optimal values of the parameters of the descent in temperature. We will successively examine four principal parameters of the annealing
program:

- the initial temperature;
- the length of the homogeneous Markov chains, i.e. the criterion for change
  of temperature stage;
- the law of decrease of the temperature;

- the criterion for program termination.

For each one of them, we will indicate initially the regulations resulting from the theory, which lead to an optimal result, but often at the cost of a prohibitive computing time. Then we mention the values obtained by the experiment.

Initial temperature

There exists a necessary condition, but insufficient, so that the optimization process does not get trapped in a local minimum. The initial temperature T_0 must be sufficiently high so that, at the end of the first stage, all the configurations can be obtained with the same probability. A suitable expression of T_0, which ensures a rate of acceptance close to 1, is the following:

$$T_0 = r \cdot \max_{ij} \Delta C_{ij}$$

with $r \gg 1$ (typically $r = 10$). In practice, in many combinatorial optimization problems, this rule is difficult to employ, because it is difficult to evaluate a priori $\max_{ij} \Delta C_{ij}$. The choice of T_0 will be able in this case to result from an experimental procedure, carried out before the process of optimization itself. During such a procedure, one calculates the evolution of the system during a limited time; one acquires some knowledge about the configuration space, from which one can determine T_0. This preliminary experiment can simply consist in calculating the average value of the variation in energy ΔC_{ij}, by maintaining the temperature to zero. Aarts and Van Laarhoven propose a more sophisticated preliminary procedure: they established an iterative formula which makes it possible to adjust the value of T_0 after each perturbation, so that the rate of acceptance is maintained constant. The authors indicate that this algorithm led to good results if the values of the cost function for the various system configurations are distributed in a sufficiently uniform way.

Length of the Markov chains (or length of the temperature stages); law of decrease of the temperature

The length of the Markov chain, which determines the length of the temperature stages, and the law of decrease of the temperature, which acts on the number of stages, are two parameters of the annealing program very closely dependent on each other and which are most critical from the point of view of the computing time involved. A first approach of the problem consists in seeking the optimal solution, by fixing the length M of the Markov chains so as to reach quasi-equilibrium, i.e. to approach equilibrium, at a short distance ϵ fixed a priori, characterized by the vector of static probability distribution $q(T)$. One obtains the following condition:

$$M > K \left(|R|^2 - 3|R| + 3 \right)$$

where K is a constant which depends on ϵ. In the majority of the combinatorial optimization problems, the total number of configurations $|R|$ is an exponential function of the number N of the variables of the system. Consequently, the preceding inequality leads to an exponential computing time, which was confirmed by experimental observations in the case of a particular form of the traveling salesman problem (the cities considered occupy all the nodes of a plane square network, which makes it possible to easily calculate the exact value of the global optimum of the cost function: the a priori knowledge of the solution is very useful to analyze the convergence of the algorithm). These experimental results also show that a considerable gain in CPU time is obtained if one agrees to deviate a little from the optimum. A deviation in the final result of only 2% compared to the optimum makes it possible to decrease the exponential computing time to a cubic time of N. This gave rise to the idea to take the theoretical investigations again, by seeking the parameters of the annealing program that ensure a deviation from the true optimum, and this, independently of the dimension of the problem considered. The starting postulate of the reasoning is as follows: for each homogeneous Markov chain generated during the process of optimization, the distribution of the states must be close to the static distribution (i.e. Boltzmann distribution, if one adopts the Metropolis rule of acceptance). This situation can be implemented on the basis of a high temperature (for which one arrives quickly at quasi-equilibrium, as indicated by the property 3). Then it is necessary to choose the rate of decrease of the temperature such that the static distributions corresponding to two successive values of T are close. This way, after each change of temperature stage, the distribution of the states approaches the new static distribution quickly, so that the length of the successive chains can be maintained small. There one can see the strong interaction that exists between the length of the Markov chains and the rate of decrease of the temperature. Let by T and T' be the temperatures of two unspecified successive stages and α be the rate of decrease of the temperature ($T' = \alpha T < T$). The condition to be satisfied can be written as:

$$\|q(T) - q(T')\| < \epsilon$$

(ϵ is a positive and small number)

This condition is equivalent to the following, which is easier to use:

$$\forall i \in I_R : \frac{1}{1+\delta} < \frac{q_i(T)}{q_i(T')} < 1 + \delta$$

(δ is also a positive and small number, called the distance parameter). It can then be shown, with the help of some approximations, that the rate of decrease of the temperature can be written as:

$$\alpha = \frac{1}{\left(1 + \frac{T \cdot \ln(1+\delta)}{3 \cdot \sigma(T)}\right)} \tag{A.1}$$

when $\sigma(T)$ is the standard deviation of the values of the cost function for the states of the Markov chain at the temperature T.

The authors moreover recommend the following choice for the length of the Markov chains:

$$M = \max_{i \in I_R} |R_i| \qquad (A.2)$$

(it is pointed out that R_i is the subset of R comprising all the configurations that can be obtained in only one movement starting from the stage i). The Markov chain formalism leads thus to an annealing program characterized by a constant length of Markov chain and a variable rate of decrease of the temperature. This result, which is based on the theory, differs from the usual empirical approach: in this last case, one adopts a variable length of stage of temperature and a constant rate α of decrease of the temperature, typically ranging between 0.90 and 0.99. It is observed however that the parameter α is not very critical in achieving the convergence of the algorithm, provided the stage of temperature lasts long enough.

Program termination criterion

Quantitative information on the progress of the optimization process can be drawn from the *entropy*, which is a natural measurement of the order of the system. This one is defined by the following expression:

$$S(T) = -\sum_{i=1}^{|R|} q_i(T) \cdot \ln(q_i(T))$$

It is shown that $S(T)$ can be written in the following form :

$$S(T) = S(T_1) - \int_T^{T_1} \frac{\sigma^2(T')}{T'^3} dT'$$

and $\sigma^2(T)$ can be easily estimated numerically using the values of the cost function, for the configurations obtained at the temperature T. A termination criterion can then be elaborated starting from the following ratio, which measures the difference between the current configuration and the optimal configuration

$$\frac{S(T) - S_0}{S_\infty - S_0}$$

where S_∞ and S_0 are defined by the relations:

$$S_\infty = \lim_{T \to \infty} S(T) = \ln|R|$$
$$S_0 = \lim_{T \to 0} S(T) = \ln|R_0|$$

One can also detect the disorder-order transition (and consequently decide to slow down cooling) by observing any steep increase in the following parameter, which is similar to the *specific heat*: $\frac{\sigma^2(T)}{T^2}$. To achieve precise numerical calculations, these criteria are applicable in practice only when the Markov chains are of sufficient length. In the contrary case, another termination criterion can be obtained starting from extrapolation, at zero temperature, of the smoothed average, $C_l(T)$, of the values of the cost function obtained during the process of optimization:

$$\left| \frac{dC_l(T)}{dT} \cdot \frac{T}{C(T_0)} \right| < \epsilon_s \tag{A.3}$$

where ϵ_s is a positive and small number, and $C(T_0)$ the average value of the cost function at the initial temperature T_0.

Remarks.

If one adopts the rate of decrease of the temperature and the termination criterion respectively defined by the relations (A.1) and (A.3), Aarts and Van Laarhoven showed the existence of an upper limit, proportional to $\ln |R|$, for the total number of temperature stages. Moreover, if the length of the Markov chains is fixed in accordance with the relation (A.2), the execution time of the annealing algorithm is proportional to the following expression:

$$\max_{i \in I_R} |R_i| \cdot \ln |R|$$

But the term $\max |R_i|$ is generally a polynomial function of the number of variables of the problem. Consequently, the annealing program consisting of the relations (A.1), (A.2), (A.3) and (A.3) allows to solve the majority of the NP-difficult problems while obtaining, in a polynomial time, a result which presents a variation of a few percent compared to the global optimum, and this, independently of the dimension of the problem considered. The preceding theoretical considerations were confirmed by the application of this annealing program to the traveling salesman and logical partitioning problems.

Modeling of the simulated annealing algorithm by inhomogeneous Markov chains

The results which we presented till now are based on the assumption of a decrease of the temperature in stages (which ensures a fast convergence of the simulated annealing algorithm, as we already mentioned before). This property makes it possible to represent the process of optimization in the form of a completed set of homogeneous Markov chains, whose asymptotic behavior can be simply described. We have seen that it results in a complete theoretical explanation of the operation of the algorithm, and development of an operational annealing program. Certain authors were interested in convergence of the simulated annealing algorithm while placing themselves within the more general framework of the theory of the inhomogeneous Markov chains. In this case, the asymptotic behavior is more delicate to study: for example, Gidas

[Gidas, 1985] shows the possibility of appearance of phenomena similar to the phase transitions. We will be satisfied here to discuss the main result of this work of primarily theoretical interest: the annealing algorithm converges towards a global optimum, with a probability equal to unity if, when time t tends towards infinity, the temperature $T(t)$ does not decrease more quickly than the expression $\frac{C}{\ln(t)}$, where C denotes a constant that is related to the depth of the "energy wells" of the problem.

B

Complete Example of Implementation of Tabu Search for the Quadratic Assignment Problem

The following procedure establishes a tabu search for the quadratic assignment problem. This procedure uses the transpositions like neighborhood. The prohibition mechanism consists in preventing two elements from having positions which they both occupied recently. The long-term memory consists in forcing a transposition that places two elements at positions that they never occupied during a large number of iterations, and this, independently of the quality of the solution that this movement carries out. This program establishing a tabu search for the quadratic assignment problem is known as *Robust taboo search* in the literature [Taillard, 1991]. For certain types of problems, it is at present one of the best methods.

A C++ program for the *Robust taboo search* method
of [Taillard, 1991].

```
#include <iostream.h>
#include <fstream.h>

const long infinite = 999999999;

typedef int*   type_vector;
typedef long** type_matrix;

/*************** L'Ecuyer random number generator ***************/
const long m = 2147483647; const long m2 = 2145483479;
const long a12= 63308; const long q12=33921; const long r12=12979;
const long a13=-183326; const long q13=11714; const long r13=2883;
const long a21= 86098; const long q21=24919; const long r21= 7417;
const long a23=-539608; const long q23= 3976; const long r23=2071;
const double invm = 4.656612873077393e-10;

long x10 = 12345, x11 = 67890, x12 = 13579,   // initialization of
     x20 = 24680, x21 = 98765, x22 = 43210;   // seeds values

double rando()
 {long h, p12, p13, p21, p23;
  h = x10/q13;  p13 = -a13*(x10-h*q13)-h*r13;
  h = x11/q12;  p12 = a12*(x11-h*q12)-h*r12;
  if (p13 < 0) p13 = p13 + m; if (p12 < 0) p12 = p12 + m;
  x10 = x11; x11 = x12; x12 = p12-p13; if (x12 < 0) x12 = x12 + m;
  h = x20/q23;  p23 = -a23*(x20-h*q23)-h*r23;
  h = x22/q21;  p21 = a21*(x22-h*q21)-h*r21;
  if (p23 < 0) p23 = p23 + m2; if (p21 < 0) p21 = p21 + m2;
  x20 = x21; x21 = x22; x22 = p21-p23; if(x22 < 0) x22 = x22 + m2;
  if (x12 < x22) h = x12 - x22 + m; else h = x12 - x22;
  if (h == 0) return(1.0); else return(h*invm);
 }

/*********** return an integer between low and high ***********/
long unif(long low, long high)
 {return(low + long(double(high - low + 1) * rando() ));}

void transpose(int & a, int & b) {long temp = a; a = b; b = temp;}

int min(long a, long b) {if (a < b) return(a); else return(b);}
```

```
/*---------------------------------------------------------*/
/*         compute the cost difference if elements i and j  */
/*            are transposed in permutation (solution) p    */
/*---------------------------------------------------------*/
long compute_delta(int n, type_matrix & a, type_matrix & b,
                   type_vector & p, int i, int j)
 {long d; int k;
  d = (a[i][i]-a[j][j])*(b[p[j]][p[j]]-b[p[i]][p[i]]) +
      (a[i][j]-a[j][i])*(b[p[j]][p[i]]-b[p[i]][p[j]]);
  for (k = 1; k <= n; k = k + 1) if (k!=i && k!=j)
    d = d + (a[k][i]-a[k][j])*(b[p[k]][p[j]]-b[p[k]][p[i]]) +
            (a[i][k]-a[j][k])*(b[p[j]][p[k]]-b[p[i]][p[k]]);
  return(d);
 }


/*---------------------------------------------------------*/
/*      Idem, but the value of delta[i][j] is supposed to   */
/*      be known before the transposition of elements r and s */
/*---------------------------------------------------------*/
long compute_delta_part(type_matrix & a, type_matrix & b,
                        type_vector & p, type_matrix & delta,
                        int i, int j, int r, int s)
 {return(delta[i][j]+(a[r][i]-a[r][j]+a[s][j]-a[s][i])*
     (b[p[s]][p[i]]-b[p[s]][p[j]]+b[p[r]][p[j]]-b[p[r]][p[i]])+
     (a[i][r]-a[j][r]+a[j][s]-a[i][s])*
     (b[p[i]][p[s]]-b[p[j]][p[s]]+b[p[j]][p[r]]-b[p[i]][p[r]]) );
 }

void tabu_search(long n,                        // problem size
                 type_matrix & a,               // flows matrix
                 type_matrix & b,               // distance matrix
                 type_vector & best_sol,        // best solution found
                 long & best_cost,              // cost of best solution
                 long min_size,                 // parameter 1 (< n^2/2)
                 long max_size,                 // parameter 2 (< n^2/2)
                 long aspiration,               // parameter 3 (> n^2/2)
                 long nr_iterations)            // number of iterations


 {type_vector p;                                // current solution
  type_matrix delta;                            // store move costs
  type_matrix tabu_list;                        // tabu status
  long current_iteration;                       // current iteration
  long current_cost;                            // current sol. value
  int i, j, k, i_retained, j_retained;          // indices
```

```
/**************** dynamic memory allocation ******************/
p = new int[n+1];
delta = new long* [n+1];
for (i = 1; i <= n; i = i+1) delta[i] = new long[n+1];
tabu_list = new long* [n+1];
for (i = 1; i <= n; i = i+1) tabu_list[i] = new long[n+1];

/************* current solution initialization ***************/
for (i = 1; i <= n; i = i + 1) p[i] = best_sol[i];

/********** initialization of current solution value **********/
/*************** and matrix of cost of moves  ****************/
current_cost = 0;
for (i = 1; i <= n; i = i + 1) for (j = 1; j <= n; j = j + 1)
 {current_cost = current_cost + a[i][j] * b[p[i]][p[j]];
  if (i < j) {delta[i][j] = compute_delta(n, a, b, p, i, j);};
 };
best_cost = current_cost;

/***************** tabu list initialization ******************/
for (i = 1; i <= n; i = i + 1) for (j = 1; j <= n; j = j+1)
   tabu_list[i][j] = -(n*i + j);

/****************** main tabu search loop *******************/
for (current_iteration = 1; current_iteration <= nr_iterations;
     current_iteration = current_iteration + 1)
 {/** find best move (i_retained, j_retained) **/

   i_retained = infinite;              // in case all moves are tabu
   long min_delta = infinite;                // retained move cost
   int autorized;                             // move not tabu?
   int aspired;                               // move forced?
   int already_aspired = false;        // in case many moves forced

   for (i = 1; i < n; i = i + 1)
     for (j = i+1; j <= n; j = j+1)
       {autorized = (tabu_list[i][p[j]] < current_iteration) ||
                    (tabu_list[j][p[i]] < current_iteration);

        aspired =
          (tabu_list[i][p[j]] < current_iteration-aspiration)||
          (tabu_list[j][p[i]] < current_iteration-aspiration)||
          (current_cost + delta[i][j] < best_cost);
```

```
    if ((aspired && !already_aspired) ||   // first move aspired
        (aspired && already_aspired &&    // many move aspired
        (delta[i][j] < min_delta) ) ||   // => take best one
       (!aspired && !already_aspired &&   // no move aspired yet
        (delta[i][j] < min_delta) && autorized))
      {i_retained = i; j_retained = j;
       min_delta = delta[i][j];
       if (aspired) {already_aspired = true;};
      };
   };

if (i_retained == infinite) cout << "All moves are tabu! \n";
else
  {/** transpose elements in pos. i_retained and j_retained **/
   transpose(p[i_retained], p[j_retained]);
                                        // update solution value
   current_cost = current_cost + delta[i_retained][j_retained];
            // forbid reverse move for a random number of iterations
   tabu_list[i_retained][p[j_retained]] =
     current_iteration + unif(min_size,max_size);
   tabu_list[j_retained][p[i_retained]] =
     current_iteration + unif(min_size,max_size);


                                   // best solution improved ?
   if (current_cost < best_cost)
     {best_cost = current_cost;
      for (k = 1; k <= n; k = k+1) best_sol[k] = p[k];
      cout << "Solution of value " << best_cost
           << " found at iter. " << current_iteration << '\n';
     };


                           // update matrix of the move costs
   for (i = 1; i < n; i = i+1) for (j = i+1; j <= n; j = j+1)
     if (i != i_retained && i != j_retained &&
         j != i_retained && j != j_retained)
       {delta[i][j] =
          compute_delta_part(a, b, p, delta,
                                 i, j, i_retained, j_retained);}
       else
       {delta[i][j] = compute_delta(n, a, b, p, i, j);};
  };

};
```

```
                                                       // free memory
  delete[] p;
  for (i=1; i <= n; i = i+1) delete[] delta[i]; delete[] delta;
  for (i=1; i <= n; i = i+1) delete[] tabu_list[i];
  delete[] tabu_list;
}                                                              // tabu

void generate_random_solution(long n, type_vector & p)
 {int i;
  for (i = 0; i <= n; i = i+1) p[i] = i;
  for (i = 1; i <  n; i = i+1) transpose(p[i], p[unif(i, n)]);
 }

int n;                                            // problem size
type_matrix a, b;                      // flows and distances matrices
type_vector solution;                      // solution (permutation)
long cost;                                        // solution cost

ifstream data_file;
char file_name[30];
int i, j;


main()
 {/************* read file name and problem size **************/
  cout << "Data file name : \n";
  cin >> file_name; cout << file_name << '\n';
  data_file.open(file_name);
  data_file >> n;

  /**************** dynamic memory allocation ****************/
  a = new long* [n+1];
  for (i = 1; i <= n; i = i+1) a[i] = new long[n+1];
  b = new long* [n+1];
  for (i = 1; i <= n; i = i+1) b[i] = new long[n+1];
  solution = new int[n+1];

  /************* read flows and distances matrices *************/
  for (i = 1; i <= n; i = i+1) for (j = 1; j <= n; j = j+1)
    data_file >> a[i][j];
  for (i = 1; i <= n; i = i+1) for (j = 1; j <= n; j = j+1)
    data_file >> b[i][j];
  data_file.close();
```

```
generate_random_solution(n, solution);
tabu_search(n, a, b,                          // problem data
            solution, cost,                   // tabu search results
            9*n/10, 11*n/10, n*n*2,                    // parameters
            1000000);                         // number of iterations

cout << "Solution found by tabu search :\n";
for (i = 1; i <= n; i = i+1) cout << solution[i] << ' ';
cout << '\n';
}
```

References

[Aarts et al., 1986] Aarts, E. H. L., De Bont, F. M. J., Habers J. H. A., and Van Laarhoven, P. J. M. (1986). A parallel statistical cooling algorithm. In *Proc. of the 3rd Annual Symposium on Theoretical Aspects of Computer Science*, volume 210 of *Lecture Notes in Computer Science*, pages 87–97.

[Aarts and Van Laarhoven, 1985] Aarts, E. H. L. and Van Laarhoven, P. J. M. (1985). Statistical cooling : a general approach to combinatorial optimisation problems. *Philips J. of Research*, 40:193–226.

[Ackley, 1987] Ackley, D. H. (1987). *A Connectionist Machine for Genetic Hill-climbing*. Kluwer.

[Ahuja et al., 1993] Ahuja, R., Magnanti, T., and Orlin, J. (1993). *Network Flows, Theory, Algorithms and Applications*. Prentice Hall.

[Altman et al., 2002] Altman, Z., Picard, J., Ben Jamaa, S., Fouresté, B., Caminada, A., Dony, T., Morlier, J., and Mourniac, S. (2002). New challenges in automatic cell planning of UMTS networks. In *IEEE International Symposium VTC*, Vancouver.

[Aluffi-Pentini et al., 1985] Aluffi-Pentini, F., Parisi, V., and Zirilli, F. (1985). Global optimization and stochastic differential equations. *J. of optimization theory and applications*, 47(1):1–16.

[Angeline, 1996] Angeline, P. J. (1996). Genetic programming's continued evolution. In Angeline, P. and Kinnear, K., editors, *Advances in Genetic Programming, vol 2*, pages 89–110. MIT Press.

[Azencott, 1992] Azencott, R., editor (1992). *Simulated Annealing : Parallelization Techniques*. John Wiley and Sons.

[Baeck et al., 2000a] Baeck, T., Fogel, D. B., and Michalewicz, Z. (2000a). *Evolutionary Computation 1: Basic Algorithms and Operators*. Institute of Physics Publishing.

[Baeck et al., 2000b] Baeck, T., Fogel, D. B., and Michalewicz, Z. (2000b). *Evolutionary Computation 2: Advanced Algorithms and Operators*. Institute of Physics Publishing.

[Baker, 1987] Baker, J. E. (1987). Reducing bias and inefficiency in the selection algorithm. In Grefenstette, J. J., editor, *Proc. 2nd Int. Conf. on Genetic Algorithms*, pages 14–21.

[Barr et al., 1995] Barr, R. S., Golden, B. L., Kelly, J. P., Resende, M. G. C., and Stewart, W. R. (1995). Designing and reporting on computational experiments with heuristic methods. *Journal of Heuristics*, 1(1):9–32.

[Beasley et al., 1993] Beasley, D., Bull, D. R., and Martin, R. R. (1993). A sequential niche technique for multimodal function optimization. *Evolutionary Computation*, 1(2):101–125.

[Beckers et al., 1992] Beckers, R., Deneubourg, J. L., and Goss, S. (1992). Trails an U-Turns in the Selection of a Path by the Ant Lasius Niger. *J. Theor. Biol.*, 159:397–415.

[Belideanu and Contjean, 1994] Belideanu, N. and Contjean, E. (1994). Introducing global constraints in chip. *Mathematical and computer Modelling*, 12:97–123.

[Ben Jamaa et al., 2003] Ben Jamaa, S., Altman, Z., Picard, J., Fourestié, B., and Mourlon, J. (2003). Manual and automatic design for UMTS networks. In *WiOpt'03: Modeling and Optimization in Mobile, Ad Hoc and Wireless Networks*, Sophia-Antipolis, France.

[Berthiau et al., 1994] Berthiau, G., Durbin, F., Haussy, J., and Siarry, P. (1994). Learning of neural networks approximating continuous functions through circuit simulator SPICE-PAC driven by Simulated Annealing. *International Journal of Electronics*, 76:437–441.

[Bertocchi and Odoardo, 1991] Bertocchi, M. and Odoardo, C. D. (1991). A stochastic algorithm for global optimization based on threshold accepting technique. In *11th European Congress on Operational Research (EURO XI)*, Aachen, Germany.

[Bessaou et al., 2000] Bessaou, M., Petrowski, A., and Siarry, P. (2000). Island model cooperating with speciation for multimodal optimization. In Schwefel, H.-P., Schoenauer, M., Deb, K., Rudolph, G., Yao, X., Lutton, E., and Merelo, J. J., editors, *Parallel Problem Solving from Nature - PPSN VI 6th International Conference*, Paris, France. Springer Verlag.

[Bessiere and Regin, 1997] Bessiere, C. and Regin, J.-C. (1997). Arc consistency for general constraint networks: preliminary results. In *Proceedings of the 15th IJCAI*, pages 398–404.

[Beyer, 2001] Beyer, H.-G. (2001). *The Theory of Evolution Strategies, Natural Computing Series*. Springer.

[Bieszczad and White, 1999] Bieszczad, A. and White, T. (1999). *The Fundamentals of Network Management*, chapter Mobile Agents for Network Management. Plenum books.

[Bilchev and Parmee, 1995] Bilchev, G. and Parmee, I. (1995). The Ant Colony Metaphor for Searching Continuous Design Spaces. *Lecture Notes in Computer Science*, 993:25–39.

[Birattari et al., 2002] Birattari, M., Di Caro, G., and Dorigo, M. (2002). Toward the Formal Foundation of Ant Programming. In Dorigo, M., Di Caro, G., and Sampels, M., editors, *Proceedings of the Third International Workshop on Ant Algorithms (ANTS'2002)*, volume 2463 of *Lecture Notes in Computer Science*, pages 188–201, Brussels, Belgium. Springer Verlag.

[Bonabeau et al., 1999] Bonabeau, E., Dorigo, M., and Theraulaz, G. (1999). *Swarm Intelligence, From Natural to Artificial Systems*. Oxford University Press.

[Bonabeau et al., 1996] Bonabeau, E., Theraulaz, G., and Deneubourg, J.-L. (1996). Quantitative Study of the Fixed Threshold Model for the Regulation of Division of Labour in Insect Societies. In *Proceedings Roy. Soc. London B*, volume 263.

[Bonabeau et al., 1998] Bonabeau, E., Theraulaz, G., and Deneubourg, J.-L. (1998). Fixed Response Thresholds and the Regulation of Division of Labor in Insect Societies. *Bulletin of Mathematical Biology*, (60):753–807.

[Bonomi and Lutton, 1984] Bonomi, E. and Lutton, J. L. (1984). The N-city travelling salesman problem, Statistical Mechanics and the Metropolis Algorithm. *SIAM Review*, 26(4):551–568.

[Brandimarte, 1992] Brandimarte, P. (1992). Neighbourhood search-based optimization algorithms for production scheduling : a survey. *Computer-Integrated Manufacturing Systems*, 5(2):167–176.

[Bullnheimer et al., 1997] Bullnheimer, B., Hartl, R., and Strauss, C. (1997). An improved ant system algorithm for the vehicle routing problem. Technical report, Institute of Management Science, Vienna, Austria.

[Bullnheimer et al., 1999] Bullnheimer, B., Hartl, R., and Strauss, C. (1999). Applying the ant system to the vehicle routing problem. In Roucairol, S., Voss, S., Martello, I., and Osman, C., editors, *Meta-heuristics : Advances and Trends in Local search Paradigms for Optimization*. Kluwer Academic Publishers, pages 285–296, USA. Kluwer Academic.

[Burkard and Fincke, 1985] Burkard, R. E. and Fincke, U. (1985). Probabilistic properties of some combinatorial optimization problems. *Discrete Applied Mathematics*, 12:21–29.

[Camazine et al., 2000] Camazine, S., Deneubourg, J., Franks, N., Sneyd, J., Theraulaz, G., and Bonabeau, E. (2000). *Self-Organization in Biological Systems*. Princeton University Press.

[Campos et al., 2000] Campos, M., Bonabeau, E., Theraulaz, G., and Deneubourg, J.-L. (2000). Dynamic Scheduling and Division of Labor in Social Insects. In *Adaptive Behavior 2000*, pages 83–96.

[Casotto et al., 1987] Casotto, A., Romea, F., and Sangiovanni-Vincentelli, A. (1987). A parallel simulated annealing algorithm for the placement of macro-cells. *IEEE Trans. on C.A.D.*, CAD-6(5):838–847.

[Cerny, 1985] Cerny, V. (1985). Thermodynamical approach to the traveling salesman problem : an efficient simulation algorithm. *J. of Optimization Theory and Applications*, 45(1):41–51.

[Charon and Hudry, 2002] Charon, I. and Hudry, O. (2002). The noising methods: a survey. In Hansen, P. and Ribeiro, C. C., editors, *Essays and Surveys in Meta-heuristics*, pages 245–261. Kluwer Academic Publishers.

[Chelouah and Siarry, 2000a] Chelouah, R. and Siarry, P. (2000a). A Continuous Genetic Algorithm Designed for the Global Optimization. *Journal of Heuristics*, 6:191–213.

[Chelouah and Siarry, 2000b] Chelouah, R. and Siarry, P. (2000b). Tabu Search Applied to Global Optimization. *European Journal of Operational Research*, 123:256–270.

[Chelouah and Siarry, 2003] Chelouah, R. and Siarry, P. (2003). Genetic and Nelder-Mead algorithms hybridized for a more accurate global optimization of continuous multiminima functions. *European Journal of Operational Research*, 148:335–348.

[Chelouah et al., 2000] Chelouah, R., Siarry, P., Berthiau, G., and De Barmon, B. (2000). An optimization method fitted for model inversion in non destructive control by eddy currents. *The European Physical Journal, Applied Physics*, 12:231–238.

[Cherruault, 1986a] Cherruault, Y. (1986a). *Mathematical modelling in Biomedicine*. D. Reidel Publishing Company.

[Cherruault, 1986b] Cherruault, Y. (1986b). *Mathematical Modelling in Biomedicine. Optimal Control of Biomedical Systems*. D. Reidel Publishing Company.

[Cherruault, 1989] Cherruault, Y. (1989). A new method for global optimization (Alienor). *Kybernetes*, 19(3):19–32.

[Choo, 2000] Choo, S.-Y. (2000). *Genetic Algorithms and Genetic Programming at Stanford 2000*, chapter Emergence of a Division of Labour in a Bee Colony, pages 98–107. Stanford Bookstore, Stanford, California.

[Christofides et al., 1979] Christófides, N., Mingozzi, A., and Toth, P. (1979). The vehicle routing problem. In Christofides, N., Mingozzi, A., Toth, P., and Sandi, C., editors, *Combinatorial Optimization*, pages 315–338. Wiley.

[Cicirello and Smith, 2001] Cicirello, V. and Smith, S. (2001). Wasp-like Agents for distributed Factory Coordination. Technical Report CMU-RI-TR-01-39, Robotics Institute, Carnegie Mellon University, Pittsburgh.

[Clarke and Wright, 1964] Clarke, G. and Wright, G. W. (1964). Scheduling of vehicles from a central depot to a number of delivery points. *Operations Research*, 12:568–581.

[Clerc et al., 2002] Clerc, G., Bessaou, M., Siarry, P., and Bastiani, P. (2002). Identification des machines synchrones par algorithme génétique ; principe et application. *Revue Internationale de Génie Electrique*, 5:485–515.

[Clerc, 2005] Clerc, M. (2005). *Particle Swarm Optimization*. Hermes Science.

[Clerc and Kennedy, 2002] Clerc, M. and Kennedy, J. (2002). The particle swarm: explosion, stability and convergence in a multi-dimensional complex space. *IEEE Transactions on Evolutionnary Computation*, 6:58–73.

[Coffin and Saltzman, 2000] Coffin, M. and Saltzman, M. J. (2000). Statistical analysis of computational tests of algorithms and heuristics. *INFORMS Journal on Computing*, 12(1):24–44.

[Cohoon et al., 1991] Cohoon, J., Hegde, S., Martin, W., and Richards, D. (1991). Distributed genetic algorithms for the floorplan design problem. *IEEE Trans. on Computer-Aided Design*, 10(4):483–492.

[Cohoon et al., 1987] Cohoon, J. P., Hedge, S. U., Martin, W. N., and Richards, D. (1987). Punctuated equilibria: A parallel genetic algorithm. In Grefenstette, J. J., editor, *Genetic algorithms and their applications : Proc. of the second Int. Conf. on Genetic Algorithms*, pages 148–154, Hillsdale, NJ. Lawrence Erlbaum Assoc.

[Collette, 2002] Collette, Y. (2002). *Contribution à l'évaluation et au perfectionnement des méthodes d'optimisation multiobjectif. Application à l'optimisation des plans de rechargement de combustible nucléaire*. PhD thesis, Université de Paris XII Val-de-Marne.

[Collette and Siarry, 2003] Collette, Y. and Siarry, P. (2003). *Multiobjective Optimization*. Springer.

[Collins et al., 1988] Collins, N. E., Eglese, R. W., and Golden, B. (1988). Simulated annealing - An annotated bibliography. *American Journal of Mathematical and Management Sciences*, 8:209–307.

[Colorni et al., 1992] Colorni, A., Dorigo, M., and Maniezzo, V. (1992). Distributed Optimization by Ant Colonies. In Varela, F. and Bourgine, P., editors, *Proceedings of ECAL'91 - First European Conference on Artificial Life*, pages 134–142, Paris, France. Elsevier Publishing.

[Conover, 1999] Conover, W. J. (1999). *Practical Nonparametric Statistics*. Wiley, Weinheim, 3rd edition.

[Courat et al., 1994] Courat, J., Raynaud, G., Mrad, I., and Siarry, P. (1994). Electronic component model minimisation based on Log Simulated Annealing. *IEEE Trans. on Circuits and Systems*, 41(12):790–795. part I.

[Courat et al., 1995] Courat, J., Raynaud, G., and Siarry, P. (1995). Extraction of the topology of equivalent circuits based on parameter statistical evolution driven by Simulated Annealing. *International Journal of Electronics*, 79:47–52.

[Courrieu, 1991] Courrieu, P. (1991). A distributed search algorithm for hard optimization. Technical Report TA-9101, Univ. de Provence, CREPCO (URA CNRS 182).

[Cramer, 1985] Cramer, N. L. (1985). A representation for the adaptive generation of simple sequential programs. In Grefenstette, J., editor, *Proc. 1st Int. Conf. on Genetic Algorithms and Their Applications*, pages 183–187.

[Creutz, 1983] Creutz, M. (1983). Microcanonical Monte Carlo simulation. *Physical Review Letters*, 50(19):1411–1414.

[Darwin, 1859] Darwin, C. (1859). *On The Origin of Species by Means of Natural Selection or the Preservation of Favored Races in the Struggle for Life*. Murray, London.

[Dasgupta, 1999] Dasgupta, D. (1999). *Artificial Immune Systems and their applications*. Springer Verlag.

[Dasgupta and Attoh-Okine, 1997] Dasgupta, D. and Attoh-Okine, N. (1997). Immune-based systems: A survey. In *Proceedings of the IEEE International Conference on Systems, Man and Cybernetics*, volume 1, pages 369–374, Orlando. IEEE Press.

[Davis, 1991] Davis, L. (1991). *Handbook of Genetic Algorithms, p. 80*. Van Nostrand Reinhold.

[De Backer et al., 2000] De Backer, B., Furnon, V., Kilby, P., Prosser, P., and Shaw, P. (2000). Solving vehicle routing problems with constraint programming et metaheuristics. *Journal of Heuristics*, 6(4).

[De Backer et al., 2002] De Backer, B., Shaw, P., and Furnon, V. (2002). A constraint programming toolkit for local search. In Voss, S. and Woodruff, D. L., editors, *Optimization Software Class Libraries*, pages 219–262. Kluwer Academic Publishers.

[De Boer, 2002] De Boer, P.-T. (2002). The Cross-Entropy Method. University of Twente. http://wwwhome.cs.utwente.nl/p̃tdeboer/ce/.

[De Castro and Von Zuben, 1999] De Castro, L. and Von Zuben, F. (1999). Artificial Immune Systems: Part I: Basic Theory and Applications. Technical Report TR-DCA 01/99, Department of Computer Engineering and Industrial Automation, School of Electrical and Computer Engineering, State University of Campinas, Brazil.

[De Castro and Von Zuben, 2000] De Castro, L. and Von Zuben, F. (2000). Artificial Immune Systems: Part II - A Survey of Applications. Technical Report DCA-RT 02/00, Department of Computer Engineering and Industrial Automation, School of Electrical and Computer Engineering, State University of Campinas, Brazil.

[De Jong, 1975] De Jong, K. A. (1975). *An Analysis of the Behavior of a Class of Genetic Adaptive Systems*. Doctoral Dissertation, University of Michigan.

[De Jong and Sarma, 1993] De Jong, K. A. and Sarma, J. (1993). Generation gaps revisited. In Whitley, L. D., editor, *Foundations of Genetic Algorithms 2*, pages 19–28. Morgan Kaufmann.

[De La Maza and Tidor, 1993] De La Maza, M. and Tidor, B. (1993). An analysis of selection procedures with particular attention paid to proportional and boltzmann selection. In Forrest, S., editor, *Proc. 5th Int. Conf. on Genetic Algorithms*, pages 124–131. Morgan Kaufmann.

[de Werra and Hertz, 1989] de Werra, D. and Hertz, A. (1989). Tabu search techniques: A tutorial and applications to neural networks. *OR Spectrum*, 11:131–141.

[De Wolf et al., 2002] De Wolf, T., Liesbeth, J., Holvoet, T., and Steegmans, E. (2002). A Nested Layered Threshold Model for Dynamic Task Allocation. In Dorigo, M., Di Caro, G., and Sampels, M., editors, *Proceedings of the Third International Workshop on Ant Algorithms (ANTS'2002)*, volume 2463 of *Lecture Notes in Computer Science*, pages 290–291, Brussels, Belgium. Springer Verlag.

[Deb, 2000] Deb, K. (2000). An efficient constraint handling method for genetic algorithms. *Computer Methods in Applied Mechanics and Engineering*, 186(2/4):311–338.

[Deb, 2001] Deb, K. (2001). *Multi-objective Optimization using Evolutionary Algorithms*. John Wiley and sons.

[Delamarre and Virot, 1998] Delamarre, D. and Virot, B. (1998). Simulated annealing algorithm : technical improvements. *Operations Research*, 32(1):43–73.

[Di Caro and Dorigo, 1997] Di Caro, G. and Dorigo, M. (1997). AntNet: A mobile agents approach to adaptive routing. Technical Report IRIDIA/97-12, IRIDIA, Université Libre de Bruxelles, Belgium.

[Di Caro and Dorigo, 1998] Di Caro, G. and Dorigo, M. (1998). AntNet: Distributed stigmergic control for communications networks. *Journal of Artificial Intelligence Research*, 9:317–365.

[Dorigo et al., 2002] Dorigo, M., Di Caro, G., and Sampels, M., editors (2002). *Proceedings of the Third International Workshop on Ant Algorithms (ANTS'2002)*, volume 2463 of *Lecture Notes in Computer Science*, Brussels, Belgium. Springer Verlag.

[Dorigo and Gambardella, 1997a] Dorigo, M. and Gambardella, L. M. (1997a). Ant Colonies for the Traveling Salesman Problem. *BioSystems*, 43:73–81.

[Dorigo and Gambardella, 1997b] Dorigo, M. and Gambardella, L. M. (1997b). Ant Colony System: A Cooperative Learning Approach to the Traveling Salesman Problem. *IEEE Trans. Evol. Comp.*, 1:53–66.

[Dorigo et al., 1996] Dorigo, M., Maniezzo, V., and Colorni, A. (1996). The Ant System: Optimization by a Colony of Cooperating Agents. *IEEE Trans. Syst. Man Cybern*, B(26):29–41.

[Dorigo and Stützle, 2003] Dorigo, M. and Stützle, T. (2003). *Handbook of Metaheuristics*, volume 57 of *International series in operations research and management science*, chapter The Ant Colony Optimization Metaheuristics: Algorithms, Applications and Advances. Kluwer Academic Publishers, Boston Hardbound.

[Dréo and Siarry, 2002] Dréo, J. and Siarry, P. (2002). A New Ant Colony Algorithm Using the Heterarchical Concept Aimed at Optimization of Multiminima Continuous Functions. In Dorigo, M., Di Caro, G., and Sampels, M., editors, *Proceedings of the Third International Workshop on Ant Algorithms (ANTS'2002)*, volume 2463 of *Lecture Notes in Computer Science*, pages 216–221, Brussels, Belgium. Springer Verlag.

[Dréo and Siarry, 2003] Dréo, J. and Siarry, P. (2003). Un algorithme de colonie de fourmis en variables continues hybridé avec un algorithme de recherche locale. In *5e Congrès de la Société Française de Recherche Opérationnelle et d'Aide à la Décision (ROADEF 2003)*, Avignon, France.

[Dueck, 1993] Dueck, G. (1993). New optimization heuristics, the great deluge and the record to record travel. *Journal of Computational physics*, 104:86–92.

[Dueck and Scheuer, 1989] Dueck, G. and Scheuer, T. (1989). Threshold accepting. Technical report, IBM Zentrum, Heidelberg, Germany.

[Duong and Faure, 1998] Duong, V. and Faure, P. (1998). On the applicability of the free-flight mode in european airspace. In *Proceedings of the 2nd usa/europe seminar*.

[Durand and White, 1991] Durand, M. and White, S. (1991). Permissible error in parallel simulated annealing. Technical report, Institut de Recherche en Informatique et Systèmes aléatoires, Rennes.

[Durand, 1996] Durand, N. (1996). *Optimisation de trajectoires pour la résolution de conflits aériens en route*. PhD thesis, Institut National Polytechnique de Toulouse.

[Durand and Alliot, 1998] Durand, N. and Alliot, J. (1998). Genetic crossover operator for partially separable functions. In *Genetic Programming*.

[Durand et al., 1996] Durand, N., Alliot, J., and Noailles, J. (1996). Collision avoidance using neural networks learned by genetic algorithms. In *Ninth International Conference on Industrial and Engineering Applications of Artificial Intelligence and Expert Systems, Fukuoka*.

[Durrenbach et al., 2003] Durrenbach, A., Fourestié, B., Renou, S., and Raoult, M. (2003). Global shadowing margins for 3G networks. In *IEEE VTC Conf.*, Korea.

[Eberhart et al., 2001] Eberhart, R., Kennedy, J., and Shi, Y. (2001). *Swarm Intelligence*. Evolutionnary Computation. Morgan Kaufmann.

[Eiben et al., 1995] Eiben, A. E., van Kemenade, C. H. M., and Kok, J. N. (1995). Orgy in the computer: Multi-parent reproduction in genetic algorithms. In Moran, F., Moreno, A., Merelo, J., and Chacon, P., editors, *Proceedings of the 3rd European Conference on Artificial Life, number 929 in LNAI*, pages 934–945. Springer-Verlag.

[Engrand and Mouney, 1998] Engrand, P. and Mouney, X. (1998). Une méthode originale d'optimisation multiobjectif. Technical Report HT-14/97/035/A, EDF-DER.

[Faigle and Kern, 1992] Faigle, U. and Kern, W. (1992). Some convergence results for probabilistic tabu search. *ORSA Journal on Computing*, 4:32–37.

[Feo and Resende, 1995] Feo, T. and Resende, M. (1995). Greedy randomized adaptive search procedure. *Journal of Global Optimization*, 42:860–878.

[Fleury, 1995] Fleury, G. (1995). Application de méthodes stochastiques inspirées du recuit simulé à des problèmes d'ordonnancement. *RAIRO A.P.I.I. (Automatique - Productique - Informatique industrielle)*, 29(4–5):445–470.

[Fogel et al., 1966] Fogel, L. J., Owens, A. J., and Walsh, M. J. (1966). *Artifical Intelligence through Simulated Evolution*. Wiley.

[Fonseca and Fleming, 1993] Fonseca, C. M. and Fleming, P. J. (1993). Genetic algorithms for multiobjective optimization: Formulation, discussion and generalization. In *Proc. of the Fifth International Conference on Genetic Algorithms*, pages 416–423. Morgan Kaufmann.

[Fraser, 1957] Fraser, A. S. (1957). Simulation of genetic systems by automatic digital computers. *Australian Journal of Biological Sciences*, 10:484–491.

[Frazzoli et al., 1999] Frazzoli, E., Mao, Z., and Feron, E. (1999). Aircraft conflict resolution via semidefinite programming. *AIAA Journal of Guidance, Control and Dynamics*.

[Fron et al., 1993] Fron, X., Maudry, B., and Tumelin, J. (1993). Arc 2000 : Automatic radar control. Technical report, Eurocontrol.

[Gambardella and Dorigo, 2000] Gambardella, L. and Dorigo, M. (2000). Ant Colony System hybridized with a new local search for the sequential ordering problem. *INFORMS Journal on Computing*, 12(3):237–255.

[Gambardella et al., 1999] Gambardella, L., Taillard, E., and Agazzi, G. (1999). *New Ideas in Optimization*, chapter MACS-VPTW: A multiple ant colony system for vehicle routing problems with time windows, pages 63–67. McGraw Hill, London, UK.

[Gambardella and Dorigo, 1995] Gambardella, L. M. and Dorigo, M. (1995). Ant-Q: A Reinforcement Learning Approach to the Travelling Salesman Problem. In *Proceedings Twelfth International Conference on Machine Learning*, volume ML-95, pages 252–260, Palo Alto. Morgan Kaufmann.

[Gämperle et al., 2002] Gämperle, R., Müller, S., and Koumoutsakos, P. (2002). *Advances in Intelligent Systems, Fuzzy Systems, Evolutionnary Computation*, chapter A Parameter Study for Differential Evolution, pages 293–298. WSEAS Press.

[Garey and Johnson, 1979] Garey, M. R. and Johnson, D. S. (1979). *Computers and Intractability: A Guide to the Theory of NP-Completeness*. W H Freeman.

[Gaspar and Collard, 1999] Gaspar, A. and Collard, P. (1999). From GAs to artificial immune systems: improving adaptation in time dependent optimization. In Angeline, P., Michalewicz, Z., Schoenauer, M., Yao, X., and Zalzala, A., editors, *Proceedings of the Congress on Evolutionnary Computation*, volume 3, pages 1859–1866, Washington D.C.

[Geman and Geman, 1984] Geman, S. and Geman, D. (1984). Stochastic relaxation, Gibbs distributions and the Bayesian restoration of images. *IEEE Trans. on Pattern Analysis and Machine Intelligence*, PAMI-6:721–741.

[Geman and Hwang, 1986] Geman, S. and Hwang, C. (1986). Diffusions for global optimization. *SIAM J. on Control and Optimization*, 24(5):1031–1043.

[Gendreau et al., 1994] Gendreau, M., Hertz, A., and Laporte, G. (1994). A tabu search heuristic for the vehicle routing problem. *Management Science*, 40:1276–1290.

[Gidas, 1985] Gidas, B. (1985). Nonstationary Markov chains and convergence of the Annealing Algorithm. *J. Statis. Phys.*, 39.

[Glover, 1977] Glover, F. (1977). Heuristics for integer programming using surrogate constraints. *Decision Sciences*, 8(1):156–166.

[Glover, 1986] Glover, F. (1986). Future paths for integer programming and links to artificial intelligence. *Computers and Operations Research*, 13:533–549.

[Glover, 1989] Glover, F. (1989). Tabu search — part I. *ORSA Journal on Computing*, 1:190–206.

[Glover, 1990] Glover, F. (1990). Tabu search — part II. *ORSA Journal on Computing*, 2:4–32.

[Glover and Laguna, 1997] Glover, F. and Laguna, M. (1997). *Tabu Search*. Kluwer Academic Publishers.

[Glover et al., 1993] Glover, F., Laguna, M., Taillard, E. D., and de Werra, D. (1993). *Tabu Search*. Number 41 in Annals of Operations Research. Baltzer Science Publisher, Basel.

[Goldberg, 1989] Goldberg, D. E. (1989). *Genetic Algorithms in Search, Optimization and Machine learning.* Addison-Wesley.

[Goldberg, 1994] Goldberg, D. E. (1994). *Algorithmes génétiques. Exploration, optimisation et apprentissage automatique.* Addison-Wesley France.

[Goldberg and Deb, 1991] Goldberg, D. E. and Deb, K. (1991). A comparison of selection schemes used in genetic algorithms. In Rawlins, G., editor, *Foundations of Genetic Algorithms*, pages 69–93. Morgan Kaufmann.

[Goldberg and Richardson, 1987] Goldberg, D. E. and Richardson, J. (1987). Genetic algorithms with sharing for multimodal function optimization. In Grefenstette, J., editor, *Proc. 2nd Int Conf. on genetic Algorithms*, pages 41–49. Hillsdale, NJ: Erlbaum.

[Gosh and Tomlin, 2000] Gosh, R. and Tomlin, C. (2000). Maneuver design for multiple aircraft conflict resolution. In *American Control Conference.*

[Goss et al., 1989] Goss, S., Aron, S., Deneubourg, J. L., and Pasteels J. M. (1989). Self-Organized Shortcuts in the Argentine Ant. *Naturwissenchaften*, 76:579–581.

[Granger, 2002] Granger, G. (2002). *Détection et résolution de conflits aériens : modélisations et analyse.* PhD thesis, Ecole Polytechnique.

[Gutjahr, 2000] Gutjahr, W. J. (2000). A graph-based Ant System and its convergence. *Future Generation Computer Systems*, 16(8):873–888.

[Gutjahr, 2002] Gutjahr, W. J. (2002). ACO algorithms with guaranted convergence to the optimal solution. *Information Processing Letters*, 82(3):873–888.

[Hadj-Alouane and Bean, 1997] Hadj-Alouane, A. B. and Bean, J. C. (1997). A genetic algorithm for the multiple-choice integer program. *Operations Research*, 45(1):92–101.

[Hajek, 1988] Hajek, B. (1988). Cooling schedules for optimal Annealing. *Math. Oper. Res.*, 13:311–329.

[Hajek and Sasaki, 1989] Hajek, B. and Sasaki, G. (1989). Simulated annealing — to cool or not. *Systems and Control Letters*, 12:443–447.

[Hanafi, 2001] Hanafi, S. (2001). On the convergence of tabu search. *Journal of Heuristics*, 7(1):47–58.

[Hansen and Mladenović, 1999] Hansen, P. and Mladenović, N. (1999). An introduction to variable neighborhood search. In Voß, S., Martello, S., Osman, I. H., and Roucairol, C., editors, *Meta-heuristics : Advances and Trends in Local Search Paradigms for Optimization*, pages 422–458. Kluwer, Dordrecht.

[Hentenyrck, 1989] Hentenyrck, P. V. (1989). *Constraint Satisfaction in Logic Programming.* The MIT Press, Cambridge, MA.

[Hérault, 1989] Hérault, L. (1989). *Réseaux de neurones récursifs pour l'optimisation combinatoire ; Application à la théorie des graphes et à la vision par ordinateur.* PhD thesis, Institut National Polytechnique de Grenoble (I.N.P.G.), Grenoble.

[Hertz and de Werra, 1987] Hertz, A. and de Werra, D. (1987). Using tabu search techniques for graph coloring. *Computing*, 39:345–351.

[Hertz and de Werra, 1991] Hertz, A. and de Werra, D. (1991). The tabu search metaheuristic: How we used it. *Annals of Mathematics and Artificial Intelligence*, 1:111–121.

[Hertz and Kobler, 2000] Hertz, A. and Kobler, D. (2000). A framework for the description of evolutionary algorithms. *European Journal of Operational Research*, 126(1):1–12.

[Holland, 1962] Holland, J. H. (1962). Outline for logical theory of adaptive systems. *J. Assoc. Comput. Mach.*, 3:297–314.

[Holland, 1992] Holland, J. H. (1992). *Adaptation in Natural and Artificial Systems, 2nd edition.* MIT Press.

[Hölldobler and Wilson, 1990] Hölldobler, B. and Wilson, E. (1990). *The Ants.* Springer Verlag.

[Holma and Toskala, 2000] Holma, H. and Toskala, A. (2000). *WCDMA for UMTS, Radio Access for Third Generation Mobile Communications.* John Wiley & Sons Ltd, England.

[Homaifar et al., 1994] Homaifar, A., Lai, S. H.-Y., and Qi, X. (1994). Constrained optimization via genetic algorithms. *Simulation,* 62(4):242–254.

[Hooker, 1995] Hooker, J. (1995). Testing heuristics : We have it all wrong. *Journal of Heuristics,* 1(1):33–42.

[Horn et al., 1994] Horn, J., Nafpliotis, N., and Goldberg, D. E. (1994). A niched pareto genetic algorithm for multiobjective optimization. In *Proc. 1st IEEE Conf. on Evolutionary Computation,* pages 82–87. IEEE Piscataway.

[Horst and Pardolos, 1995] Horst, R. and Pardolos, P. M., editors (1995). *Handbook of Global Optimization.* Kluwer Academic Publishers.

[Horst and Tuy, 1995] Horst, R. and Tuy, H. (1995). *Global Optimization, Deterministic Approaches.* Springler.

[Hutchinson, 1957] Hutchinson, G. E. (1957). Concluding remarks, population studies: Animal ecology and demography. In *Cold Spring Harbor Symposia on Quantitative Biology 22,* pages 415–427.

[ILOG, 2002a] ILOG (2002a). *ILOG Dispatcher Reference Manual, Version 3.3.* ILOG S.A.

[ILOG, 2002b] ILOG (2002b). *ILOG Solver Reference Manual, Version 5.3.* ILOG S.A.

[Jeffcoat and Bulfin, 1993] Jeffcoat, D. and Bulfin, R. (1993). Simulated annealing for resource-constrained scheduling. *European Journal of Operational Research,* 70(1):43–51.

[Jennings, 1996] Jennings, N. R. (1996). Coordination Techniques for Distributed Artificial Intelligence. In G. M. P. O'Hare and N. R. Jennings, editor, *Foundations of Distributed Artificial Intelligence,* pages 187–210. John Wiley & Sons.

[Jimenez and Marzal, 1999] Jimenez, V. M. and Marzal, A. (1999). Computing the k shortest paths: A new algorithm and an experimental comparison. In Vitter, J. S. and Zaroliagis, C. D., editors, *Proc. 3rd Worksh. Algorithm Engineering,* number 1668 in Lecture Notes in Computer Science, pages 15–29. Springer-Verlag.

[Johnson et al., 1989] Johnson, D., Aragon, C., McGeoch, L., and Schevon, C. (1989). Optimization by simulated annealing : an experimental evaluation - Part I (Graph partitioning). *Opns. Res.,* 37(6):865–892.

[Johnson et al., 1991] Johnson, D., Aragon, C., McGeoch, L., and Schevon, C. (1991). Optimization by simulated annealing : an experimental evaluation - Part II (Graph coloring and number partitioning). *Opns. Res.,* 39(3):378–406.

[Johnson et al., 1992] Johnson, D., Aragon, C., McGeoch, L., and Schevon, C. (1992). Optimization by simulated annealing : an experimental evaluation - Part III (The travelling salesman problem). *Opns. Res.*

[Joines and Houck, 1994] Joines, J. and Houck, C. (1994). On the use of non-stationary penalty functions to solve nonlinear constrained optimization problems with gas. In Michalewicz, Z., Schaffer, J. D., Schwefel, H.-P., Fogel, D. B., and Kitano, H., editors, *Proceedings of the First IEEE International Conference on Evolutionary Computation,* pages 579–584. IEEE Press.

[Kennedy, 2000] Kennedy, J. (2000). Stereotyping: Improving particle Swarm performances with cluster analysis. In *Proceedings of the 2000 Congress on Evolutionnary Computation*, pages 1507–1512, Piscataway. IEEE Service Center.

[Kennedy and Eberhart, 1995] Kennedy, J. and Eberhart, R. C. (1995). Particle swarm optimization. In *Proc. IEEE Int'l. Conf. on Neural Networks*, volume IV, pages 1942–1948, Piscataway, NJ: IEEE Service Center.

[Kilby et al., 1997] Kilby, P., Shaw, P., and Prosser, P. (1997). Guided local search for the vehicle routing problem. In *In Proceedings of the 2nd International Conference on Meta-heuristics*.

[Kirkpatrick et al., 1983] Kirkpatrick, S., Gelatt, C., and Vecchi, M. (1983). Optimization by simulated annealing. *Science*, 220(4598):671–680.

[Kirkpatrick and Toulouse, 1985] Kirkpatrick, S. and Toulouse, G. (1985). Configuration space analysis of travelling salesman problems. *J. Physique*, 46:1277–1292.

[Koza, 1992] Koza, J. R. (1992). *Genetic Programming: on the Programming of Computers by Means of Natural Selection*. MIT Press.

[Koza, 1994] Koza, J. R. (1994). *Genetic Programming II: Automatic Discovery of Reusable Programs*. MIT Press.

[Kravitz and Rutenbar, 1987] Kravitz, S. and Rutenbar, R. (1987). Placement by simulated annealing on a multiprocessor. *IEEE Trans. on Computer Aided Design*, CAD-6:534–549.

[Lampinen, 2001] Lampinen, J. (2001). A Bibliography of Differential Evolution Algorithm. Technical report, Lappeenranta University of Technology, Department of Information Technology, Laboratory of Information Processing.

[Larranaga and Lozano, 2002] Larranaga, P. and Lozano, J. (2002). *Estimation of Distribution Algorithms, A New Tool for Evolutionnary Computation*. Genetic Algorithms and Evolutionnary Computation. Kluwer Academic Publishers.

[Leriche et al., 1995] Leriche, R. G., Knopf-Lenoir, C., and Haftka, R. T. (1995). A segregated genetic algorithm for constrained structural optimization. In Eshelman, L. J., editor, *Proc. of the Sixth Int. Conf. on Genetic Algorithms*, pages 558–565, San Francisco, CA. Morgan Kaufmann.

[Ling et al., 2002] Ling, C., Jie, S., Ling, Q., and Hongjian, C. (2002). A Method for Solving Optimization Problems in Continuous Space Using Ant Colony Algorithm. In Dorigo, M., Di Caro, G., and Sampels, M., editors, *Proceedings of the Third International Workshop on Ant Algorithms (ANTS'2002)*, volume 2463 of *Lecture Notes in Computer Science*, pages 288–289, Brussels, Belgium. Springer Verlag.

[Mackworth, 1977] Mackworth, A.-K. (1977). Consistency in networks of relations. *Artificial Intelligence*, 8:99–118.

[Mahfoud, 1992] Mahfoud, S. W. (1992). Crowding and preselection revisited. In Manner, R. and Manderick, B., editors, *Proc. of Parallel Problem Solving from Nature*, pages 27–36. Elsevier.

[Mathias and Whitley, 1992] Mathias, K. and Whitley, D. (1992). Genetic operators, the fitness landscape and the traveling salesman problem. In Manner, R. and Manderick, B., editors, *Parallel Problem Solving from nature, 2*, pages 221–230. Elsevier Science Publishers.

[Mathur et al., 2000] Mathur, M., Karale, S. B., Priye, S., Jyaraman, V. K., and Kulkarni, B. D. (2000). Ant Colony Approach to Continuous Function Optimization. *Ind. Eng. Chem. Res.*, 39:3814–3822.

[Mautor and Michelon, 1997] Mautor, T. and Michelon, P. (1997). Mimausa : a new hybrid method combining exact solution and local search. In 2nd *International Conference on Metaheuristics*, page 15, Sophia-Antipolis, France.

[Médioni, 1998] Médioni, F. (1998). *Méthodes d'optimisation pour l'évitement aérien : systèmes centralisés, systèmes embarqués.* PhD thesis, Ecole Polytechnique.

[Merkle et al., 2000] Merkle, D., Middendorf, M., and Schmek, H. (2000). Ant Colony Optimization for resource-constrained project scheduling. In *Proceedings of the Genetic and Evolutionnary Computation Conference (GECCO-2000)*, pages 839–900, San Francisco, CA. Morgan Kaufmann Publishers.

[Metropolis et al., 1953] Metropolis, N., Rosenbluth, A., Rosenbluth, M., Teller, A., and Teller, E. (1953). Equation of state calculations by fast computing machines. *J. Chem. Phys.*, 21:1087–1090.

[Meunier et al., 2001] Meunier, H., Bachelet, V., Talbi, E., and Caminada, A. (2001). A multi-objective genetic approach applied to cellular network design. In *ALGOTEL' 2001*, France.

[Michalewicz, 1996] Michalewicz, Z. (1996). *Genetic Algorithms + Data Structures = Evolution Programs, 3rd rev. and extended ed.* Springer Verlag.

[Michalewicz and Janikow, 1994] Michalewicz, Z. and Janikow, C. Z. (1994). Genocop: A genetic algorithm for numerical optimization problems with linear constraints. *Communications of the ACM*.

[Michalewicz and Nazhiyath, 1995] Michalewicz, Z. and Nazhiyath, G. (1995). Genocop III : A co-evolutionary algorithm for numerical optimization problems with nonlinear constraints. In *Proceedings of the Second IEEE Conference on Evolutionary Computation*, pages 647–651.

[Miclo, 1991] Miclo, L. (1991). *Evolution de l'énergie libre. Applications à l'étude de la convergence des algorithmes du recuit simulé.* PhD thesis, Université de Paris 6.

[Monmarché et al., 1999] Monmarché, N., Ramat, E., Dromel, G., Slimane, M., and Venturini, G. (1999). On the similarities between AS, BSC and PBIL: toward the birth of a newmeta-heuristics. E3i 215, Université de Tours.

[Monmarché et al., 2000] Monmarché, N., Venturini, G., and Slimane, M. (2000). On how Pachycondyla apicalis ants suggest a new search algorithm. *Future Generation Computer Systems*, 16:937–946.

[Montana, 1995] Montana, D. J. (1995). Strongly typed genetic programming. *Evolutionary Computation*, 3(2):199–230.

[Montana and Davis, 1989] Montana, D. J. and Davis, L. (1989). Training feedforward neural networks using genetic algorithms. In Sridharan, N., editor, *Proc. 11th. Int. Joint Conf. on Artificial Intelligence*, pages 762–767.

[Moscato, 1999] Moscato, P. (1999). Memetic algorithms : A short introduction. In Corne, D., Glover, F., and Dorigo, M., editors, *New Ideas in Optimisation*, pages 219–235. McGraw-Hill, Londres.

[Mühlenbein et al., 1988] Mühlenbein, H., Gorges-Schleuter, M., and Krämer, O. (1988). Evolution algorithms in combinatorial optimization. *Parallel Computing*, 7:65–88.

[Mühlenbein and Paaß, 1996] Mühlenbein, H. and Paaß, G. (1996). From recombination of genes to the estimation of distributions I. Binary parameters. *Lecture Notes in Computer Science 1411: Parallel Problem Solving from Nature*, PPSN IV:178–187.

[Musser et al., 1993] Musser, K., Dhingra, J., and Blankenship, G. (1993). Optimization based job shop scheduling. *IEEE Trans. on Automatic Control*, 38(5):808–813.

[Niedringhaus, 1989] Niedringhaus, W. (1989). A mathematical formulation for planning automated aircraft separation for AERA3. Technical report, FAA. DOT/FAA/DS-89/20.

[Nomura and Shimohara, 2001] Nomura, T. and Shimohara, K. (2001). An analysis of two-parent recombinations for real-valued chromosomes in an infinite population. *Evolutionary Computation*, 9(3):283–308.

[Nouyan, 2002] Nouyan, S. (2002). Agent-Based Approach to Dynamic task Allocation. In Dorigo, M., Di Caro, G., and Sampels, M., editors, *Proceedings of the Third International Workshop on Ant Algorithms (ANTS'2002)*, volume 2463 of *Lecture Notes in Computer Science*, pages 28–39, Brussels, Belgium. Springer Verlag.

[Osman and Laporte, 1995] Osman, I. H. and Laporte, G. (1995). Routing problems: A bibliography. *Annals of Operations Research*, 61:227–262.

[Panta, 2002] Panta, L. (2002). *Modeling Transportation Problems Using Concepts of Swarm Intelligence and Soft Computing*. PhD thesis, Virginia Tech, Blacksburg, Virginia.

[Petrowski, 1996] Petrowski, A. (1996). A clearing procedure as a niching method for genetic algorithms. In *IEEE 3rd International Conference on Evolutionary Computation (ICEC'96)*, pages 798–803.

[Petrowski and Girod Genet, 1999] Petrowski, A. and Girod Genet, M. (1999). A classification tree for speciation. In *Congress on Evolutionary Computation (CEC99)*, pages 204–211. IEEE Piscataway.

[Pham and Karaboga, 2000] Pham, D. and Karaboga, D. (2000). *Intelligent optimisation techniques. Genetic Algorithms, Tabu Search, Simulated Annealing and Neural Networks*. Springer.

[Pirlot, 1992] Pirlot, M. (1992). General local search heuristics in Combinatorial Optimization : a tutorial. *Belgian Journal of Operations Research and Computer Science*, 32(1–2):7–67.

[Rammal et al., 1986] Rammal, R., Toulouse, G., and Virasoro, M. (1986). Ultrametricity for Physicists. *Reviews of Modern Physics*, 58(3):765–788.

[Rechenberg, 1965] Rechenberg, I. (1965). *Cybernetic Solution Path of an Experimental Problem*. Royal Aircraft Establishment Library Translation.

[Rechenberg, 1973] Rechenberg, I. (1973). *Evolutionsstrategie: Optimierung technischer Systeme nach Prinzipien der biologischen Evolution*. Frommann-Holzboog, Stuttgart.

[Reeves, 1995] Reeves, C., editor (1995). *Modern Heuristic Techniques for Combinatorial Problems*. Advances topics in computer science. Mc Graw-Hill.

[Regin, 1994] Regin, J.-C. (1994). A filtering algorithm for constraints of difference in CSPs. In *Proceedings of the 11th AAAI*. The MIT Press.

[Regin, 1996] Regin, J.-C. (1996). Generalized arc consistency for global cardinality constraint. In *Proceedings of the 13th AAAI*. The MIT Press.

[Rego and Roucairol, 1996] Rego, C. and Roucairol, C. (1996). A parallel tabu search algorithm using ejection chains for the VRP. In Osman, I. and Kelly, J., editors, *Meta-Heuristic : Theory & Applications*, pages 253–295. Kluwer, Dordrecht.

[Renders and Flasse, 1996] Renders, J. and Flasse, S. (1996). Hybrid methods using genetic algorithms for global optimization. *IEEE Trans. on Systems, Man, and Cybernetics — Part B: Cybernetics*, 26(2).

[Resende, 2000] Resende, M. (2000). Greedy randomized adaptive search procedures (GRASP). Technical Report TR 98.41.1, AT&T Labs-Research.

[Rinnooy Kan and Timmer, 1987a] Rinnooy Kan, A. and Timmer, G. (1987a). Stochastic global optimization methods — Part I: Clustering methods. *Mathematical Programming*, 39:27–56.

[Rinnooy Kan and Timmer, 1987b] Rinnooy Kan, A. and Timmer, G. (1987b). Stochastic global optimization methods — Part II : Multi level methods. *Mathematical Programming*, 39:57–78.

[Rochat and Semet, 1994] Rochat, Y. and Semet, F. (1994). A tabu search approach for delivering pet food and flour in Switzerland. *Journal of the Operational Research Society*, 45:1233–1246.

[Rochat and Taillard, 1995] Rochat, Y. and Taillard, E. D. (1995). Probabilistic diversification and intensification in local search for vehicle routing. *Journal of Heuristics*, 1(1):147–167.

[Rogalsky and Derksen, 2000] Rogalsky, T. and Derksen, R. (2000). Hybridization of Differential Evolution for Aerodynamic Design. In *8th Annual Conference of the Computational Fluid Dynamics Society of Canada*, Montréal.

[Rogers and Prügel-Bennett, 1999] Rogers, A. and Prügel-Bennett, A. (1999). Genetic drift in genetic algorithm selection schemes. *IEEE Transactions on Evolutionary Computation*, 3(4):298–303.

[Rossi, 2000] Rossi, F. (2000). Constraint logic programming. In *Proceedings of the ERCIM/Compulog Net workshop on constraints, LNAI 1865*. Et.

[Roussel-Ragot, 1990] Roussel-Ragot, P. (1990). *La méthode du recuit simulé : accélération et parallélisation*. PhD thesis, Université de Paris 6.

[Roussel-Ragot et al., 1990] Roussel-Ragot, P., Siarry, P., and Dreyfus, G. (1990). A problem independent parallel implementation of simulated annealing: Models and experiments. In *IEEE Trans. CADICS*, volume 9, pages 827–835.

[Rubinstein, 1997] Rubinstein, R. (1997). Optimization of computer simulation models with rare events. *European Journal of Operations Research*, 99:89–112.

[Rubinstein, 1999] Rubinstein, R. (1999). The simulated entropy method for combinatorial and continuous optimization. *Methodology and Computing in Applied Probability*, 2:127–190.

[Rubinstein, 2001] Rubinstein, R. (2001). *Stochastic Optimization: Algorithms and Applications*, chapter Combinatorial optimization, cross-entropy, ants and rare events, pages 304–358. Kluwer Academic Publishing.

[Rudolph, 1994] Rudolph, G. (1994). Convergence analysis of canonical genetic algorithms. *IEEE Trans. on Neural Networks*, 5(1):96–101.

[Rudolph, 1996] Rudolph, G. (1996). Convergence of evolutionary algorithms in general search spaces. In *Proc. IEEE Int. Conf. on Evolutionary Computation 96*, pages 50–54. IEEE Press.

[Runarsson and Yao, 2000] Runarsson, T. P. and Yao, X. (2000). Stochastic ranking for constrained evolutionary optimization. *IEEE Transactions on Evolutionary Computation*, 4(3):274–283.

[Saït and Youssef, 1999] Saït, S. and Youssef, H. (1999). Iterative computer algorithms with applications in engineering. *IEEE Computer Society Press*.

[Sareni and Krahenbuhl, 1998] Sareni, B. and Krahenbuhl, L. (1998). Fitness sharing and niching methods revisited. *IEEE Transactions on Evolutionary Computation*, 2(3):97–106.

[Schoenauer and Michalewicz, 1996] Schoenauer, M. and Michalewicz, Z. (1996). Evolutionary computation at the edge of feasibility. In Voigt, H.-M., Ebeling, W., Rechenberg, I., and Schwefel, H.-P., editors, *Parallel Problem Solving from Nature – PPSN IV*, pages 245–254, Berlin. Springer.

[Schoenauer and Xanthakis, 1993] Schoenauer, M. and Xanthakis, S. (1993). Constrained GA optimization. In Forrest, S., editor, *Proceedings of the Fifth International Conference on Genetic Algorithms*, pages 573–580, San Mateo, CA. Morgan Kaufmann.

[Schoonderwoerd et al., 1996] Schoonderwoerd, R., Holland, O., Brutent, J., and Rothkrantz, L. (1996). Ant-based load balancing in telecommunication networks. *Adaptive Behavior*, 5(2):169–207.

[Schwefel, 1981] Schwefel, H.-P. (1981). *Numerical Optimization of Computer Models*. Wiley.

[Sechen, 1988] Sechen, C. (1988). *VLSI placement and global routing using simulated annealing*. Kluwer Academic Publishers.

[Shaw, 1998] Shaw, P. (1998). Using constraint programming and local search methods to solve vehicle routing problems. Technical report, ILOG S.A., Gentilly, France.

[Shi and Eberhart, 1998] Shi, Y. and Eberhart, R. (1998). A modified particle swarm optimizer. In *Proceedings of the IEEE International Conference on Evolutionnary Computation*, pages 69–73, Piscataway. IEEE Press.

[Shi and Eberhart, 1999] Shi, Y. and Eberhart, R. (1999). Empirical study of particle swarm optimization. In *Proceedings of the 1999 Congress on Evolutionnary Computation*, pages 1945–1950, Piscataway. IEEE Service Center.

[Siarry, 1986] Siarry, P. (1986). *La méthode du recuit simulé : application à la conception de circuits électroniques*. PhD thesis, Université de Paris 6.

[Siarry, 1994] Siarry, P. (1994). La méthode du recuit simulé en électronique. Adaptation et accélération. Comparaison avec d'autres méthodes d'optimisation. Application dans d'autres domaines. Habilitation à diriger les recherches en sciences physiques, Université de Paris-Sud (Orsay).

[Siarry, 1995] Siarry, P. (1995). La méthode du recuit simulé : théorie et applications. *RAIRO A.P.I.I. (Automatique - Productique - Informatique industrielle)*, 29(4–5):535–561.

[Siarry et al., 1987] Siarry, P., Bergonzi, L., and Dreyfus, G. (1987). Thermodynamic optimization of block placement. *IEEE Trans. on Computer Aided Design*, CAD-6(2):211–221.

[Siarry and Berthiau, 1997] Siarry, P. and Berthiau, G. (1997). Fitting of tabu search to optimize functions of continuous variables. *International Journal for Numerical Methods in Engineering*, 40:2449–2457.

[Siarry and Dreyfus, 1984] Siarry, P. and Dreyfus, G. (1984). An application of physical methods to the computer aided design of electronic circuits. *J. Physique Lettres*, 45:L39–L48.

[Siarry and Dreyfus, 1989] Siarry, P. and Dreyfus, G. (1989). *La méthode du recuit simulé : théorie et applications*. ESPCI – IDSET, 10 rue Vauquelin, 75005 Paris.

[Sizun, 2003] Sizun, H. (2003). *La propagation des Ondes Radioélectriques*. Springer-Verlag, Paris.

[Solla et al., 1986] Solla, S., Sorkin, G., and White, S. (1986). Configuration space analysis for optimization problems. In Bienenstock, E., Fogelman Soulie, F., and Weisbuch, G., editors, *Disordered Systems and Biological Organization*, pages 283–292, New York. Springer-Verlag.

[Solomon, 1987] Solomon, M. M. (1987). Algorithms for the vehicle routing and scheduling problem with time window constraints. *Operations Research*, 35:254–265.

[Sorkin, 1991] Sorkin, G. B. (1991). Efficient Simulated Annealing on Fractal Energy Landscapes. *Algorithmica*, 6:367–418.

[Spears, 1994] Spears, W. M. (1994). Simple subpopulation schemes. In *Proc. Conf. on evolutionary Programming*, pages 296–307. World Scientific.

[Srinivas and Deb, 1994] Srinivas, N. and Deb, K. (1994). Multiobjective function optimization using nondominated sorting genetic algorithms. *Evolutionary Computation*, 2(3):221–248.

[Storn, 1997] Storn, R. (1997). Differential Evolution – A Simple and Efficient Heuristic Strategy for Global Optimization over Continuous Spaces. *Journal of Global Optimization*, 11:341–359.

[Storn and Price, 1995] Storn, R. and Price, K. (1995). Differential Evolution – A simple and efficient adaptive scheme for global optimization over continuous spaces. Technical Report TR95 -012, International Computer Science Institute, Berkeley, California.

[Stützle and Hoos, 1997] Stützle, T. and Hoos, H. (1997). Improvements on the Ant System: Introducing MAX-MIN Ant System. In *Proceedings International Conference on Artificial Neural Networks and Genetic Algorithms*, Vienna. Springer-Verlag.

[Stützle and Hoos, 2000] Stützle, T. and Hoos, H. (2000). MAX-MIN Ant System. *Future Generation Computer System*, 16:889–914.

[Surry et al., 1995] Surry, P. D., Radcliffe, N. J., and Boyd, I. D. (1995). A Multi-Objective Approach to Constrained Optimisation of Gas Supply Networks : The COMOGA Method. In Fogarty, T. C., editor, *Evolutionary Computing. AISB Workshop. Selected Papers*, Lecture Notes in Computer Science, pages 166–180, Sheffield, U.K. Springer-Verlag.

[Taillard, 1990] Taillard, E. D. (1990). Some efficient heuristic methods for the flow shop sequencing problem. *European Journal of Operational Research*, 47(1):65–74.

[Taillard, 1991] Taillard, E. D. (1991). Robust taboo search for the quadratic assignment problem. *Parallel Computing*, 17:443–455.

[Taillard, 1993] Taillard, E. D. (1993). Parallel iterative search methods for vehicle routing problems. *Networks*, 23:661–673.

[Taillard, 1994] Taillard, E. D. (1994). Parallel taboo search techniques for the job shop scheduling problem. *ORSA Journal on Computing*, 6(2):108–117.

[Taillard, 1995] Taillard, E. D. (1995). Comparison of iterative searches for the quadratic assignment problem. *Location science*, 3(2):87–105.

[Taillard, 1998] Taillard, E. D. (1998). *Programmation à mémoire adaptative et algorithmes pseudo-gloutons : nouvelles perspectives pour les méta-heuristiques*. Thèse d'habilitation à diriger des recherches, Université de Versailles, France.

[Taillard, 2002] Taillard, E. D. (2002). Principes d'implémentation des méta-heuristiques. In Teghem, J. and Pirlot, M., editors, *Optimisation approchée en recherche opérationnelle*, pages 57–79. Lavoisier, Paris. Note : STAMP software is available at http://ina.eivd.ch/taillard.

[Taillard, 2003a] Taillard, E. D. (2003a). Heuristic methods for large centroid clustering problems. *Journal of Heuristics*, 9(1):51–73.

[Taillard, 2003b] Taillard, E. D. (2003b). A statistical test for comparing success rates. In *Metaheuristic international conference MIC'03*, Kyoto, Japan.

[Taillard et al., 1997] Taillard, E. D., Badeau, P., Guertin, F., Gendreau, M., and Potvin, J. (1997). A tabu search heuristic for the vehicle routing problem with soft time windows. *Transportation science*, 31(2):170–186.

[Taillard et al., 1998] Taillard, E. D., Gambardella, L. M., Gendreau, M., and Potvin, J.-Y. (1998). Adaptive Memory Programming: A Unified View of Meta-Heuristics. *European Journal of Operational Research*, 135(1):1–16.

[Taillard and Voß, 2002] Taillard, E. D. and Voß, S. (2002). POPMUSIC — partial optimization meta-heuristic under special intensification conditions. In Ribeiro, C. and Hansen, P., editors, *Essays and surveys in metaheuristics*, pages 613–629. Kluwer, Dordrecht.

[Talbi, 2002] Talbi, E. (2002). A taxonomy of hybrid metaheuristics. *Journal of Heuristics*, 8:541–564.

[Teghem and Pirlot, 2002] Teghem, J. and Pirlot, M. (2002). *Optimisation approchée en recherche opérationnelle. Recherches locales, réseaux neuronaux et satisfaction de contraintes*. Hermès.

[Theraulaz et al., 1998] Theraulaz, G., Bonabeau, E., and Deneubourg, J.-L. (1998). Response Threshold Reinforcement and Division of Labour in Insect Societies. In *Proceedings of the Royal Society of London B*, volume 265, pages 327–335.

[Tsang, 1993] Tsang, E. (1993). *Foundations of Constraint Satisfaction*. Academic Press.

[Tsui and Liu, 2003] Tsui, K. and Liu, J. (2003). Multiagent Diffusion and Distributed Optimization. In *Proceedings of the Second International Joint Conference on Autonomous Agents & Multiagent Systems*.

[Ulungu et al., 1999] Ulungu, E., Teghem, J., Fortemps, P., and Tuyttens, D. (1999). MOSA method : a tool for solving multiobjective combinatorial optimization problems. *Journal of Multicriteria Decision Analysis*, 8:221–236.

[Van Laarhoven and Aarts, 1987] Van Laarhoven, P. and Aarts, E. (1987). *Simulated annealing : theory and applications*. D. Reidel Publishing Company, Dordrecht (The Netherlands).

[Van Laarhoven et al., 1992] Van Laarhoven, P., Aarts, E., and Lenstra, J. (1992). Job-shop scheduling by simulated annealing. *Op. Res.*, 40:113–125.

[Vecchi and Kirkpatrick, 1983] Vecchi, M. and Kirkpatrick, S. (1983). Global wiring by simulated annealing. *IEEE Trans. on C.A.D.*, CAD-2(4):215–222.

[Voß and Woodruff, 2002] Voß, S. and Woodruff, D. L. (2002). *Optimization Software Class Libraries*. OR/CS Interfaces Series. Kluwer, Dordrecht.

[Weiss, 1999] Weiss, G. (1999). *Multiagent Systems*. MIT Press.

[White et al., 1998] White, T., Pagurek, B., and Oppacher, F. (1998). Connection Management using Adaptive Agents. In *Proceedings of the International Conference on Parallel and Distributed Processing Techniques and Applications (PDPTA'98)*, pages 802–809.

[Wilson, 1984] Wilson, E. (1984). The relation between Caste Ratios and Division of Labour in the Ant Genus Pheidole (Hymenoptera: Formicidae). *Behav. Ecol. Sociobiol.*, 16:89–98.

[Wilson and Hölldobler, 1988] Wilson, E. and Hölldobler, B. (1988). Dense Heterarchy and mass communication as the basis of organization in ant colonies. *Trend in Ecology and Evolution*, 3:65–68.

[Wittner and Helvik, 2002] Wittner, O. and Helvik, B. (2002). Cross-Entropy Guided Ant-Like Agents Finding Cyclic Paths in Scarely Meshed Networks. In Dorigo, M., Di Caro, G., and Sampels, M., editors, *Proceedings of the Third International Workshop on Ant Algorithms (ANTS'2002)*, volume 2463 of *Lecture Notes in Computer Science*, pages 121–134, Brussels, Belgium. Springer Verlag.

[Wodrich and Bilchev, 1997] Wodrich, M. and Bilchev, G. (1997). Cooperative distributed search: the ant's way. *Control and Cybernetics*, 26(3).

[Wong et al., 1988] Wong, D., Leong, H., and Liu, C. (1988). *Simulated annealing for VLSI design*. Kluwer Academic Publishers.

[Woodruff and Zemel, 1993] Woodruff, D. L. and Zemel, E. (1993). Hashing vectors for tabu search. In Glover, G., Laguna, M., Taillard, E. D., and de Werra, D., editors, *Tabu Search*, number 41 in Annals of Operations Research, pages 123–137, Basel, Switzerland. Baltzer.

[Wright, 1932] Wright, S. (1932). The roles of mutation, inbreeding, crossbreeeding and selection in evolution. In Jones, D. F., editor, *Proc. of the Sixth International Congress on Genetics*, pages 356–366.

[Xu and Kelly, 1996] Xu, J. and Kelly, J. P. (1996). A network flow-based tabu search heuristic for the vehicle routing problem. *Transportation Science*, 30(4):379–393.

[Yao and Liu, 1996] Yao, X. and Liu, Y. (1996). Fast evolutionary programming. In Fogel, L. J., Angeline, P. J., and Bäck, T., editors, *Proc. 5th Ann. Conf. on Evolutionary Programming*, pages 451–460. MIT Press.

[Yin and Germay, 1993a] Yin, X. and Germay, N. (1993a). A fast genetic algorithm with sharing scheme using cluster analysis methods in multimodal function optimization. In R.F.Albrecht, C. R. and Steele, N., editors, *In proceedings of the Artificial Neural Nets and Genetic Algorithm International Conference, Insbruck Austria*. Springer-Verlag.

[Yin and Germay, 1993b] Yin, X. and Germay, N. (1993b). A fast genetic algorithm with sharing scheme using cluster methods in multimodal function optimization. In Albrecht, R. F., Reeves, C. R., and Steele, N. C., editors, *Proc. of the International Conference on Artificial Neural Nets and Genetic Algorithms*, pages 450–457. Springer-Verlag.

[Zeghal, 1994] Zeghal, K. (1994). *Vers une théorie de la coordination d'actions, application à la navigation aérienne*. PhD thesis, Universite Paris VI.

[Zitzler and Thiele, 1999] Zitzler, E. and Thiele, L. (1999). Multiobjective evolutionary algorithms : A comparative case study and the strength pareto approach. *IEEE Trans. on Evolutionary Computation*, 3(4):257–271.

Index

3-opt, 133

ACO, 135
Adaptive Memory Programming, 235,
    236, 239, 247
ADF, 115
AIS, 172
algorithms
    evolutionary, 75
    genetic, 76, 118, 252, 277
allele, 82
amplification, 165
annealing, 5
    microcanonic, 155
    simulated, 6, 51, 208
    simulated logarithmic curve, 41
annealing scheme, 30
Ant Colony System, 132, 316
Ant Programming, 149
Ant System, 130, 316
    elitism, 131
Ant-Q, 132
Antigens, 172
AntNet, 147
API, 145
artificial immune systems, 172
aspiration, 47, 66, 71
asynchronous, 34
automatically defined functions, 115
azimuth, 263

CACO, 140
candidate list, 56

capacitated vehicle routing problem,
    225
capacity, 259, 269
CE, 170
cell breathing, 260
cellular network, 252
change in temperature stage, 45
channels of communication, 142
    direct, 143
    integrity, 142
    memory, 142
    range, 142
    stigmergic, 143
CIAC, 141
clearing procedure, 201
cloning, 173
clusters, 166, 205, 280
combinatorial explosion, 2
communication
    ant colonies, 141
    process, 124
comparison
    heuristics, 240
    multiple, 244
    optimization methods, 243, 248
    STAMP software, 244
    success rates, 241
competitive exclusion, 197
complex neighborhood, 230
configuration space, 28
conflict in route, 277
constrained evolutionary optimization,
    216

constraint programming, 233
constraints, 1, 216, 316
Continuous Ant Colony Algorithm, 140
Continuous Interacting Ant Colony, 141
convergence, 27
  of tabu search, 66
  premature, 82
coverage, 259
Cross-Entropy method, 170
crossover, 78, 95
  arithmetic, 104, 222
  contracting, 103
  lethal, 96
  linear BLX-α, 104
  operator, 237
  rate, 95
  voluminal BLX-α, 102
cycle, 51, 57, 59, 62, 68

darwinist operator, 77
decentralization, 13
decentralized control, 126
decision variables, 4
decomposing problems, 231, 248
decrease of the temperature, 26, 45, 334
dense heterarchy, 126
detection of the conflicts, 283
deterministic crowding, 199, 200
differential evolution, 173
distributed problems, 14
diversification, 10, 72, 137
dominance of Pareto, 207
dynamic optimization, 3

ecological niche, 196
EDA, 166
ejection chains, 230, 239
elitism, 91, 131, 202, 214
emergence, 124
energy, 6
entropy, 336
Estimation of Distribution Algorithms, 166
Evolution Strategies, 76, 105
evolutionary computation, 76
exploitation, 137
exploration, 137
exponential scaling, 87
extensions, 3

feasible individual, 216
feedback, 124, 165, 172
fitness function, 77, 92
flexibility, 13
fluctuations, 124
fractal properties, 29

GBAS, 139
generations, 77
genetic
  algorithms, 76, 97, 118
  drift, 82, 199, 202
  programming, 113
GENOCOP, 222
GENOCOP III, 220
genotype, 98, 118
Graph-Based Ant System, 139
GRASP, 169
Gray code, 100
Greedy Randomized Adaptive Search Procedure, 169

Hamming cliffs, 99
HCIAC, 144
heuristic, 2
horizon effect, 286
hybrid methods, 15
hybridization, 138
hyper-mutation, 172

image processing, 42
indirect interactions, 138
inertia weight, 165
intensification, 10, 72, 137
interferences, 256
island model, 205
iterative amelioration, 4

landscape
  energy, 5
  fitness, 94
layout of electronic circuits, 36
linear scaling, 86
list of limited candidates, 169
local search, 82, 231
lymphocytes, 172

macro-diversity, 260
maneuvers for avoidance, 283
Markov chain, 26, 331
  homogeneous, 332

inhomogeneous, 337
Markovian field, 42
Max-Min Ant System, 133
memory
    adaptive, 137, 239, 247
    adaptive programming, 235
    behavioral, 223
    establishment, 339
    long-term, 10, 69, 71, 72
    population, 236
    short-term, 10, 57, 59, 72
    trails, 236
    type, 63
method
    descent, 4
    penalization of the fitness function,
        217
    Alienor, 160
    artificial immune systems, 172
    classical, 4
    Cross-Entropy, 170
    crowding, 199
    death penalty, 217
    differential evolution, 173
    distributed search, 159
    estimation of distribution, 166
    GRASP, 169
    great deluge, 157
    hybrid, 3, 15
    kangaroo, 43
    microcanonic annealing, 155
    noising, 159
    particle swam optimization, 162
    repair, 220
    roulette, 83
    sharing, 196
    simulated diffusion, 154
    social insects, 175
    stochastic universal sampling, 83
    threshold, 157
    travel of record in record, 157
methodology, 239
migration, 204
minimal coupling of points, 36
MOGA, 213
move, 52, 64
    aspired, 66
    candidate, 56
    evaluation, 54, 55

forced, 71
inverse, 59
inversion, 53
penalized, 69
reverse, 63
tabu, 59–61
transposition, 53
movement, 4
multimodal, 3, 15, 196
multiobjective, 3, 15
multiobjective optimization, 206
mutation, 78, 97, 99, 104, 117, 267
    bit-flip, 99
    boundary, 222
    deterministic, 99
    gaussian, 105
    non uniform, 222
    operator, 237
    self-adaptive, 106
    uniform, 104, 222
mutation rate, 97

neighborhood, 51, 52, 228, 239
    evaluation, 54
    on a permutation, 51, 53
    vehicle routing, 229
net of safeguard, 278
niching, 196
    sequential, 196
NPGA, 214
NSGA, 213

operator
    crossover, 78, 95, 287
    darwinist, 77
    mutation, 78, 97, 236
    recombination, 78
    replacement, 77, 90
    search, 78
    selection, 77, 81
    variation, 78, 93
optimization
    difficult, 1
    global, 2
    multimodal, 196
    multiobjective, 206

Pachycondyla apicalis, 145
parallel implementations, 3

parallelism, 138
  intrinsic, 13
parallelization, 15, 32
parameter
  tuning, 60, 63, 68, 70, 240, 241
Pareto front, 207
Pareto optimal, 15
Pareto-optimal set, 207
parse trees, 113
Particle Swarm Optimization, 162
partitioning of graph, 36
path relinking, 72, 239
penalties
  adaptive, 219
  dynamic, 218
  static, 218
penalty, 69, 227, 235
  death, 217
permutation problem, 53
perturbation, 107
phenotype, 98, 118
pheromonal trails, 127
pheromone, 13, 127
POPMUSIC, 231–233, 238, 239, 248
  parameter, 232
population, 77
power of the pilot channel, 263
probability distribution, 166
problem
  quadratic assignment, 339
  decomposition, 231
  modeling, 227
  permutation, 108
  quadratic assignment, 49
  traveling salesman, 49, 108, 129
programming
  constraint, 307
  evolutionary, 76
  genetic, 113
  linear, 16
PSO, 162

quadratic assignment, 36, 60
  and tabu search, 62, 64, 71, 244, 339
  definition, 49
  example, 50, 64
  neighborhood, 55
quenching, 6

radio planning, 261
radius
  niche, 197, 202
  of restriction, 96
real time management, 284
recombination, 78
  intermediary, 104
regulate
  acceptance, 45
repair methods, 220
replacement, 77, 90
  elitist, 91
  generational, 90
  steady state, 90
representation, 78, 93
  binary, 97
  ordinal, 109
  paths, 109
  real, 101
  sequence, 109
robustness, 13
roulette wheel selection method, 83
routing, 147
rule
  acceptance, 30
  of 1/5, 106
RWS, 84

Scatter Search, 236
search
  local, 51, 138
  tabu, 47, 51, 52, 66
Segregated Genetic Algorithms, 219
selection
  intensity, 81
  operator, 77, 81, 237
  pressure, 81, 85
  proportional, 82–84
  rank based, 88
  tournament, 88
  truncation, 90
self-organization, 14, 124, 164, 172, 175
separability, 287
sexual reproduction, 78
shared fitness, 197
SPEA, 214
speciation, 196, 203
  by clustering, 205
  island model, 205

label based, 204
specific heat, 336
spin glasses, 5
STAMP, 244
statistical test
    Mann-Whitney, 243
    proportion, 241
stigmergy, 13, 126, 137
stochastic, 2
strategic oscillations, 72
subpopulation, 196, 203
SUS, 84
swarm, 162
swarm intelligence, 139, 148

tabu, 60, 61, 63
    duration, 70
tabu (see under "search"), 47
tabu condition
    duration of, 60
tabu list, 8, 57, 59
    basic, 47
    hashing tables, 57
    length, 61
    random, 61–63
    size, 60

type, 63, 67
tag-bits, 204
takeover time, 81
tandem-running, 146
temperature, 6
    initial, 44
tilt, 263
tournament, 88
    deterministic, 89
    domination, 214
    stochastic, 90
trade-off surface, 15, 207
traffic simulator, 284
trails of pheromone, 136
traveling salesman, 2, 35, 108
TSP, 108, 129
type of antenna, 263

ultrametricity, 29
uncertainty, 281

variables, 15
variations, 78
vehicle routing, 67, 225, 230, 233, 307
vocabulary building, 72, 238